The Culture of Science
in France, 1700–1900

Professor Robert Fox

Robert Fox

The Culture of Science
in France, 1700–1900

Taylor & Francis Group

LONDON AND NEW YORK

First published 1992 by Variorum, Ashgate Publishing

Published 2017 by Routledge
2 Park Square, Milton Park, Abingdon, Oxon, OX14 4RN
52 Vanderbilt Avenue, New York, NY 10017

*Routledge is an imprint of the Taylor & Francis Group,
an informa business*

A CIP catalogue record for this book is available
from the British Library and the
US Library of Congress.

ISBN 13: 978-0-86078-339-8 (hbk)

COLLECTED STUDIES SERIES CS381

CONTENTS

This book contains xiii + 335 pages

PUBLISHER'S NOTE

The articles in this volume, as in all others in the Collected Studies Series, have not been given a new, continuous pagination. In order to avoid confusion, and to facilitate their use where these same studies have been referred to elsewhere, the original pagination has been maintained wherever possible.

Each article has been given a Roman number in order of appearance, as listed in the Contents. This number is repeated on each page and quoted in the index entries.

PREFACE

I have never been happy with the divisions that have often been thought to separate the internalist and externalist approaches to the history of science. It is with some satisfaction, therefore, that I have brought together in this volume articles that invade both of these supposedly disparate worlds. Under the heading "Patronage and institutions", I have assembled six texts concerned with the institutional structures of science in nineteenth-century France; and in a second section, entitled "The physics of heat", I have included an equal number that reflect my long-standing interest in the theories of imponderable fluids that were accepted and eventually abandoned in the late eighteenth and early nineteenth centuries. Despite this two-fold structure, the points of contact between the items in the two sections are numerous. They reinforce my conviction that the history of scientific ideas can never be detached from a serious consideration of the context from which they sprang and that social histories of science that find no place for ideas and the practices of the laboratory or the field tell only a partial story.

As the volume shows, most of my work as an historian has been directed to France. It was the literature of France that first fascinated me at school, and it has been one of the privileges of my career to have been able to extend that interest to the other realms of French culture treated here. From my earliest days in research, nearly thirty years ago, I was struck by the diversity of the contexts in which the French pursued science and learned culture in the nineteenth century. The notion that the revolutionary and Napoleonic régimes succeeded in strengthening the already powerful centralizing tendencies inherited from Colbert clearly has its core of truth: no scientist from the time of the Directory to the Third Republic could ignore the authority vested in ministers and the bureaucracy of government, and many (like Laplace and other patrons and leaders of schools) found the power of the centre a valuable tool in their personal careers. But to suppose that the course of French science was guided exclusively by the regulations and patronage emanating from the great ministries and institutions of the capital is a grave mistake.

Several of the contributions reproduced in *The culture of science, 1770–1900* reinforce this point. The article on "Scientific enterprise and the patronage of research" (I) and my discussion of Laplacian physics

(IX), for example, show how meticulously the opportunities inherent in the rich institutional structure of early-nineteenth-century French science had to be exploited, in order to create a successful research school of the kind that Laplace and Berthollet established at Arcueil. In the far humbler contexts discussed in the papers on the provincial and national *sociétés savantes* (II, III, and IV) and on the cultural world of the booming textile-town of Mulhouse (V), the importance of individual initiative is brought out even more starkly. In those contexts, science proceeded largely in isolation from the centre. The opportunities to which it responded were predominantly local ones, and attempts at ministerial intervention were often viewed with suspicion, in particular by the Norman antiquarian and naturalist Arcisse de Caumont, who established a national network of *savants* dedicated to the fostering of a determinedly provincial world of science and learning. The fact that Caumont found the struggle so arduous and that his annual Congrès scientifiques and other fiercely autonomous organizations collapsed so quickly after his death in 1873 point to the inevitability of the eventual victory of the centre over the periphery and of "official" science over the science of the independent devotee. But the length of Caumont's campaign (it lasted almost half a century) suggests that the victory was not an easy one and that the contrast that is traditionally made between France and Britain, with regard to the relative importance of state patronage and locally focussed individualism, may have been somewhat over-drawn. It is certainly my sense that the interactions between the regional economy and the institutions for promoting and disseminating science and technology in Mulhouse were not conspicuously different from those that evolved during the nineteenth century in many an industrial town in Britain.

Although the papers in the section entitled "The physics of heat" broach questions of a different order, I have been mindful even here of the diversity of French styles in physics. The point is made clearly enough in item VII by J.H. Lambert's decidedly idiosyncratic theory of heat (if a resident of Mulhouse in Lambert's day can legitimately be classed as French) and by the work of Joseph Mollet and his colleagues in Lyons on the ignition of tinder by the rapid compression of air — observations that gave rise to the curious fire-making device known as the fire piston (X). Yet I do concede that in the early nineteenth century, the history of French physical science at its highest levels was essentially a Parisian story, dominated by the towering figures of Laplace and his neighbour at Arcueil, Berthollet. The quality of the circle that Laplace and Berthollet gathered around them helped to make the First Empire a golden age for the physical sciences in France, although my account of the rise and fall of Laplacian physics (IX) points to the dangers inherent in the power of two great patrons to perpetuate doctrines, based in this case on short-range intermolecular forces, which by 1815 were widely regarded by younger scientists as suspect. Dulong and Petit, the discovery of whose law I discuss in paper VIII, were typical members of that younger generation, both in the high quality of their work and in their critical attitude towards the theories on which they had been raised. Like Arago, Fresnel, and Gay-Lussac, among others, they helped to give French science a new brilliance in the decade after Waterloo, as the old Laplacian principles yielded to new theories of heat and light and chemical atomism.

One contemporary who did not share in the heady excitement of physics and chemistry in the Académie des Sciences, the Ecole Polytechnique, and the Sorbonne and the other main seats of power after 1815, was Sadi Carnot. The son of a regicide and a diffident young man who never took to the military career that awaited him in peace-time, Carnot moved on the fringes of Parisian scientific life during the Bourbon Restoration. It was a world in which the education of working men for the new industrial age mattered more than the esoteric debates of the academicians. In this world, Charles Dupin, whose travels during the Empire and the Restoration I discuss in "From Corfu to Caledonia" (VI), was a powerful figure, typical in that his activities spanned science, engineering, and politics, as well as the new departures in technical education that he helped to realize, from 1819, at the Conservatoire Royal des Arts et Métiers. But Carnot's contacts were with less prominent figures.

The most influential of these figures for Carnot was Nicolas Clément, the first professor of industrial chemistry at the Conservatoire. In "Watt's expansive principle" (XII), I argue that Carnot's work on the theory of the heat engine owed some of its most distinctive elements, in particular the notion of the adiabatic phase in his cycle, to ideas already developed and expounded, rather hazily, by Clément. In a similar vein, in "The challenge of a new technology" (XI), I identify Carnot's problem, that of determining the conditions for securing the maximum output from a heat engine, as one that had engaged physicists and engineers in France for a decade by the time he published his *Réflexions sur la puissance motrice du feu* in 1824. This tradition of studies that were ignored in their own time, as they have been ignored ever since by historians, helps us to see the *Réflexions* as part of a capacious but obscure literature whose neglect by contemporaries is all the more understandable since none of it offered serious guidance to practising engineers.

Looking back over these articles, I feel generally heartened in my conviction that the study of science in a national context is a

xii

worth-while exercise and that there are advantages as well as snares for scholars working on countries other than their own. In the year of our newly united Europe, I am grateful to Variorum for the opportunity of making this point, however obliquely. I also thank the various copyright holders without whose agreement the volume would not have come to be.

ROBERT FOX

Oxford,
February 1992

ACKNOWLEDGEMENTS

I am grateful to the following individuals and organizations, for giving permission for the reproduction of articles for which they hold the copyright: Mrs Gillian Anderson, Managing Editor of *Minerva*, 19 Nottingham Road, London SW17 7EA (for Study I); Mr Richard Fisher, Editorial Manager, Social Science Publications, Cambridge University Press (II); the editor of *Historical Reflections/Réflexions historiques* (III); the President, Vice-President and General Secretary of the Société Zoologique de France (IV); Dr J.H. Brooke, Honorary Editor, *The British Journal for the History of Science*, on behalf of the British Society for the History of Science (V, VIII); Kluwer Academic Publishers, Dordrecht (VI); Editions Ophrys, Gap and Monsieur Gilbert Salmob, Directeur du Centre de Recherches et d'Etudes Rhénanes, Université de Haute Alsace, Mulhouse (VII); Princeton University Press, Princeton, New Jersey (IX); Dr R.C. Post, Editor of *Technology and Culture*, on behalf of the Society for the History of Technology (X); Monsieur G. Lilamand, Directeur du CNRS Publications (XI); Ms Helen Gardner, on behalf of the Royal Society (XII).

I

Scientific Enterprise and the Patronage
of Research in France 1800-70

In the mid-1860s French scientists came, quite suddenly, to see the state
of scientific research in their country as a national scandal. In 1868, in
Le budget de la science, Louis Pasteur wrote passionately of the " suffer-
ing " and " penury " of the deprived community for which he considered
himself a spokesman.[1] With no financial support for their experiments
and with access to only the most rudimentary laboratory facilities, the
scientists of France were like soldiers without arms, sacrificing health and
personal fortune in their dedication to research. The remedy, for Pasteur,
was clear: the financial provision for research laboratories—" temples of
the future, of wealth, and well-being " [2]—had, quite simply, to be drasti-
cally increased. In all this Pasteur was by no means alone. Indeed, his
criticism was only one product of a campaign of protest and often harsh
recrimination which involved nearly all of the country's leading scientists,
as well as many other intellectuals, and which culminated in the remarkable meeting of the Académie des sciences on 6 March, 1871, just five
weeks after the end of the Franco-Prussian war, at which member after
member, including Henri Sainte-Claire Deville, Jean-Baptiste Dumas, and
Charles Hermite, rose to speak bitterly of the parlous state of French
science.[3]

In nearly all the complaints of the 1860s there was a strong common
theme: admiration for the successes of the German research schools and
the peculiarly German ideals of *Wissenschaft* and university-based research
united the scientists of France and their sympathisers,[4] as it united so

[1] Pasteur, Louis, *Le budget de la science* (Paris: Gauthier-Villars, 1868), p. 5.

[2] *Ibid.,* p. 4.

[3] *Comptes rendus hebdomadaires des séances de l'Académie des Sciences,* LXXII (1871),
pp. 231–239 and 261–269.

[4] See, for example, Renan, Ernest, " L'instruction supérieure en France. Son histoire
et son avenir ", *Revue des deux mondes,* 2e période, LI (1864), especially pp. 80 and 86–89.
Although my subsequent references are to the article as it appeared in the *Revue des deux
mondes,* it may also be consulted in Psichari, Henriette (ed.), *Oeuvres complètes de Ernest
Renan,* 10 vols. (Paris: Calmann-Lévy, 1947–1961), Vol. I, pp. 69–97. Admiration for
Germany is also evident in Pouchet, Georges, " L'enseignement supérieur des sciences en
Allemagne ", *Revue des deux mondes,* 2e période, LXXXIII (1869), especially pp. 442–443,
and Papillon, Fernand, " Les laboratoires en France et à l'étranger ", *ibid.,* 2e période,
XCIV (1871), especially pp. 596–600 and 605–606. There were frequent references to the
success of German science in the statements made at the Académie des sciences on 6 March,
1871. See also Pasteur, L., *op. cit.,* p. 4, and Pasteur's article of 1871 in the Lyons
newspaper *Salut public,* of which passages are reproduced in Vallery-Radot, René, *La vie
de Pasteur* (Paris: Hachette, 1900), pp. 278–280. For a more general study of French
attitudes towards German science, see Paul, Harry W., *The Sorcerer's Apprentice: The
French Scientist's Image of German Science 1840–1919* (Gainesville: University of Florida
Press, 1972).

many of the educational reformers of Britain at precisely the same time.[5] French studies of German practices in scientific research and education abounded in this period, and, almost without exception, they took the form of eulogies. This was true equally of the abundant periodical literature [6] and of more formal reports, such as those on German medical education by Sigismond Jaccoud and Paul Lorain, published in 1864 and 1868 respectively,[7] and that of Adolphe Wurtz on German university laboratories, published in 1870.[8]

In support of their view that greatly increased financial support for their work was urgently needed, the French scientists had ample evidence. Learned papers, books, and highly trained students flowed in abundance from the research schools of universities like Giessen, Berlin, Heidelberg, Göttingen, and Bonn, and in some cases had been doing so for over 20 years. Although the point was rarely conceded by the protesting scientists, the excellence of German science had even been recognised, albeit slowly and with great caution, in certain government circles. It was at the request of the Minister of Education, Gustave Rouland, that Jaccoud undertook his investigation of the German medical faculties in 1863,[9] and from 1863 to 1869 the Ministry of Education was headed by Victor Duruy, a man who had imbibed a lasting admiration for German scholarship through his own historical research and, more particularly, through his youthful contact with Jules Michelet, his professor at the École normale in the early 1830s.[10] Not surprisingly, it was Duruy who, in 1868, commissioned the report by Wurtz,[11] which was perhaps the most telling of all the indictments of the conditions for laboratory research in France.

Now it could be objected that the statements of such men as Pasteur and Wurtz are not reliable evidence since, like most French scientists in the late 1860s, they had a strong personal interest in encouraging government intervention in science on a large scale and they were quite openly

---

[5] See, for example, Huxley, Thomas Henry, " A Liberal Education; and where to find it " (1868), in his *Lay Sermons, Addresses and Reviews* (London: Macmillan, 1895), pp. 43–44; Arnold, Matthew, *Schools and Universities on the Continent* (London: Macmillan, 1868), pp. 222–232; Playfair, Lyon, *On Teaching Universities and Examining Boards* (Edinburgh: Edmonston and Douglas, 1872), pp. 23–24 and 30–31; and, on Mark Pattison's conversion to German ideas of the university, Sparrow, John, *Mark Pattison and the Idea of a University* (Cambridge: Cambridge University Press, 1967), pp. 105–149.

[6] See, for example, Renan, Ernest, *op. cit.*, pp. 80 and 86–89; Pouchet, George, *op. cit.*, pp. 442–443; and Papillon, Fernand, *op. cit.*, pp. 596–600 and 605–606.

[7] Jaccoud, Sigismond, *De l'organisation des facultés de médecine en Allemagne* (Paris: Delahaye, 1864), and Lorain, Paul, *De la réforme des études médicales par les laboratoires* (Paris: J. Bonaventure, 1868).

[8] Wurtz, Charles Adolphe, *Les hautes études pratiques dans les universités allemandes* (Paris: Imprimerie Impériale, 1870).

[9] Jaccoud, S., *op. cit.*, *passim*.

[10] On Michelet's admiration for Germany, which was shared in the 1820s and 1830s by Romantics like Victor Hugo, Gérard de Nerval, and Jean Jacques Ampère, see Monod, Gabriel, *La vie et la pensée de Jules Michelet 1798–1852*, 2 vols. (Paris: Champion, 1923), Vol. I, pp. 145–169. Duruy most handsomely acknowledged his debt to Michelet in a letter of 22 April, 1884, probably to Monod, reproduced in Monod, G., *op. cit.*, Vol. I, p. 257.

[11] Wurtz, C. A., *op. cit.*, pp. 1 and 3.

involved in polemical activity to that end.[12] However, in this paper I shall not question the opinion, shared by Pasteur, Wurtz, and the great majority of their fellow scientists, that on the eve of the Franco-Prussian war the provision for laboratory research in France was lamentable by the contemporary standards in German universities. What I shall question, with special reference to the physical sciences—though it is difficult to treat them in isolation—is whether the commonly drawn contrast between the apparently flourishing state of scientific research under Napoleon and its condition in the 1860s can be explained in terms of government attitudes, in the way that both Pasteur and Wurtz, for example, suggested.[13] I shall argue that even during the First Empire, governmental patronage of science was rarely enlightened and only intermittently generous and that the indifference which was complained of so bitterly in the 1860s, far from being something new, had in reality been characteristic of the attitudes of governments throughout the century. Of course, the degree of indifference varied. For example, the governments of the Second Empire— especially while Duruy was Minister of Education— were more responsive to the needs of scientific research than were those of the restored Bourbons. But even when nineteenth-century French governments did act as the patrons of research, they normally did so in response to external pressures, such as the personal approaches that were made to Napoleon III by Claude Bernard and Henri Sainte-Claire Deville,[14] and they acted on their own initiative only on the rare occasions when a piece of research promised a solution to some urgent problem.[15]

So the task, as I see it, is to ascertain why from the 1820s the demands made on governments and the institutions of science were so slight, whereas before 1815 men of high intellectual and social standing, like the mathematician and physicist Pierre Simon de Laplace and the chemist Claude Louis Berthollet, were content to devote themselves to research and to the far from easy task of acquiring good facilities both for their own work and for that of their disciples. With this as my problem, I shall pay special attention to certain attitudes towards research which

---

[12] I discuss the difficulties encountered by Pasteur and Wurtz in their personal research later in this paper. In the late 1860s Pasteur, as administrator and director of scientific studies at the École normale, was trying to promote research there and to free the institution from its traditional preoccupation with teaching. Wurtz was engaged on the investigation of research facilities in German universities; see Wurtz, C. A., op. cit.

[13] See Pasteur's contribution to the *Salut public*, quoted in Vallery-Radot, R., op. cit., p. 279, and Wurtz, C. A., op. cit., pp. 6 and 12–13.

[14] See Olmsted, James Montrose Duncan and Olmsted, E. H., *Claude Bernard and the Experimental Method in Medicine* (London, New York, and Toronto: Abelard-Schuman, 1952), pp. 126–127, and Liard, Louis, *L'enseignement supérieur en France 1789–1889*, 2 vols. (Paris: Armand Colin, 1888–1894), Vol. II, p. 273. On the campaign by a group of scientists including Bernard, Victor Regnault, and Jean-Baptiste Dumas, which led to the appointment of Marcellin Berthelot to the newly established posts at the Collège de France, first as a lecturer in 1863, then as a professor of organic chemistry in 1865, see Velluz, Léon, *Vie de Berthelot* (Paris: Plon, 1964), pp. 63–69.

[15] Two such occasions, referred to later in this paper, occurred when Dulong and, later, Regnault were commissioned to perform research concerned with the operation of steam-engines.

came to be held by French scientists during the second quarter of the nineteenth century. Although I do not claim that these new attitudes were alone responsible for France's loss of the supremacy in scientific research which she enjoyed in Europe during the early years of the nineteenth century, I do believe that they contributed greatly, and in a way which has not been recognised before, to the French failure to respond to the growing German challenge of the 1830s and 1840s. In so far as my paper directs attention to the attitudes of scientists themselves, and assigns to their paymasters a more passive role than is customary, I hope that it will also serve a more general purpose in pointing to the importance of personal motivation in nineteenth-century French science. For it is my belief that it was individual initiative, rather than government sponsorship, which brought French science its successes early in the century, and it was the lack of such initiative which brought French science to its unhappy state during the Second Empire.

Individual Initiative and Public Patronage under Napoleon

The excellence of French science in the age of Napoleon has rarely been doubted.[16] It was an age in which not only Laplace and Berthollet but also Jean-Baptiste Biot, Siméon Denis Poisson, Étienne Malus, Joseph Louis Gay-Lussac, Louis Jacques Thenard, Georges Cuvier, Joseph Fourier, and Jean-Baptiste Delambre, as well as members of an older generation like Gaspard Monge, Lazare Carnot, and Antoine François de Fourcroy, were all at the height of their powers. It was an age of great discoveries, notably Gay-Lussac's law of combining volumes and Malus's discovery of the polarisation of light, and of major experimental investigations like the work of Gay-Lussac and Thenard on chlorine and the alkali metals and the study of the specific heats of gases undertaken by Delaroche and Bérard for the Institut's prize competition of 1812. Among the books published were the first four volumes of Laplace's *Traité de mécanique céleste* (1799–1805) and Berthollet's *Essai de statique chimique* (1803); and the many distinguished contributions which appeared in the journals of the period included all Malus's writings on mathematical physics and some of the most important of the papers of Laplace, Biot, and Poisson.

For these successes Napoleon himself has received much credit. He was eulogised in his own lifetime, most fulsomely perhaps in the reports on the state of science which Delambre and Cuvier presented to the Emperor

---

[16] However, for a particularly unfavourable view of Napoleonic science see [Moll, Gerrit], *On the Alleged Decline of Science in England. By a Foreigner* (London: T. and T. Boosey, 1831), pp. 6–7, 10–16, 18–19, and 21–33; also Moll's letters to Michael Faraday, dated 24 December, 1830, and 13 November, 1831, in Williams, L. Pearce (ed.), *The Selected Correspondence of Michael Faraday*, 2 vols. (Cambridge: Cambridge University Press, 1971), Vol. I, pp. 187–189 and 204–208. For more recent criticism of Napoleonic science see Williams, L. P., " Science, Education and Napoleon I ", *Isis*, XLVII (1956), pp. 369–382, and Hahn, Roger, *The Anatomy of a Scientific Institution. The Paris Academy of Sciences, 1666–1803* (Berkeley, Los Angeles, and London: University of California Press, 1971), pp. 310–312.

in 1808,[17] and later, in 1830, both Babbage and Brewster portrayed him as a model patron.[18] By his own membership of the First Class of the Institute (he was elected to the mechanics section in 1797), the personal favour which he showed to scientists, his provision of prizes for work in voltaic electricity, his encouragement of industry, and his celebrated gift of a huge electric pile for research at the École polytechnique, he gave some justification for these favourable opinions.[19] But, even when he was granting favours, considerations of utility or national prestige were never far from Napoleon's mind.[20] And, in any case, Napoleonic science was not entirely the creation of Napoleon himself, as critics such as the Dutch observer Gerrit Moll, professor of physics at Utrecht, were quick to point out.[21] Most of the young men who distinguished themselves at the Institute or in the circle of Laplace and Berthollet were products of the École polytechnique, created in 1794 some five years before Napoleon became First Consul; and other scientific institutions, such as the École des ponts et chaussées, the Écoles des mines, the Muséum d'histoire naturelle, and the First Class of the Institut, were relics of the Ancien Régime.[22] Moreover, when Napoleon did introduce measures relevant to science they showed little sympathy for science as an intellectual activity. For example, when in 1804 he subjected the École polytechnique to a military regime, he did so with complete disregard for the interests of science and for the protests of Monge, Berthollet, and Fourcroy.[23] The decision that henceforth students should wear uniform and live in barracks was in itself not too harmful, although the increasingly military tone may well have affected recruitment [24]; but the drastic reduction in the financial provision for research and laboratories was a measure of striking insensi-

[17] Delambre, Jean-Baptiste Jean, *Rapport historique sur les progrès des sciences mathématiques depuis 1789, et sur leur état actuel* (Paris: Imprimerie Impériale, 1810), especially pp. 1-3 and 40-42, and Cuvier, Jean L. N. F., *Rapport historique sur les progrès des sciences naturelles . . .* (Paris: Imprimerie Impériale, 1810), especially pp. 389-394. Both reports were commissioned by Napoleon himself.

[18] Babbage, Charles, *Reflections on the Decline of Science in England and on Some of its Causes* (London: B. Fellowes and J. Booth, 1830), p. 26, and [Brewster, David], " Decline of science in England ", *Quarterly Review*, XLIII (1830), pp. 315-316.

[19] On Napoleon's scientific interests and his contributions as a patron of science, see Crosland, Maurice P., *The Society of Arcueil. A View of French Science at the Time of Napoleon I* (London: Heinemann, 1967), pp. 4-55.

[20] It is particularly difficult to see his patronage of electrical research as anything but an attempt to gain for France some of the prestige that the work of Volta and Davy had earned for Italian and English science.

[21] Moll, G., *op. cit.*, pp. 18-19. The large measure of continuity between eighteenth-century Science and the science of the First Empire has been pointed out more recently in Ben-David, Joseph, " The Rise and Decline of France as a Scientific Centre ", *Minerva*, VIII, 2 (April 1970), pp. 160-179, especially pp. 165-172.

[22] However, the Muséum d'histoire naturelle (previously the Jardin du Roi) and the First Class of the Institut (previously the Académie royale des sciences) had been reorganised and renamed in 1793 and 1795 respectively.

[23] Pinet, Gaston, *Histoire de l'École polytechnique* (Paris: Baudry, 1887), pp. 48-55.

[24] The possible effect on recruitment is suggested by L. P. Williams, " Science, education and Napoleon I " . . . pp. 372-373. However, on some entrants to the École polytechnique who were attracted by the prospect of a military career, see Crosland, M. P., *op. cit.*, pp. 83-87. A more serious impediment to recruitment after 1804 was the withdrawal of the grant to students and the imposition of a charge of 800 francs per annum.

tivity,[25] and it set the tone for the remaining years of the Empire, in which the interests of science at the École polytechnique, the Institut, and, from 1808, in the newly founded Imperial University were consistently made subservient to practical demands.[26]

In short, there was a darker side of French science under Napoleon. Against the brilliance of many of the young *polytechniciens* of the early years, like Biot, Poisson, Dulong, Arago, and Petit, we have to set the fact that they achieved their distinction in physics despite the poor teaching in the subject which they received as students under Jean Henri Hassenfratz, who was a thoroughly incompetent professor of physics from 1794 to 1815.[27] And in admiring the extraordinarily high quality of the winning entries in the Institut's prize competitions, especially those of 1807 (by Fourier), 1809 (by Malus), and 1812 (by Delaroche and Bérard), we should not overlook the crisis of 1809 when two meetings of the First Class of the Institut had to be abandoned because of a lack of communications.[28]

Had French science depended solely on the patronage of Napoleon or on the provision made in the institutions of science, it is difficult to see how it could have flourished as it did during the First Empire. Hence, if we are to account for its successes, I believe we must look not to official or government patronage but rather to the personal motives of the individual scientists. In particular, as far as the physical sciences are concerned, we must look to Laplace and Berthollet, the two men who dominated French physics and chemistry throughout the years of Napoleon's rule. For in my opinion the source of most of the achievements for which Napoleonic science is justly celebrated lies in the ambition of these two men in pursuit of a joint programme of work, which they conceived as the fulfilment of a research tradition the origins of which lie in the Newtonian science of the eighteenth century. In an almost literal sense, they sought to be the Newtons of their age and they planned their research to this end.

I shall summarise what I call the " Laplacian " programme very briefly.[29] The goal which Laplace and Berthollet set themselves was the explanation

²⁵ See Hahn, R., *op. cit.*, p. 311.

²⁶ See Williams, L. P., " Science, education and Napoleon I ". . . . pp. 376–379, and Crosland, M. P., *op. cit.*, pp. 40–45 and 155–158. On the wretched state of the science faculties in the Napoleonic university see Liard, L., *op. cit.*, Vol. II, pp. 105–124 and 136–138. Examining for the *licence* and *baccalauréat* and, to a lesser extent, teaching, were seen as the proper functions of the faculties of science and of letters. Research had no place in them.

²⁷ See Fox, Robert, *The Caloric Theory of Gases from Lavoisier to Regnault* (Oxford: Clarendon Press, 1971), pp. 231–232; also Arago's recollections of Hassenfratz in his " Histoire de ma jeunesse ", in *Oeuvres complètes de François Arago*, 17 vols. (Paris: Gide; and Leipzig: Weigel, 1854–1862), Vol. I, pp. 12–13.

²⁸ Crosland, M. P., *op. cit.*, pp. 158–161, and Hahn, R., *op. cit.*, pp. 306–307.

²⁹ I have given more detailed accounts of the programme elsewhere, and full references for the summary which follow may be found in these other sources. See Fox, Robert, " The Laplacian Programme for Physics ", *Boletin de la Academia Nacional de Ciencias de la Republica Argentina*, XLVIII (1970), pp. 429–437, and " The Rise and Fall of Laplacian Physics ", *Historical Studies in the Physical Sciences*, IV (in press).

of all physical and chemical phenomena in terms of attractive and repulsive forces which they assumed to exist between the particles of matter, whether the matter in question was that of ordinary ponderable substances or one of the imponderable fluids of heat, light, electricity, or magnetism. In his earlier writings, such as the *Exposition du système du monde* (1796),[30] Laplace gave only a hint of what by 1809 had become an explicit research programme.[31] However, there is little doubt that the programme was conceived clearly enough by 1805. In that year, encouraged no doubt by the support he had gained in Berthollet's writings on chemical affinity,[32] he devoted lengthy sections of the fourth volume of the *Mécanique céleste* to mathematical treatments of optical refraction and capillary action in terms of the short-range attractive forces first postulated by Newton.[33] And after 1805 most of his own researches, and those of the disciples who worked with him and Berthollet in Berthollet's private laboratory at Arcueil, were directed to the fulfilment of the programme.

The willingness of Laplace and Berthollet to spend so much of their admittedly considerable senatorial income in order to provide research facilities at Arcueil is the most obvious mark of their dedication to the programme.[34] But it is equally revealing to observe the way in which they used their influence in manipulating for their own ends facilities which were not directly under their control. They secured their disciples' leisure and freedom for research by placing them in teaching and other posts at various institutions in Paris [35]; and at the Institut they achieved a dual purpose by proposing subjects for the annual prize competitions which would yield

[30] In this book Laplace looked forward to the time when not only optical refraction and capillary action (already treated in the Laplacian manner by eighteenth-century Newtonians) but also the cohesion of bodies, their crystalline properties, and even chemical reactions, would be explained in terms of the attractive forces exerted by the ultimate particles of matter. See Laplace, P. S., *Exposition du système du monde*, 2 vols. (Paris: Imprimerie du Cercle-Social, an IV [1796]), Vol. II, pp. 196–198.

[31] The programme is stated in Laplace, P. S., "Mémoire sur les mouvemens de la lumière dans les milieux diaphanes", *Mémoires de la classe des sciences mathématiques et scientifiques de l'Institut de France*, X (1809), pp. 329 and 338.

[32] Berthollet's writings on chemical affinity include his *Recherches sur les lois de l'affinité* (Paris: Baudouin, an IX [1801]), as well as the *Essai de statique chimique*, 2 vols. (Paris: Firmin Didot, an XI [1803]).

[33] Laplace, P. S., *Traité de mécanique céleste*, 5 vols. (Paris: Duprat; Courcier; Bachelier, 1799–1825), Vol. IV (1805), pp. 231–281. Two separately paginated supplements, added in 1806 and 1807, contained his theory of capillary action.

[34] On the incomes of Berthollet and Laplace, which exceeded 50,000 francs annually during the First Empire, see Crosland, M. P., *op. cit.*, pp. 69–74. The incomes should be compared with the 6,000 francs paid to the permanent secretaries of the First Class of the Institut and to professors at the École polytechnique. According to a letter to Macvey Napier, cited in Morrell, J. B., "Science and Scottish University Reform: Edinburgh in 1826", *The British Journal for the History of Science*, VI (1972–1973), p. 51, John Leslie was impressed to find Berthollet and Laplace with annual incomes of between £5,000 and £6,000 each when he visited Paris in 1814. At the current rate of exchange, this suggests incomes in each case of well over 100,000 francs per annum. On the income of Berthollet after the Bourbon Restoration, see Crosland, M. P., *op. cit.*, pp. 398–401.

[35] For example, the influence of Laplace helped Biot, Arago, and Poisson to obtain appointments at the Bureau des longitudes. And, to varying extents, the backing of Arcueil furthered the careers of Thenard, Gay-Lussac, Poisson, Arago, and Malus at the École polytechnique, of Biot at the Collège de France, and of Dulong at the École normale. See Crosland, M. P., *op. cit.*, pp. 190–221.

both success, and a prize of 3,000 francs, for one of their protégés, and an answer to some problem raised by their programme. Malus was someone who benefited in an obvious way from patronage of this kind, for it seems certain that the prize competition for a mathematical study of double refraction was set, in December 1807, not only in the hope that Laplace's earlier treatment of refraction would be extended to embrace double refraction as well, but also with Malus specifically in mind. Predictably Malus, who treated the problem in the classic Laplacian manner, won the prize [36]; so, in 1812, did Delaroche and Bérard, whose experiments on the specific heats of gases, performed at Arcueil, upheld the position favoured by Laplace on a technical issue in the caloric theory.[37]

By the end of the First Empire, Laplacian physics (though arguably not Berthollet's chemistry) was, to all appearances, in an unassailable position. It had stimulated work of outstanding quality; its principles, most notably the theories of imponderable fluids, were taught as standard doctrine in science courses at all levels; and its exponents dominated not only teaching but also research. Yet, however remote such a possibility may have appeared early in 1815, change sufficiently drastic to overthrow both the content and the structure of Napoleonic physical science was imminent.

The Restoration

There is abundant evidence that in the years immediately after the downfall of Napoleon French cultural life underwent a radical change. François Guizot, Edgar Quinet, and Alphonse de Lamartine, among others, all convey the feeling of exhilaration and intellectual liberation which was so common in the period, despite the sinister activities of the Congrégation, and which led to so many brilliant achievements in literature and the arts. For Guizot the Restoration came like spring after a long, hard winter,[38] while Quinet, in retrospect, described the Empire as an intellectual " desert ".[39]

Not surprisingly, most men of science did not react to the Restoration with quite the unbridled joy which so many of their literary contemporaries displayed. For example, they could scarcely accept Lamartine's view of mathematics as the shackles which had bound human thought under Napoleon but which were now, to his relief, broken.[40] However, Lamartine

[36] The prize-winning paper was " Théorie de la double réfraction " *Mémoires présentés à l'Institut . . . par divers savans . . . Sciences mathématiques et physiques*, II (1811), pp. 303–508.
[37] Their entry was published as " Mémoire sur la détermination de la chaleur spécifique des différens gaz ", *Annales de chimie et de physique*, LXXXV (1813), pp. 72–110 and 113–182. On the 1812 competition see Fox, R., *Caloric Theory of Gases. . . .* pp. 131–150.
[38] See Bertier de Sauvigny, Guillaume André de, *La restauration*, new ed. (Paris: Flammarion, 1955), p. 328.
[39] Quinet, Edgar, *Histoire de mes idées. Autobiographie*, 7th ed. (Paris: Hachette, 1895), Vol. XV of the *Oeuvres complètes d'Edgar Quinet*, p. 241.
[40] Lamartine, Alphonse M. L. de P. de, *Des destinées de la poésie* (1834), in *Oeuvres complètes de Lamartine publiées et inédites*, 41 vols. (Paris: Chez l'auteur, 1860–1866), Vol. I, pp. 30–32.

was correct in perceiving that a cultural change had taken place, and, by winning the love of the attractive wife of the aged physicist, J. A. C. Charles, and making her the beloved Elvire of the *Méditations*,[41] he symbolised the nature of the change in a curiously appropriate way. Science could no longer command the special respect which it had enjoyed under Napoleon, and success in the literary salons, on the stage or political platform, or in the revived political press, rather than at the laboratory bench, became increasingly the typical ambition of gifted young men. It was entirely characteristic of this period, and symptomatic of a considerable " generation gap ", that two such important young literary figures as Victor Hugo and Quinet should have turned to literature in the first years of the Restoration after being destined for the École polytechnique by their fathers.[42]

Yet Lamartine overstated his case, for, despite a certain loss of prestige, the Restoration did little harm to science. Some politically suspect scientists, it is true, became the victims of reaction. But while Monge, Guyton de Morveau, Lazare Carnot, J. N. P. Hachette, Louis Poinsot, and Louis Benjamin Francoeur were punished for their earlier services to the revolution and Napoleon by being removed from positions in teaching or in the Académie des sciences, the majority of scientists reconciled themselves without difficulty to the monarchy,[43] and the institutions of science were left to function much as they had done in the last years of the Empire. In fact, when changes did occur, they were usually for the better, as at the École polytechnique, where demilitarisation and the replacement of the former Jacobin Hassenfratz by Petit stimulated a (temporary) rise in academic standards.[44] But in all too many cases the opportunity for change was missed and the weaknesses of Napoleonic science were perpetuated. This was especially unfortunate in the case of the faculties, which at the Restoration were ripe for reorganisation. The harm done by their isolation from one another, and by their financial impoverishment, was fully recognised by Royer-Collard and Guizot as they worked for the strengthening of higher education in the provinces during the 11 months of the First Restoration. But the resulting legislation of 17 February, 1815, which would have created, in place of the monolithic Imperial University, 17 regional universities, each consisting of a group of faculties and possessing a large measure of independence in administration and teaching, was never

[41] On this episode see, for example, Bertrand, Louis, *Lamartine* (Paris: Arthème Fayard, 1940), pp. 170–172.

[42] See, in addition to the standard biographical sources, Quinet, E., *op. cit.*, pp. 208–216, 225–232, and 239–247.

[43] None, except perhaps Cuvier, did so more easily than Laplace. However, his adoption of the prevailing illiberal attitudes was to do him great harm, as I point out later in this section.

[44] On demilitarisation, which was revoked in 1830, see Pinet, G., *op. cit.*, pp. 93–103 and 421–430. The replacement of Hassenfratz is discussed in Fox, R., *Caloric Theory of Gases* . . . pp. 231–232.

put into effect.[45] For by 1 March Napoleon had landed at Fréjus, and by the time the Hundred Days were over the faculties were fortunate to survive in any form at all. The new bill of 15 August, 1815, merely confirmed the old imperial structure,[46] and the faculties resumed their modest roles as examining bodies and as the purveyors of low-level instruction—roles which they were to fulfil for the rest of the century.

So in many respects continuity rather than change characterised the transition from the Empire to the Restoration, as far as science was concerned. But I believe that in some less obvious ways the decade 1815–1825 did mark a turning-point. In the history of scientific thought, as in so many intellectual activities, these were years of great change, years in which science shared some of the upheavals that are more commonly associated with the emergence of the new Romantic movement in literature, music, and art. There were challenges to established beliefs in medicine and mathematics [47]; and in the physical sciences, between 1815 and 1820, the Laplacian programme came to be discredited in the eyes of all but a few increasingly isolated diehards, headed by Laplace himself and the most loyal of his disciples, Poisson.[48]

The discrediting of the Laplacian orthodoxy, and of Berthollet's chemistry, have a special and obvious importance for our understanding of French science during the Restoration. From the point of view of the " internal " history of science, the best known casualty of the anti-Laplacian movement of 1815–20 was the corpuscular theory of light, which Augustin Fresnel, with the backing of François Arago, ruthlessly undermined in his celebrated early papers on diffraction and double refraction. But scarcely less important were the attacks on the material theory of heat—the caloric theory—by Alexis Thérèse Petit and Pierre Louis Dulong, and the rapid acceptance of the chemical atomic theory, which implied clearly the rejection of Berthollet's chemistry of affinities and molecular forces. However, for the purposes of this paper, it is more important to see the turning away from Laplace and his school as one aspect of the decline of

45 Liard, L., *op. cit.*, Vol. II, pp. 125–136 and 181–183. Guizot's preference was for no more than four regional universities, though he fully shared Royer-Collard's desire for decentralisation.

46 However, some strengthening was attempted by the closure of 17 faculties of letters and three science faculties, at Besançon, Lyons, and Metz. See Liard, L., *op. cit.*, Vol. II, pp. 136–138.

47 On medicine see Ackerknecht, Erwin H., *Medicine at the Paris Hospital 1794–1848* (Baltimore: Johns Hopkins University Press, 1967), pp. 61–80. On the declining influence of Lagrange and the rise of Fourier see Grattan-Guinness, Ivor, *The Development of the Foundations of Mathematical Analysis from Euler to Riemann* (Cambridge, Mass., and London: MIT Press, 1970), pp. 1–45, and the same author's *Joseph Fourier 1768–1830. A Survey of his Life and Work based on a Critical Edition of his Monograph on the Propagation of Heat, presented to the Institut de France in 1807* (Cambridge, Mass., and London: MIT Press, 1972), pp. 441–490.

48 For much fuller studies of the discrediting of Laplacian physics and Berthollet's chemistry, see Fox, R., *Caloric Theory of Gases. . . .* pp. 227–248 and 270–280, and my papers " The Laplacian Programme for Physics ", pp. 435–436, and " The Rise and Fall of Laplacian Physics ". Full references for the account which follows will be found in these sources.

Arcueil as a source of influence and patronage. For during the first half of the nineteenth century there was no change in the course and style of French scientific research more drastic than that which accompanied this decline.

From 1815, when advancing age and a diminished income were impeding the activity of Berthollet,[49] little research was performed at Arcueil. And Laplace, with his programme discredited on scientific grounds and his personal standing, especially in the eyes of younger scientists, adversely affected by his pliability on political matters and his illiberal views—notably on the freedom of the press—found it increasingly difficult to exert his old influence.[50] Power quickly passed into the hands of a new anti-Laplacian generation with Fresnel, Arago, Dulong, and Petit as its leading members and with Joseph Fourier as its detached though sympathetic elder statesman. By 1816, Laplacian influence in the leading scientific journal, the *Annales de chimie et de physique*, was greatly reduced when Arago was made one of the two joint editors (with Gay-Lussac [51]), and at the École polytechnique and the Société philomathique the critics of Laplace and his school became rapidly more powerful. At the Académie des sciences also the Laplacians lost control, the decisive event there being the victory of Fourier over Biot, who was still loyal to Laplace, in the election of 1822 for the post of permanent secretary for the mathematical sciences. From that point the science of Laplace was doomed, as Biot himself acknowledged by holding aloof from the scientific community of Paris until the 1830s.

Of course, the decline of a research school is not in itself remarkable. But what is remarkable and significant in this case, and in need of explanation, is the fact that there emerged no school to take the place of the school at Arcueil. In attempting to explain this, I shall continue to pay special attention not to official policies towards science, which changed little after the Restoration, but rather to the attitudes of individual scientists. In particular, I shall try to show how the interests of scientists under the Bourbon and Orléans monarchies were diverted by their adoption of a mode of life incompatible with a deep commitment to research.

The New Style of Science

When Amaury, the hero of Sainte-Beuve's novel *Volupté*, returned about 1820 to a France that he had not seen for some three years, he

---

[49] By February 1816, Berthollet's income was reduced to 24,000 francs. See Crosland, M. P., *op. cit.*, pp. 398–401.

[50] Laplace's growing unpopularity is well conveyed by the critical comments in the articles on him in Rabbe, Alphonse, *et al.* (eds.), *Biographie universelle et portative des contemporains*, 5 vols. (Paris: Levrault, 1834), Vol. III, pp. 151–153, and Hoefer, Johann Christian Ferdinand, *Nouvelle biographie générale*, 46 vols. (Paris: Firmin Didot, 1855–66), Vol. XXIX, cols. 533–534.

[51] Although he was not prominent in the revolt against the ideas of Berthollet and Laplace, Gay-Lussac was decidedly sympathetic to the new theories, especially to Fresnel's wave theory of light.

observed with astonishment the change which had come over French life. He wrote:

> What struck me most when I first returned from America . . . was that, after the Empire and the excess of military force which had been prevalent in that period, people had gone over to excess in words, to a wasteful and elaborate use of declamation, imagery, and promises, and to an equally blind confidence in these new weapons.[52]

In pointing to the new fashion for declamation and the consequent striving for oratorical effect, Sainte-Beuve was making an observation of characteristic astuteness which complements the recollections of Lamartine, Guizot, and Quinet referred to in the previous section. That he was correct in identifying a new spirit abroad in France in the early years of the Restoration is apparent in the flowering of political oratory, which continued throughout the reigns of Louis XVIII and Charles X, with such brilliant debaters as the Comte de Villèle, Royer-Collard, and Benjamin Constant dominating the Chamber of Deputies,[53] and the point is supported also by the success of the many religious preachers of the period.[54] But the correctness of Sainte-Beuve's observation cannot be better illustrated than by the ability of Victor Cousin, Guizot, and Villemain to attract such huge and admiring crowds to their lectures at the Sorbonne on, respectively, philosophy, history, and French literature,[55] and we may be sure that Sainte-Beuve himself knew of the scenes of enthusiasm with which these three men were received, especially by the young. Increasingly, success in higher education came to be measured by success at the lectern, rather than in research and scholarship, and Cousin, Guizot, and Villemain were only the most brilliant of numerous teachers in the faculties and other institutions who aspired to popular acclaim as lecturers. Some of these teachers were scarcely less charismatic than Cousin, Guizot, and Villemain themselves; and this was true even of lecturers in science, who in this respect have not received the attention they deserve.

In science, as in other subjects, success was more readily achieved in Paris, with its large educated public, than in the provinces, where audiences for lectures in the faculties, when they could be assembled at all, were

---

[52] Sainte-Beuve, Charles Augustin, *Volupté*, 2 vols. (Paris: Eugène Renduel, 1834), Vol. II, p. 285.

[53] This flowering of political oratory, which men of the stature of Guizot and Thiers maintained throughout the Orléans monarchy, though with somewhat diminished brilliance, has been widely recognised. For a contemporary study, see "Timon" [L. M. de La Haye, vicomte de Cormenin], *Études sur les orateurs parlementaires* (Paris: La Nouvelle Minerve, 1836).

[54] However, in this case it should be noted, first, that the abbé de Frayssinous, who was perhaps the most successful preacher under the restored Bourbons, had already had a considerable following at Saint-Sulpice during the First Empire, and, secondly, that the finest achievements in nineteenth-century religious oratory belong to a later period, from the mid-1830s, when Ravignan, Dupanloup, and, most brilliant of all, Lacordaire enjoyed their greatest renown.

[55] On these lectures, see Johnson, Douglas, *Guizot: Aspects of French History 1787–1874* (London: Routledge & Kegan Paul, 1963), pp. 118–119 and 121–122.

commonly sparse and nearly always ill-informed [56]; even modest successes, like that of Pasteur, who had audiences of 250 at the faculty of science in Lille in the 1850s,[57] were unusual. Of course, some lecturers, like the gauche and inarticulate Ampère, were incapable of attracting a large audience even in the most favourable circumstances, in Paris or anywhere else.[58] But Thenard, Biot, and Gay-Lussac, for example, were so successful at the Sorbonne that the young Charles de Rémusat, whose educational background at the Lycée Napoléon had been almost exclusively linguistic and philosophical, attended their lectures, in preference to those of Cousin and Villemain, shortly after the Restoration.[59] By comparison with Villemain's lectures, which he regarded as no more than an " entertainment ", Thenard's exposition of chemistry was enthralling. As Thenard spoke, according to Rémusat, " It seemed . . . as though the veil of nature was being raised before my eyes. I felt that I was gazing for the first time on a world in which, until then, I had lived as though surrounded by magical mysteries or rather by senseless prodigies." [60]

Dumas was another attractive lecturer who, when Pasteur attended his lectures in the 1840s, regularly had audiences of between 700 and 800, or even 1,000, at the Sorbonne.[61] And even a less celebrated figure like the physicist C. S. M. Pouillet was able to attract an audience of over 800 when he performed some experiments of Faraday, presumably his demonstration of electromagnetic induction, at the Sorbonne in 1832.[62]

The new style of declamatory science was most evident in the faculties, where audiences could be won, and to a large extent had to be won, by the attractiveness of the lecturer. Although payment in the faculties was

[56] On the audiences in the provincial faculties, see Liard, L., *op. cit.*, Vol. II, pp. 136–138 and 276–282. The problems of teaching in a provincial faculty are well illustrated by the experiences of Antoine Augustin Cournot, a conscientious and able teacher who in 1834 began lecturing at the faculty of science in Lyons. Although a large audience was present for the first month of the course—apparently attracted by the novelty of the newly founded faculty—attendance had fallen to about a dozen by the end of the year; see Cournot, Antoine Augustin, *Souvenirs (1760–1860)*, ed. Botinelli, E. P. (Paris: Hachette, 1913), p. 156. Victor Duruy's report of 1868 on higher education suggests that in the 1860s the composition of the audiences in the science faculties was a somewhat greater problem than their size. For example, in 1865 only 94 out of the 2,232 who attended lectures in the country's 16 science faculties were registered students; the remainder were for the most part casual students whose commitment to serious study was slight. See *Ministère de l'Instruction Publique. Statistique de l'enseignement supérieur. 1865–1868* (Paris: Imprimerie Impériale, 1868), pp. xxxii–xxxiii.

[57] See Pasteur's letter to his father, 19 December, 1855, in Vallery-Radot, Pasteur, *Pasteur. Correspondance 1840–1895*, 4 vols. (Paris: Flammarion, 1940–1951), Vol. I, pp. 383–384.

[58] On Ampère's difficulties as a lecturer see Launay, Louis de, *Le grand Ampère* (Paris: Perrin, 1925), pp. 189–191.

[59] Rémusat, Charles François Marie de, *Mémoires de ma vie*, 3 vols. (Paris: Plon, 1958–1960), Vol. I, pp. 241–244.

[60] *Ibid.*, Vol. I, p. 242.

[61] See Pasteur, Louis, " Souvenirs intimes ", *Le centenaire de l'École normale 1795–1895* (Paris: Hachette, 1895), p. 478, and the letter from Pasteur to his parents, 9 December, 1842, in Vallery-Radot, P., *Pasteur. Correspondance . . .* Vol. I, p. 81. It was some years later that Pasteur reported that audiences of over 1,000 were normal; see the letter to his father, 25 February, 1854, *ibid.*, Vol. I, p. 328.

[62] See the letter from J. N. P. Hachette to Michael Faraday, 9 July 1832, in Williams, L. P., *Selected Correspondence. . . .* Vol. I, p. 229.

not determined by the size of the audience, as it was in many European universities at this time, notably in Germany and Scotland,[63] self-esteem and the successes of their literary and political contemporaries seem to have provided the scientists with a sufficient incentive to attract the *grand public*. Trivialisation of the subject-matter was a constant danger, and scientific lectures in the faculties came to be criticised for the low level at which they were pitched.[64] Such exceptions as the lectures of Dumas at the Sorbonne, which had sufficient intellectual content to stimulate the interests of Pasteur and Sainte-Claire Deville in chemistry,[65] were rare. But well prepared popular lectures were preferable to many given in the institutions devoted to preparation for a professional qualification. At the École polytechnique, for instance, where syllabuses between the 1820s and 1850s showed a remarkable degree of inflexibility and disregard for new ideas and discoveries,[66] the quality of the teaching appears to have deteriorated markedly. In the early 1820s Biot, who had known the École polytechnique in its great days, wrote of it " descending every day to a state of uniform mediocrity, which engenders neither resistance nor noise "[67]; and 10 years later he wrote of the suspicion of intellectual activity which had been apparent in government policies since the beginning of the century and which had been manifested, to the serious detriment of science, both in the closure of the *écoles centrales* (now replaced by the more classically oriented *lycées*) and in the increasingly mundane and specialised character of the studies at the École polytechnique.[68] It is true

---

[63] However, the number of registered students was one factor that determined the *éventuel*, paid in addition to the basic salary to professors in the faculties; see *Statistique de l'enseignement supérieur* . . . pp. 342–343. On the class-fee system in German and Scottish universities see Paulsen, Friedrich, *The German Universities: Their Character and Historical Development*, trans. Perry, W. D. (New York: Macmillan, 1895), pp. 136–140, and Morrell, J. B., *op. cit.*, pp. 48–56. In the opinion of Biot, who was generally critical of the state of science in France at the time, it was one of the strengths of the French system that professors were not paid directly by their students; see his review of Babbage's *Reflections on the Decline of Science in England*, in the *Journal des savants* (January, 1831), p. 46. However, this advantage seems scarcely to have diminished the French lecturers' concern to impress their audiences.

[64] Renan, E., *op. cit.*, pp. 81–86, and Monod, Gabriel, *De la possibilité d'une réforme de l'enseignement supérieur* (Paris: Ernest Leroux, 1876), pp. 26–27. Renan was critical not only of the faculties but also of other institutions, such as the Collège de France and the Muséum d'histoire naturelle, which (unlike the École polytechnique, for example) were not concerned with giving a professional training.

[65] On these lectures see the comments of Désiré Gernez and Pasteur in *Le centenaire de l'École normale*, *op. cit.*, pp. 408 and 458.

[66] For example, it was only after 1850 that references to " electric fluid " disappeared from the annually published syllabuses of the École polytechnique. See Fox, R., " The rise and fall of Laplacian physics" . . . f.n. 142.

[67] Letter from Biot to Thomas Chalmers, *c.* 1822, in Hanna, William, *Memoirs of the Life and Writings of Thomas Chalmers, D.D. LL.D.*, 4 vols. (Edinburgh and London: Thomas Constable, 1849–1852), Vol. II, p. 15.

[68] See his review of Babbage's *Reflections*, in the *Journal des savants* (January 1831), p. 47. The specialised nature of the education at the École polytechnique was also complained of by Théodore Olivier in his *Mémoires de géométrie descriptive, théorique et appliquée* (Paris: Carillan-Goeury and Dalmont, 1851), pp. vi–xviii, though with a somewhat different thrust. Olivier's complaint was that ever since the reorganisation of the École in 1816, under the influence of Laplace, Poisson, and Cauchy, there had been an undue preoccupation with abstract mathematics, in particular with algebra, to the detriment of education in both technical subjects and the experimental sciences.

that such criticisms appear to have had little effect on recruitment, for the careers open to *polytechniciens* were, then as now, attractive ones [69]; but its continuing effectiveness in the training of professional engineers, and in the consolidation of a social and intellectual elite,[70] cannot obscure the fact that the École polytechnique had declined as a centre of innovation both in teaching and research by the middle of the century.

A growing preoccupation with the wider audiences for science, at the expense of dedication to original research, is apparent also at the Académie des sciences. Indeed, it was one of the most serious charges directed against Arago in a virulent attack by the Italian-born mathematician, Guglielmo Libri, in 1840, that, in the 10 years since his election as a permanent secretary of the Académie, Arago had consistently sacrificed accuracy and content in the interests of oratorical effect, especially in his official *éloges* of deceased members.[71] Moreover, he had aggravated the fault and had led other academicians to seek public acclaim by encouraging the publication of the weekly *Comptes rendus* of the activities of the Académie, which began to appear in 1835.[72] Biot, a member of the older generation accustomed to closed sessions to which visitors were admitted only by special invitation and whose activities were surrounded by an air of secrecy, viewed the new publication with equal distaste, and he too protested—as bitterly as he had done some years earlier when representatives of the press began to be admitted to meetings.[73] Yet Biot and Libri were not surprised by the new publicity which was accorded to the Académie and sought by Arago; both men saw it simply as part of the vulgarity that had come to afflict French science. In fact, the scientists who were so criticised by Biot and Libri were simply following the dominant style of French intellectual activity between the Empires, a style which had its most glittering successes in the rise of the Romantic movement in literature and the arts but which, as we have seen, could readily be adapted to the needs of science, at least of science in its popular form.

[69] On the continuing demand for places at the École polytechnique, see Artz, Frederick Binkerd, *The Development of Technical Education in France, 1500–1850* (Cambridge, Mass., and London: MIT Press, 1966), pp. 232–233.

[70] I use the word " consolidation " here since the great majority of students at the École polytechnique between 1815 and 1848 were the sons of men who had been successful in military or administrative careers or in the professions, while fewer than 1 in 7 were the sons of businessmen and industrialists, and scarcely any came from the *classes populaires*. See Daumard, Adeline, " Les élèves de l'École polytechnique de 1815 à 1848 ", *Revue d'histoire moderne et contemporaine*, V (1958), pp. 226–234.

[71] [Libri, Guglielmo B.I.T.], " Lettres à un Américain sur l'état des sciences en France. I. L'Institut ", *Revue des deux mondes*, 4th ser., XXI (1840), pp. 802–807. Arago was also criticised in this article for his oratorical performances in his lectures at the Observatory in Paris. It is interesting that these lectures by Arago, with those of Cuvier at the Muséum d'histoire naturelle and the Collège de France, were the very ones which had so enthralled the duchesse de Duras some 20 years earlier; see Villemain, A. F., *Souvenirs contemporains d'histoire et de littérature*, 2 vols. (Paris: Didier, 1854–1855), Vol. I, pp. 467–468.

[72] Libri, G. B. I. T., *op. cit.*, p. 795.

[73] See his comments in the *Journal des savants* for February 1837 (pp. 78–84) and November 1842 (pp. 642–661), reproduced in Biot, Jean-Baptiste, *Mélanges scientifiques et littéraires*, 3 vols. (Paris: Michel Lévy, 1858), Vol. II, pp. 257–292.

The activities of the Restoration salons provide further evidence of the extent to which the scientists of the day followed the prevailing cultural pattern. Although the evidence now available is less than conclusive, there seems little doubt that scientists became increasingly prominent in salon society after 1815. Of course, science had been by no means a stranger in the salons of earlier periods. In the eighteenth-century salons, such works as Buffon's *Histoire naturelle* had been much discussed, and the cabinets of natural curiosities and the private laboratories of that period had probably done as much to sustain polite conversation as to promote serious research.[74] During the First Empire, too, science had not been ignored in cultivated society; it was under Napoleon, for example, that the house of Madame de Rumford, the widow of Lavoisier and the unhappy wife of Count Rumford, became a fashionable centre for gatherings of men and women in Paris with scientific interests.[75] But it was as clear to the leading scientists of the Empire as it had been in the eighteenth century that the salons could never serve science in the way that they patently served the interests of politics and literature; and it is remarkable, there- fore, that after 1815, at a time when the dependence of scientific research on expensive laboratory facilities grew rapidly, so many leading men of French science were apparently beguiled by the resurgence of salon activity which followed the Restoration. Cuvier's weekly soirées were perhaps the most celebrated,[76] and it seems that no one was more assiduous than he and Alexander von Humboldt in attending the salons of others.[77] But, between the Empires, Cuvier and Humboldt were by no means the only scientists who were prominent in the Parisian salons; Maine de Biran, Baron Férussac, and Jean-Baptiste Dumas [78] were all at the centre of regular gatherings, and many others, including Arago, Fourier, and the engineer G. J. Christian, were active in salon society. Although the " scientific " groups never acquired the renown of, say, Charles Nodier's brilliant gathering of early Romantics at the Arsenal,[79] just as Madame de

[74] Taton, René (ed.), *Enseignement et diffusion des sciences en France au XVIIIe siècle* (Paris: Hermann, 1964), pp. 619–712.

[75] For a first-hand account of Madame de Rumford's salon see Rémusat, C. F. M. de, *op. cit.*, Vol. I, pp. 174–175.

[76] Cuvier's salon is particularly well described in the correspondence of the young Charles Lyell. See, for example, Lyell's letters of July and August 1823, February 1829, and October 1830, in Lyell, Katherine Mary, *Life, Letters and Journals of Sir Charles Lyell, Bart.*, 2 vols. (London: John Murray, 1881), Vol. I, pp. 125–128, 134–140, 248–251, and 306–307.

[77] Cuvier appears to have excelled in the salons and was a special favourite of the duchesse de Duras, at whose salon he mixed freely with ambassadors, men of government, and courtiers. Humboldt (" as indefatigable in society as in science ", as Villemain put it) was also much admired at the Duras salon, as he was at the salon of Madame de Montcalm, sister of the duc de Richelieu; see Villemain, A. F., *op. cit.*, Vol. I, pp. 465–467 and 477.

[78] On the circles of these three, see, respectively, Gouhier, Henri (ed.), *Maine de Biran. Journal*, 3 vols. (Neuchâtel: Éditions de la Baconnière, 1954–1957); Lyell, K. M., *op. cit.*, Vol. I, p. 303; and Vallery-Radot, P., *Pasteur. Correspondance* . . . Vol. I, p. 152.

[79] On Nodier's salon, see, for example, Salomon, Michel, *Charles Nodier et le groupe romantique d'après des documents inédits* (Paris: Perrin, 1908), pp. 116–224.

Rumford's circle never matched Madame de Récamier's, they were by no means ephemeral and were attractive to many men of distinction, like Cousin, Guizot, and Royer-Collard—who had no specialist interest in science.

It was clearly in their non-specialist character that the weaknesses of these groups, as centres of creative scientific activity, lay. At Cuvier's salon, discussion in the 1820s ranged over politics and such topics as Scott's latest novel and the Marquis Las Cases's account of Napoleon on St. Helena, while science was largely ignored [80]; and we know that Humboldt, like Cuvier, was quite as ready to discuss politics or the arts as science.[81] So even if the " scientific " salons of the 1820s, 1830s, and 1840s did not offer such fripperies as dancing and cards, as Nodier's and certain others did, they were still very different from the small, select gatherings which had met, for the purposes of serious discussion and corporate research, in Berthollet's house at Arcueil. Although most of the discussion in the Restoration salons has naturally passed unrecorded, there is no evidence that they stimulated research or that they were ever intended to do so.

Of course, the public character of so much Restoration science was not harmful in every respect. Substantial, well delivered lectures could benefit science by arousing the interest of able young men; the influence of Dumas's lectures on Pasteur and Sainte-Claire Deville, already referred to, is a good illustration. And the discussion of science in polite society, among literary men, philosophers, and politicians, helped to create a broad, popular interest of which there is much evidence in Balzac's *Comédie humaine* and in the novels of Stendahl.[82] But such benefits were slight, for generally the large lecture theatre and the non-specialist salon were not conducive to a deep commitment to research. The wit and rhetorical skills which they fostered had had no place at Arcueil, just as they had no place in the chemical laboratory which Liebig founded at Giessen in 1824 or in the other German university laboratories for which Giessen served as a model.

Penury and Indifference

In the quest for recognition by the lay public, it was serious research which suffered most. If we accept that a new, public style of science did

---

[80] It was Cuvier himself who set the tone in this respect: "Not a word could I get on natural history" was Lyell's complaint in 1829, and in the following year Cuvier was persuaded to discuss fossil anatomy only because he was "not in spirits about political affairs". See Lyell's letters to his sister and to Gideon Mantell, 23 February, 1829, and 10 October, 1830, in Lyell, K. M., *op. cit.*, Vol. I, 249 and 307. See Villemain, A. F., *op. cit.*, p. 467, on the breadth of Cuvier's conversation in the salons.

[81] Villemain, A. F., *op. cit.*, Vol. I, pp. 465–466 and 477.

[82] The scientific basis for the structure and conception of the *Comédie humaine* is described explicitly in Balzac's general preface to the series (dated July 1842), in *Oeuvres complètes de Honoré de Balzac*, 40 vols. (Paris: Louis Conard, 1926–1963), Vol. I, pp. xxv–xxxviii. Stendahl's science has been the subject of a detailed study by Jean Théodoridès; see his *Stendahl du côté de la science* (Aran: Éditions du Grand Chêne, 1972).

emerge after 1815, it is certainly not surprising that French scientists until the mid-1860s were generally so content with conditions for research which we now recognise to have been quite inadequate, when judged by the standards of the research facilities available in so many German universities by the middle of the century. Yet the paucity of public complaints [83] should not be allowed to obscure the fact that, at least until the 1850s, there was almost no institutional provision in France either for the personal work of professors or for that of advanced students wishing to perform research under their supervision. The situation was well summarised by Pasteur's assistant E. Duclaux. Referring to conditions which he observed at the École normale but which would have been found in virtually any institution of higher education about the middle of the century, he wrote:

> . . . the professors had at their disposal only the premises and funds allocated to their chair. No private laboratory; nothing resembling what today we should call a research laboratory. When they wanted to work, they exercised their wits; they skimped on the essentials in order to have something left over; they set themselves up as best they could in the quiet corner of a lecture room, always ready to clear away their apparatus at lecture times. [84]

In reading this account, it should be borne in mind that, from the time the École normale moved from the Collège du Plessis to its new premises in the rue d'Ulm in 1847, the laboratories there were considered to be the best available anywhere in France. [85] The teaching laboratories were comparatively spacious, and from 1851, when he arrived there from Besançon to take the chair of chemistry, Sainte-Claire Deville worked hard for an improvement in the facilities for the research of professors. [86] But even the energetic Deville had very modest success, and it was not until 1866 that he acquired the only laboratory which, according to Fernand Papillon, could even remotely be compared with those in the German universities. [87]

In the other institutions of Paris conditions for research were even less favourable than those at the École normale. The " little closet, a few metres square " at the Collège de France, where François Magendie performed his great physiological researches between 1830 and 1855, was not only unsuitable for the purposes of research, it was also damp and unhealthy, so much so that even the rats found it difficult to survive! There, if anywhere, Claude Bernard's description of French laboratories as

---

[83] I stress " public " here, since certain individuals, among them Pasteur, did complain privately about their conditions before the 1860s. See, for example, Pasteur's letters of 20 January, 1851, to the Minister of Education, and 12 and 14 December, 1854, to the dean of the faculty of science in Strasbourg and the Minister of Education respectively, in Vallery-Radot, P., *Pasteur. Correspondance* . . . Vol. I, pp. 217–218 and 351–355. On some other complaints dating from as early as the 1830s, see Paul, H. W., *op. cit.*, pp. 5–7. Despite this evidence, the fact remains that the complaints only became widespread, public, and concerted in the mid-1860s.

[84] Duclaux, E., " Le laboratoire de M. Pasteur ", in *Le centenaire de l'École normale* . . . p. 458.

[85] Papillon, F., *op. cit.*, pp. 601–602.

[86] On the difficulties and achievements of Deville at the École normale see Gay, Jules, *Henri Sainte-Claire Deville. Sa vie et ses travaux* (Paris: Gauthier-Villars, 1889), pp. 10–16.

[87] Papillon, F., *op. cit.*, pp. 601–602.

" the graves " of scientists had a special relevance.[88] Yet it was in this very same laboratory that Bernard himself worked, first as assistant to Magendie and then, from 1855, as his successor as professor of medicine.[89] The facilities for the important work in organic chemistry which Marcellin Berthelot performed in the " insignificant " laboratories of the École de pharmacie were equally unsatisfactory.[90] And nowhere was the scandal greater than at the ancient and revered Sorbonne. There the inadequacy of the buildings was acknowledged by the laying of the foundation stone for new premises to house all the faculties in 1855, but, in the absence of the necessary funds, it was only in 1895 that the science faculty was eventually able to move from the cramped quarters in the old Sorbonne which it had occupied since 1821.[91] In the old buildings, facilities for laboratory work had been derisory, and the establishment of modest new laboratories for physics, physiology, and chemistry in the late 1860s did little to improve conditions which were almost as bad for teaching as for research. From 1874 to 1881, as professor of organic chemistry, Adolphe Wurtz still had to prepare all his demonstration apparatus in the École de médecine and then transport it to the Sorbonne for his lectures, and it was only in 1881 that he was given a small room, vacated through the death of Deville, for the preparation of his demonstrations.[92]

Such experiences, together with accounts of Flourens's miserable *cabanon* at the Jardin des plantes and of laboratories at the École polytechnique which, even by 1871, had not changed since the days of Gay-Lussac's early triumphs some 50 years earlier,[93] all confirm that Pasteur was almost literally correct when he wrote, in 1868, that the educational budget did not contain one penny allocated for laboratory research.[94] In such circumstances the question of taking in students scarcely arose, although Wurtz, who was far more aware than most Frenchmen of the effectiveness of the German traditions of research, did seek to copy the German model in the 1850s and 1860s by accepting pupils in his far from spacious laboratory at the École de médecine.[95]

[88] *Ibid.*, p. 60, and Pasteur, L., *Le budget de la science* . . . p. 6. In support of Bernard's claim, Pasteur asserted that the professor of chemistry in the science faculty at Lyons, Amant Bineau, had died prematurely because of the conditions in the " cellar " which served as his laboratory; *ibid.*, p. 7.

[89] However, as a result of his personal approach to Napoleon III, Bernard was able to effect some improvement in his conditions.

[90] Papillon, F., *op. cit.*, p. 601.

[91] On the facilities at the Sorbonne, *ibid.*, pp. 601 and 604; Pasteur, L., *Le budget de la science* . . . p. 6; Caullery, Maurice, " La Faculté des Sciences ", in Durkheim, Émile, *et al.*, *La vie universitaire à Paris* (Paris: Armand Colin, 1918), pp. 51–55.

[92] Friedel, Charles, " Notice sur la vie et les travaux de Charles-Adolphe Wurtz ", *Bulletin de la Société chimique de Paris*, new ser., XLIII (1885), p. xx. I am grateful to my colleague, Dr. J. H. Brooke, for drawing my attention to this article.

[93] Papillon, F., *op. cit.*, p. 602.

[94] Pasteur, L., *Le budget de la science*. . . . p. 8.

[95] Friedel, C., *op. cit.*, pp. xii–xvii. Wurtz's standards and style of work, like Gerhardt's and Regnault's, were coloured by his experiences in Liebig's laboratory at Giessen, where he was a pupil in the 1840s.

If the facilities for research were poor in Paris, they were better, at least in the case of certain laboratories, than those available in the provinces, where the science faculties had failed utterly to become centres of learning. Indeed, during the Bourbon and Orléans monarchies their condition even deteriorated, and this despite the fact that they never lacked influential friends who wished to see them strengthened academically. For example, Paul-François Dubois, the joint founder of *Le Globe*, deputy for Loire-Inférieure, and from 1840 to 1850 director of the École normale, argued strongly for vigorous and adequately endowed provincial universities in the report on the budget for education which was presented in the Chamber of Deputies in 1836,[96] while in his eight months as Minister of Education in 1840 Victor Cousin set about grouping the isolated faculties into a small number of well financed universities in the main provincial centres.[97] But Dubois's views, like Guizot's 20 years before and Jean-Baptiste Dumas's during the Second Empire,[98] were not heeded, and by the time he left office Cousin had done no more than obtain the permission of the Chamber of Deputies to establish the new science faculty which he needed at Rennes for his proposed Breton university. So it was the very different philosophy of such extreme exponents of decentralisation as Villemain and Salvandy that prevailed from 1815 and ensured that faculties remained isolated from one another and poorly equipped until Liard's reorganisation of 1896.[99]

The situation was only aggravated by the creation of several new faculties in the 1830s and 1840s.[100] By the time Victor Duruy began his investigation in 1865, the state of the provincial faculties, whether of law, medicine, letters, theology, or science, was for the most part deplorable. Well housed science faculties, such as those at Caen, Nancy, and Clermont-Ferrand, were exceptions; elsewhere cramped and unhealthy conditions in poorly adapted premises were normal, and such laboratories as did exist were inadequate.[101] There was little money for the purchase of equipment and books [102] and no provision whatsoever for the private research of profes-

---

[96] Liard, L., *op. cit.*, Vol. II, pp. 183–185 and 190.

[97] For Cousin's own account, see his " Huit mois au Ministère de l'Instruction Publique ". *Revue des deux mondes*, 4th ser., XXV (1841), pp. 387–390. For a statement of his educational philosophy, see his comments in the *Revue de Paris*, XXX (1831), pp. 184–185.

[98] On Guizot see Liard, L., *op. cit.*, Vol. II, pp. 125–136 and 181–183, and my comment in f.n. 45. On Dumas, see Rohr, Jean, *Victor Duruy, ministre de Napoléon III. Essai sur la politique de l'instruction publique au temps de l'Empire libéral* (Paris: Pichon and Durand-Auzias, 1967), pp. 90–91.

[99] Liard, L., *op. cit.*, Vol. II, pp. 188–199.

[100] *Ibid.*, Vol. II, pp. 188–189. Three of the new faculties, at Lyons, Bordeaux, and Besançon, were science faculties, bringing the total to 10 by 1848. The new faculties were created largely to meet the growing need for examiners for the *baccalauréat*; *ibid.*, Vol. II, pp. 190–199.

[101] *Ibid.*, Vol. II, pp. 271–276, and *Statistique de l'enseignement supérieur. . . .* pp. 15–17, 56–59, and 416–429. On Duruy's investigation and his findings see also Rohr, J., *op. cit.*, pp. 90–109.

[102] Provincial science faculties had no more than 1,800 francs annually to cover expenses incurred in teaching and research. Often there were no funds at all for the purchase of books.

sors. Intellectual stagnation was a natural corollary of this material neglect, and it is not surprising that such a gifted and ambitious young man as Charles Gerhardt, whose education and family background in Strasbourg had brought him into close contact with Germany and who had known the delights of working with Liebig at Giessen, was so dissatisfied with the research facilities available to him as a professor at Montpellier in the 1840s.[103]

When conditions in the institutions were so bad, interest in research was not easily stimulated. But, unlike so many critics in the 1860s, I believe that the poor conditions should not be seen as a primary cause of France's modest record in research about the middle of the century. Rather they were a symptom of the indifference of French scientists themselves, who, as I have already observed, were so rarely moved to protest, at least publicly. In some cases the complacency of the French scientists took a positive form. A. J. Balard, who held chairs of chemistry successively at the École normale and the Collège de France between the 1830s and the 1870s, took a real pride in the handicaps under which he worked. And Marcellin Berthelot, while admitting the great difficulties encountered in undertaking serious research in France, declared his preference for the French system at the time of a visit to Heidelberg in 1858. In his view, the advantages of well endowed laboratories and *Lehrfreiheit* were outweighed by the low income of the *Privatdozenten* and the heavy burden of routine laboratory teaching imposed on German professors.[104] Both Berthelot and, in his less distinguished way, Balard showed by their own research that the French system could be made to work. But, unfortunately, these two men had a commitment greater than that of many of their contemporaries, who, far from being dedicated to research, seem to have been ever-ready to assume responsibilities which took them away from their laboratories.

Some of the activities which lured French scientists from research between the Empires were academic. Especially for teachers in the provincial faculties, with meagre basic salaries of 4,000 francs per annum and few serious students, examining was attractive as a source of additional income; paid at the rate of 5 and later 7 francs for each candidate in the *baccalauréat*, one week of examining could yield over 300 francs.[105] In

---

[103] Grimaux, Édouard and Gerhardt, Charles, *Charles Gerhardt. Sa vie, son oeuvre, sa correspondance 1816–1856* (Paris: Masson, 1900), pp. 13–29, 55–57, 59–61, and 93–94.

[104] See his letter to Renan, quoted in Velluz, L., *op. cit.*, pp. 45–46. Berthelot remained cautious in his criticism of the facilities for science in France, even after the campaign of protest in the 1860s (in which he did not participate) and the military defeat of 1870–1871. As late as 1882 he wrote to Michel Bréal: " Granted, it serves some purpose to stress, with the Chambers and the public, the condition of German professors, in order to improve that of French professors. But we should not be deceived by this comparison, in which the deficiencies of our own system are contrasted with the outstanding aspects of that of our neighbours ". The passage from the letter to Bréal is quoted in Paul, H. W., *op. cit.*, p. 10.

[105] Prost, Antoine, *L'enseignement en France 1800–1967* (Paris: Armand Colin, 1968), p. 227. Another supplement to the basic salary was the *éventuel*, a sum determined by the budget of the faculty in question and the number of registered students. Since the number of such students in science faculties was low—especially when compared with student

Paris, an even greater snare was *cumul*. Although it had been common enough under Napoleon I,[106] *cumul* seems to have been practised even more extensively after the Restoration, in science as in other disciplines. Of the great scientific *cumulards* the best known was perhaps Dumas, who had teaching duties at the Athénée, the École centrale des arts et manufactures, the Sorbonne, the École polytechnique, the École de médecine, and the Collège de France—though admittedly not *all* at the same time. But even Dumas, who excelled in the multiplicity of the responsibilities which he accepted, was not without his rivals; indeed, there were very few of his contemporaries in science who did not hold at least two teaching posts.[107] In doing so they not only jeopardised their research but also, as we can see from complaints that were made against them, did much harm to the career prospects of younger men.[108]

Dumas's career is especially interesting, for his decision to abandon science for politics in the exciting days of 1848 points to another, non-academic outlet for the talents of scientists. Like *cumul*, politics and public life in general had been attractive to French scientists and mathematicians before 1815; during the First Empire Monge, Chaptal, Fourcroy, and, for six unhappy weeks, Laplace had held government posts. But from the 1820s, as political activity grew in interest and intensity, scientists, like Ampère and Biot, who did not take positions in politics or in academic or government administration, became increasingly uncommon, and instances of successful scientists who sacrificed their research for public life, like Arago, Thenard, and Gay-Lussac, were correspondingly more frequent.

The diversion of interest from research in France would not have been

numbers in the faculties of law—the *éventuel* was small, exceeding 2,000 francs in only three of the country's 16 science faculties in 1865. In none of the 11 law faculties of France was the *éventuel* less than 2,150 francs. See *Statistique de l'enseignement supérieur . . .* pp. 342–343.

[106] At this time young family men, like Biot and Dulong, would have found it difficult to live on the income from only one post. On the *cumul* of scientists during the First Empire, see Crosland, M. P., *op. cit.*, pp. 224–226.

[107] See, in addition to the usual biographical sources, Crosland, M. P., *op. cit.*, pp. 226–231. Gay-Lussac, who was particularly enterprising, acted as industrial consultant as well as taking on a wide range of teaching duties. The combined salaries of Thenard, who defended the system of *cumul* against the attacks of Gerhardt in 1848, came to 30,000 francs p.a. for a few years. This sum far exceeded the 13,000–14,000 francs which, according to Thenard, a successful academic *cumulard* might expect to earn between his mid-thirties and his fifties. It is clear from his reply to Gerhardt that Thenard considered *cumul* to be normal and proper, given that salaries in teaching compared most unfavourably with the salaries of 15,000–18,000 francs which were commonly paid in government administration; see Thenard, Paul, *Un grand Français. Le chimiste Thenard* (Dijon: Jobard: 1950), pp. 206–209.

[108] Indignation against *cumul* seems to have reached a peak in 1848, when a petition on the subject was presented to the provisional government by a group of scientists and doctors. No one protested more vigorously than Charles Gerhardt, who saw *cumul* as the chief cause of his failure to obtain a teaching post in Paris; see Grimaux, E. and Gerhardt, C., *op. cit.*, pp. 172–179. As the naturalist Quatrefages observed, an unfortunate consequence of *cumul* was that appointments to the most important chairs tended to be made rather late in life; see Quatrefages de Bréau, Jean Louis Armand de, " De l'enseignement scientifique en France ", *Revue des deux mondes*, 5th ser., XXII (1848), pp. 501–502.

so remarkable and damaging had it not become widespread just at a time, in the 1830s and 1840s, when research in the German universities was beginning to flourish. And how ironical this was, for the German achievements owed much to French science. For example, no one did more to publicise the case for experimental research and exact science in a Germany dominated by *Naturphilosophie* than did Alexander von Humbo!dt, whose observations were based almost entirely on his experiences as one of the most prominent members of the Arcueil circle during the First Empire.[109] It was through working in Gay-Lussac's laboratory in 1824 that Liebig was inspired by the ideal of exact experimental science for which he provided such a favourable institutional setting at Giessen.[110] But, as research became increasingly integrated in education in the German universities,[111] the French maintained their rigid distinction between the conveying of established truths and the acquisition of new knowledge, and, at least in the institutions for higher education in science, largely ignored the latter. While French lecturers strove for oratorical effect, German professors developed the severe style of which Ernest Renan and George Pouchet thought so highly.[112] And even when the effectiveness of German research and research schools was beyond dispute and when students throughout Europe were flocking to the German universities, the French remained unimpressed—so much so that, of the 169 matriculated students who came from outside Germany to work in Liebig's laboratory at Giessen between 1829 and 1850, only 22 came from France, and of these no fewer than 14 came from Alsace, where German influences were unusually strong.[113] Frenchmen willing to express admiration for German research and scholarship, in the way that Victor Cousin did as a result of the tours in Germany and Holland which he made after the Restoration, were decidedly in a minority until the 1860s.[114]

Awareness of Governmental Neglect

The complacency of the French lasted far into the Second Empire, until, almost without warning, in the mid-1860s, public complaints about the

---

[109] See Crosland, M. P., *op. cit.*, pp. 104–113.

[110] On the importance of Liebig's French experiences for his later work, see the autobiographical passage quoted in Merz, John Theodore, *A History of European Thought in the Nineteenth Century*, 4 vols. (Edinburgh and London: Blackwood, 1896–1914), Vol. I, pp. 190–191.

[111] See, on this process, Paulsen, F., *op. cit.*, pp. 79–88.

[112] See Renan, E., *op. cit.*, p. 83 and Pouchet, G., *op. cit.*, pp. 443–446.

[113] Wankmüller, Armin, " Ausländische Studierende der Pharmazie und Chemie bei Liebig in Giessen ", *Tübinger Apothekengeschichtliche Abhandlungen*, Heft 15 (Stuttgart, 1966), pp. 8–9 and 11–16. I am indebted to Dr. W. H. Brock for this reference.

[114] For Cousin's view of German university life see his " Lettres à M. le comte de Montalivet, Ministre de l'Instruction Publique et des Cultes, sur l'état de l'instruction publique en Allemagne ", *Revue de Paris*, XXIX (1831), pp. 15–37, 74–107, and 209–240; *ibid.*, XX (1831), pp. 101–117 and 164–186; also the later versions of these articles, published as *Rapport sur l'état de l'instruction publique dans quelques pays de l'Allemagne, et particulièrement en Prusse* (Paris: Levrault, 1833); 3rd edn., 2 vols. (Paris: Pitois-

state of the country's science became both frequent and bitter. The complaints were nearly always presented in the form of an explicit and unfavourable comparison with Germany, and this is not surprising since by that time the French had had occasion to reflect not only on the effectiveness of German higher education, but also on the growing political and military power of Prussia, especially after her crushing victory over Austria in 1866. In such circumstances, it was natural that the case for a greater provision for scientific research should normally be argued with reference to the national interest. Renan's much-quoted assertion that it was the science of Germany which had triumphed at Sadowa was probably in the minds of many of the critics,[115] and after the tragedy of the Franco-Prussian war both Sainte-Claire Deville and Pasteur put at least some of the blame for France's defeat on her neglect of science.[116]

The long-standing lack of government support for research was most frequently blamed in the late 1860s, and there can be no denying that the complaints had substance. For although French governments had not neglected research entirely—the Bureau des longitudes is a good example of an official research institution—the motives for official sponsorship had always been, understandably, utilitarian.[117] For instance, the only special research projects in physics to be financed by government funds between 1815 and 1870 were the two investigations of the properties of steam and gases which Dulong, between 1823 and 1829, and Regnault, between 1840 and 1870, were commissioned to conduct.[118] In both cases the patronage was lavish enough to allow the performance of experiments other than those that were asked for, and results of some theoretical importance were obtained.[119] But the purpose was clearly and explicitly to provide information that could be put to immediate use, whether (as in the case of Dulong's investigation) it was in order to enhance the safety of steam-engines, or (as in the case of Regnault's) it was to increase their efficiency.

Levrault, 1840). This form of admiration for scholarship and research and the institutions which fostered them should not be confused with the Romantic interest in German literature and philosophy which was common at this time.

[115] Wurtz, for example, argued strongly for the connection between the intellectual attainments of a nation and its political and military strength. See Wurtz, C. A., *op. cit.*, pp. 12–13.

[116] For Deville's comment, see *Comptes rendus hebdomadaires des séances de l'Académie des sciences*, LXXII (1871), pp. 237–238. For Pasteur's, see Vallery-Radot, R., *La vie de Pasteur* . . . pp. 277–280.

[117] Research institutions which failed to show their usefulness in the way that the Bureau did were, by comparison, sadly neglected. The Muséum d'histoire naturelle, where a state of extreme delapidation was exposed by a government committee of enquiry in 1858–1859, is a good illustration of this; see Du Camp, Maxime, " Les écoles à Paris ", *Revue des deux mondes*, 2e période, CIII (1873), pp. 823–825. Although an enquiry in 1863 had confirmed the findings of the earlier committee, showing that the gardens, specimens, and live animals were still in a deplorable condition, there had been little improvement by 1872, when Du Camp made his critical observations.

[118] On this research see Fox, R., *Caloric Theory of Gases.* . . . pp. 249–250 and 297–303.

[119] Regnault was even able to take in students at his laboratory in the Collège de France. See *Mémoires de l'Académie des sciences*, XXVI (1862), p. ix. The young William Thomson worked there in 1845.

466

If cases of government intervention in support of research were few between 1815 and 1870, the reason was that science so rarely appeared to be useful. But also science appeared to do little harm, and when, very occasionally, measures unfavourable to science were taken, they were the result of short-term political considerations rather than of a consistently hostile policy. This was certainly true of the discrediting of mathematicians and scientists that took place immediately after the Bourbon Restoration, for science at that time was all too easily charged with being a product of the revolutionary and Napoleonic periods.[120] Indeed, the elite of scientific apprentices, the *polytechniciens*, had aroused the justified suspicion of royalist sympathisers by fighting ardently in the defence of Paris in 1814 and by rallying joyfully to Napoleon during the Hundred Days.[121] So it was only to be expected that after the Restoration science would suffer and that scientists would be scrutinised with special care.

By the 1820s, a much greater threat to the political and religious ideals of the Restoration seemed to lie in the eclectic philosophy of Cousin and in the prevailing anti-clericalism of the teachers in secondary and higher education.[122] As lay and clerical officials recognised the new threat, so the radical associations of science began to be forgotten. It is entirely characteristic of these changing attitudes that the government and the newly created Grand Master of the University, Monseigneur Frayssinous, demonstrated their suspicion of subversive intellectual activity by closing the École normale in 1822, while the specialist institutions of science went almost unscathed.[123] To a government concerned to maintain loyalty to the Bourbons and the church, science appeared neutral, and, once the early years of the Restoration were past, it appears to have lost its neutrality only when it became involved in movements for popular education, which in the 1820s aroused as much fear among the legitimists of France as they did among the Tories of Britain.[124]

Of the targets for suspicion of this kind, the most obvious was the Conservatoire des arts et métiers, founded in 1794 as an industrial museum

[120] I have already referred to some of the scientists and mathematicians of the Restoration who lost their positions in teaching and the Académie des sciences for political reasons.

[121] Pinet, G., *op. cit.*, pp. 73–92.

[122] On the conflicts between the administrators and the predominantly liberal, anti-clerical teaching body—among whom, in Paris, at least, the *normaliens* were the most influential, see Gerbod, Paul, " La vie universitaire à Paris sous la Restauration de 1820 à 1830 ", *Revue d'histoire moderne et contemporaine*, XIII (1966), pp. 5–48. The suspension of Cousin from his chair at the Sorbonne between 1820 and 1828 was characteristic of this period, as was the determination with which the church gained control of the new *agrégation* in philosophy in 1825.

[123] On this closure, which lasted until the École normale was re-opened in 1826 as the École préparatoire, see Liard, L., *op. cit.*, Vol. II, pp. 164–166, and *Le centenaire de l'École normale* . . . pp. 222–228. The only purely scientific institution to suffer at this time was the École de médecine in Paris, where courses were suspended for several months in March 1822; see Liard, L., *op. cit.*, Vol. II, pp. 162–163.

[124] On the fear of the Mechanics' Institutes in England, see Cardwell, Donald S. L., *The Organisation of Science in England*, 2nd edn. (London: Heinemann, 1972), pp. 41–42, and the anonymous attempt to allay the fears of "A Country Gentleman", in the *Edinburgh Review*, XLV (1826–1827), pp. 189–199.

but far more important since 1819 as an institution offering part-time education in technological subjects to working men. Because of its concern with technology and industry, the Conservatoire had liberal and Saint-Simonian associations of which none of the reactionary governments of the 1820s could possibly approve. Suspicion was greatly heightened by the fact that, as far as its teaching function was concerned, it was the creation of the same Charles Dupin who, almost alone, had defended Lazare Carnot when Carnot was exiled from France shortly after the downfall of Napoleon [125] and who, especially during the ministry of the duc de Richelieu (1815–1818), had done so much to arouse the hostility of ministers, in whose eyes his brilliant, first-hand investigation of the industrial, commercial, and military strength of Britain was subversive and unpatriotic.[126] The professors of the Conservatoire included not only Dupin but also Jean-Baptiste Say, professor of industrial economy and, like Dupin, no friend of reactionary governments.[127] They were therefore unusually vulnerable to official censure and they were perhaps fortunate to suffer no more than mild harassment by the police and a withdrawal of permission to give their lectures in the evenings.[128] But the important point is that the motives for the harsh treatment of the Conservatoire were political and only incidentally concerned with science itself.

So after the Restoration science proceeded against a background of government indifference rather than of special favour or hostility. Modest provision was made for what government officials regarded as essential needs, which were invariably those of the unambitious teacher rather than those of the research scientist aspiring to an international reputation. It is understandable that the twenty-one-year-old Pasteur, fresh from Besançon, was impressed by the facilities available to him as a student at the École

[125] On this episode, see Lacaine, A. Victor and Laurent, H. Charles, *Biographies et nécrologies des hommes marquants du XIXe siècle*, 7 vols. (Paris: À la Direction, 1844–1850), Vol. IV, p. 279, and Bertrand, Joseph, *Éloges académiques* (Paris: Hachette, 1890), pp. 234–236. Dupin's sympathy for those who suffered at the Restoration was demonstrated also by his *Essai historique sur les services et les travaux scientifiques de Gaspard Monge* (Paris: Bachelier, 1819), in which he championed Monge, his teacher at the École polytechnique from 1801 to 1803.

[126] On Dupin's tours of Britain and the official impediments to his work that he encountered in France, see especially Lacaine, A. V. and Laurent, H. C., *op. cit.*, Vol. IV, pp. 280–284. Although Dupin had sought permission to visit Britain since the fall of the Empire, it was only in the summer of 1816 that permission was granted. For some three years thereafter Dupin was suspect, and his earliest published accounts of his observations, notably his *Mémoires sur la marine et les ponts et chaussées de France et d'Angleterre* (Paris: Bachelier, 1818), were temporarily withdrawn from naval and military libraries. It was, significantly, during the ministry of the mildly liberal duc de Decazes that the Conservatoire began to offer lectures. A study of the Conservatoire's early years which pays special attention to the political background is my article "Education for a New Age: the Conservatoire des arts et métiers, 1815–1830", in Cardwell, Donald S. L. (ed.), *University of Manchester Institute of Science and Technology. Commemorative Essays, 1824–1974* (Manchester: Manchester University Press, to appear in 1974).

[127] On the police surveillance of Say and Dupin, who were both considered to be extremely suspect, see Dreyfus, Ferdinand, *Un philanthrope d'autrefois. La Rochefoucauld-Liancourt 1747–1827* (Paris: Plon, 1903), pp. 415–416.

[128] On the difficulties of the Conservatoire, see Dupin, F. P. Charles, *Forces productives et commerciales de la France*, 2 vols. (Paris: Bachelier, 1827), Vol. II, pp. 232–233, and Fox, R., "Education for a New Age . . .".

normale in 1843 [129]—but it is equally understandable that, by 1851, when he was teaching in the science faculty at Strasbourg (generally considered to have been the best of the provincial science faculties in the nineteenth century), he felt greatly constrained by an annual budget for chemistry of between 1,300 and 1,400 francs, of which 400 francs went immediately to pay the laboratory assistant.[130] At Strasbourg, as elsewhere, there was no provision for research.

It is important to stress that poor conditions and lack of encouragement did not make research impossible. The teaching load was not intolerable in any of the institutions of higher education, although the practice of *cumul* could easily make it appear so [131]; C. M. Despretz, for many years professor of physics at the lycée Henri IV in Paris, showed that serious research could even be undertaken by professors in *lycées*. In all such work, however, strong individual initiative and dedication were indispensable, and, especially for teachers in the faculties, it was always tempting to abandon laboratory work, which brought no monetary reward and often little prestige or advancement, in favour of extra teaching or examining, which was at least lucrative.[132] In these circumstances, it was only to be expected that those who excelled in research were often men who possessed not only exceptional personal ambition and ability but also private means. Claude Bernard, good entrepreneur that he was, gained his support first by contracting a judicious marriage and later by making a personal approach to Napoleon III.[133] More often the source of finance was a man's own family, as in the cases of Berthelot, Sainte-Claire Deville, and Hippolyte Fizeau, all of whom had fathers able to support them for some years in their important early researches.[134]

Those who were less fortunate had to finance their research from their earned incomes, and when we consider the personal sacrifice which this

[129] See his letter of 11 November, 1843, to his parents, in Vallery-Radot, P., *Pasteur. Correspondance* . . . Vol. I, pp. 92–93.

[130] See his letter of 20 January, 1851, to the Minister of Education (Parieu), *ibid.*, Vol. I, pp. 217–218.

[131] Dulong was someone who suffered in this way. See Fox, R., *Caloric Theory of Gases* . . . pp. 255–256.

[132] In Quatrefages de Bréau, J. L. A. de, *op. cit.*, pp. 500–501, it is stated that the temptation was especially great in the provincial science faculties, where professorial incomes were normally lower than in the faculties of law or medicine. Certainly law and medicine provided more opportunities for extra-mural earnings, but Duruy's enquiry of 1865 showed that the incomes of professors in the provincial faculties of law, medicine, science, and letters did not differ greatly, nearly all of them lying between 5,000 and 7,000 francs; see *Statistique de l'enseignement supérieur* . . . pp. 342–343.

[133] Olmsted, J. M. D. and Olmsted, E. H., *op. cit.*, pp. 46–47 and 126–127.

[134] On Berthelot's background, see the article on him, by M. P. Crosland, in Gillispie, Charles Coulston (ed.), *Dictionary of Scientific Biography* (New York: Charles Scribner's Sons, 1970 in progress), Vol. II, p. 63. On Sainte-Claire Deville, see Gernez's comments in *Le centenaire de l'École normale* . . . p. 409. In his early work Fizeau was supported by his father, a professor at the École de médecine in Paris. Although he performed some of his most important work at the Paris Observatory, he never held a permanent full-time post and he continued to finance most of his research privately throughout his life. See Cornu, Marie Alfred, "Notice sur l'oeuvre scientifique de H. Fizeau", *Annuaire pour l'an 1898*, *publié par le Bureau des longitudes* (Paris: 1898), especially pp. C.2–3.

entailed, French achievements in research between the 1820s and 1860s seem remarkable. One unidentified foreign visitor to a Parisian laboratory about the middle of the century wrote: " I used to respect your work; it appeared great. Now that I know the material resources at your disposal, I wonder at it." [135] In the great majority of French laboratories admiration would have been fully justified. Nowhere was this truer than in the private laboratory in the rue Cuvier where Dumas performed the research which did so much to establish the science of organic chemistry in the 1830s and 1840s, and where from 1832 to 1848 he offered one of the few courses of laboratory instruction available in France between the Empires.[136] Other private laboratories, such as those of J. B. J. D. Boussingault, Fizeau, J. T. Pelouze, Léon Foucault (in his own residence in the rue d'Assas), and the physiologist E. J. Marey (in the attic of an abandoned theatre), were generally more modest,[137] and, since they offered no systematic courses of instruction, students were very rarely admitted to them. The difficulties of the young Thomas Andrews, professor of chemistry at Queen's College, Belfast, from 1849 to 1879, whose vain search for a course of laboratory instruction in chemistry ended in Paris only when he made a personal contact with Dumas,[138] would have been all too common for a visitor to France about 1830.

Thus by the middle of the century scientific research was by no means flourishing in France—a fact which was especially evident in the physical sciences, where the development of thermodynamics, the kinetic theory of gases, and electro-magnetism was proceeding in the almost total absence of French contributions. However, without the enterprise and dedication of a few individuals like Dumas, Pasteur, and Wurtz, the situation would certainly have been very much worse. Naturally this leads us to the question why those who did perform serious research at this time made the sacrifices which were necessary in order to do so. The most pressing motive, at least for younger men, seems to have been the desire for professional advancement, usually in the form of an appointment to one of the famous and comparatively lucrative teaching posts in Paris. As ambitious men like Pasteur knew well, the preoccupation with teaching in the institutions for higher education in no way diminished the importance of research as a means of demonstrating one's competence to an appointing committee. The mere desire for renown in the national and international scientific community was another incentive which might have been expected to weigh heavily, but, in keeping with my analysis of the new style of science after

[135] Quoted in Pasteur, L., *Le budget de la science.* . . . p. 10, and Papillon, F., *op. cit.*, p. 609.

[136] Wurtz was among those who worked in Dumas's laboratory in the 1840s. On him and other students who attended about the same time, see Friedel, C., *op. cit.*, p. ix.

[137] See Papillon, F., *op. cit.*, pp. 594, 604 and 607.

[138] *The Scientific Papers of the late Thomas Andrews, M.D., F.R.S.* (London: Macmillan, 1889), pp. xi-xii. Dumas had only one other pupil, an American, in his laboratory at the time.

the Restoration, I see it as one of the saddest aspects of nineteenth-century French science that this particular incentive was of slight importance. As the chief evidence for this, I would point once again to the frequency with which scientists like Arago, Gay-Lussac, Thenard, Dumas, Berthelot, and Joseph Bertrand achieved eminence by their record in research but then, rather than building a research school in the German manner, used their eminence in a second career unrelated to research or, in many cases, even to science.

I believe, then, that there can be no doubt concerning the great importance of the determination and initiative of a few individuals in nineteenth-century French science. The point is reinforced if we examine developments in scientific and technological education after the Restoration, for these were no less dependent on private enterprise than was research. For example, at the Conservatoire des arts et métiers the appointment of three professors and the establishment of courses of public lectures in 1819 followed an approach by Charles Dupin to the duc Decazes.[139] Similarly the founding of the École centrale des arts et manufactures and the Association polytechnique in 1829 and 1830 resulted from the activities of small groups of, respectively, capitalists and graduates of the École polytechnique.[140] Even the government reforms of the 1860s, notably the founding of the research-oriented École pratique des hautes études in 1868, could never have come about but for the personal convictions of Victor Duruy, whose policies as Minister of Education from 1863 to 1869 reflected his whole-hearted support for the scientists' demands for improved research facilities.[141]

By the time he left the Ministry, Duruy had gone some way towards fulfilling his self-imposed task of providing government support for research on a totally new scale. The Sorbonne, for example, had new laboratories for physics, chemistry, and physiology, all established with Duruy's backing,[142] and the École pratique des hautes études, although no more than an administrative union of teachers and existing facilities for research, did at least strengthen the claims of laboratories and their users in the struggle for resources. But even while the energetic Duruy was minister, his schemes were always hindered by the lack of funds, and the disasters which followed in 1870–1871 so inhibited reform after his departure from the Ministry that many of the complaints of the 1860s, especially those concerning the faculties, continued to be voiced throughout the rest of the

---

139 On the establishment of lectures at the Conservatoire see, for example, *Ministère de l'Éducation Nationale. Cent-cinquante ans de haut enseignement technique au Conservatoire national des arts et métiers* (Paris: Conservatoire national des arts et métiers, 1970), p. 21.

140 See Comberousse, Charles de, *Histoire de l'École centrale des arts et manufactures depuis sa fondation jusqu'à ce jour* (Paris: Gauthier-Villars, 1879), pp. 10–36, and *Histoire de l'Association polytechnique et du développement de l'instruction populaire en France* (Paris: Imprimerie et Librairie Centrales des Chemins de Fer, 1880), pp. 33–40.

141 On Duruy's role in the founding of the École pratique see Liard, L., *op. cit.*, Vol. II, pp. 286–295, and Rohr, J., *op. cit.*, pp. 116–121.

142 Papillon, F., *op. cit.*, pp. 596, 601, 604, and 607.

century. It is perhaps not surprising that Gabriel Monod should have still been complaining bitterly of the state of the faculties in the 1870s, in particular of their isolation from one another and the popular nature of their teaching;[143] but what is striking and significant is that 30 years later the historian Ferdinand Lot could still describe the faculties, now reorganised into 15 universities, in a way which suggests that there had been little improvement in their condition, either materially or intellectually.[144]

Even in the early 1900s, it seems, the need for the individual scientist to create his own opportunities in the face of financial deprivation and an almost total absence of official encouragement was scarcely diminished.

Conclusion

In this paper I have examined the way in which an ideal of corporate, planned research first flourished, in Napoleonic France, and was then abandoned in the 50 years following the overthrow of Napoleon. I have tried to show that France's successes in scientific research under Napoleon and her comparative quiescence from the 1820s cannot be explained adequately in terms of either the generosity or the parsimony of government sponsorship. Of course, government initiative in, say, the 1840s, when the superiority of the facilities in German laboratories was becoming so very marked, would have been of great benefit, and it is tempting to censure successive ministers of education for their failure to act. But why should a government have taken such an initiative at that time? Even in Germany governments made their provision for research largely in response to external pressure, such as Liebig for one never ceased to exert[145]; the same was true *a fortiori* in Britain, where the individualist tradition in scholarship was strongest.[146] What France by contrast patently lacked was a scientific community ready to assert its demands persistently and publicly on behalf of science.

Since I believe that the role of individual initiative has not received the attention it deserves from historians of French science, I have discussed in some detail the highly personal motives which led Laplace and Berthollet to act as patrons in the way they did. Clearly their success as patrons owed

---

[143] Monod, G., *De la possibilité d'une réforme de l'enseignement supérieur, passim.*

[144] Lot, Ferdinand, *De la situation faite à l'enseignement supérieur en France*, 2 vols. (Paris: Cahiers de la Quinzaine, 1906). Similarly in his *Les hautes études pratiques dans les universités d'Allemagne et d'Autriche* (Paris: Masson, 1882), pp. 1–3, Wurtz wrote of a number of laudable projects, such as the rebuilding of the science faculties in Lyons and Bordeaux, but he made it clear that, with the notable exception of the enlargement of the faculty of medicine in Paris, little progress had been made since his earlier report in 1870.

[145] The importance of Liebig's enterprise is brought out well in Morrell, J. B., "The Chemist-Breeders: the Research Schools of Liebig and Thomas Thomson", *Ambix*, XIX (1972), pp. 1–46.

[146] See Morrell, J. B., "Individualism and the Structure of British Science in 1830", *Historical Studies in the Physical Sciences*, III (1971), pp. 183–204.

a great deal to facilities which were dependent on government funds. But for the training at the École polytechnique, where would they have obtained the supply of recruits for their school? And without the financial support and prospects of good careers which came of victory in the competitions of the Académie, how could they have hoped to engage gifted young men in long and expensive research? Yet, for all this, I believe that the achievements of Laplace and Berthollet, though obviously dependent upon the facilities which were available under Napoleon, owed far more to the skill and determination with which the facilities were utilised. The importance of individual initiative becomes still clearer when we make the comparison with the years after the collapse of the Arcueil school. In these years, despite a large measure of institutional continuity with the period before 1815, there were no schools of research in France, no one willing to take on the mantle of the masters of Arcueil.[147]

In attempting to explain this striking contrast I have paid special attention to a change in the style of French intellectual life after the Restoration. In my view, this change, above all else, was responsible for the ease with which the interests of even established scientists were diverted to popular lecturing and public life and for the preoccupation with teaching and success in examinations, such as the highly esteemed *agrégation*, which characterised institutions which might otherwise have become centres of research.[148] In putting this case I do not claim to have considered every possible cause of the decline of the research ideal in France. How, for example, are we to gauge the effect of the premature deaths of such men as Petit, in 1820, and Fresnel, in 1827, who might conceivably have filled the gap left by the great Napoleonic patrons? And what importance should be attached to the size of personal fortunes? Certainly incomes had some effect, at least on the style of patronage, for after the Restoration, when an annual income of 30,000 francs was exceptional and could only be earned at the expense of time for personal research, and when in any case laboratory research was becoming more costly and complex, it was out of the question for any French scientist privately to offer facilities which

---

[147] There were still, of course, traditions of research, notably in the work of Gabriel Lamé and J. M. C. Duhamel, which owed much to Fourier. And such "elder statesmen" of French science as Fourier, Ampère, and Arago continued to help younger men. The mathematician C. F. Sturm, for example, was greatly helped in his career through influence of this kind; see Speziali, Pierre, *Charles-François Sturm (1803–1855). Documents inédits* (Paris: Palais de la découverte, 1964). But the unity and close supervision that were characteristic of the Arcueil school were absent.

[148] For Renan, the École normale and the *agrégation* were the chief culprits in diverting the attentions of France's intellectual elite from original research. *Agrégés*, educated by "lengthy exercises in the skills of rhetoric", dominated secondary and higher education both in administration and teaching, and between 1821 and 1842 well over one third of all the *agrégés* were also *normaliens*. See Renan, Ernest, *op. cit.*, pp. 84–86, and Gerbod, Paul, *La condition universitaire en France au XIXe siècle* (Paris: Presses Universitaires de France, 1965), pp. 62–64. For Lot, writing in 1906, the *agrégation* was the "intellectual scourge" of France, which both discouraged research and failed to select those who would become the best teachers; see Lot, F., *op. cit.*, Vol. II, pp. 148 and 162–166.

were even remotely comparable with those which Berthollet and Laplace had offered at Arcueil.[149]

Perhaps also we should take an internalist view and consider the possibility that the eminence of such men as Fourier, Cauchy, Navier, and Lamé in mathematical physics diverted attention to mathematical problems and made laboratory research and its shortcomings appear unimportant.[150] A lack of enthusiasm for research may also have been one unhappy consequence of the centralisation of French higher education, which gave no encouragement to the invigorating spirit of rivalry which was so strong in the German universities.[151] And, finally, we might consider the effect of secondary education, which between the Empires was dominated by an abiding preoccupation with the classics, to the undoubted detriment of science.[152] At the very least, the classical bias of the *lycées* probably contributed to a diversion of able students from science and hence to a weakening of the scientific community.[153]

Even if my characterisation of the public, rhetorical style of post-Restoration science, with its accompanying attitudes towards research, is accepted, how are we to explain the readiness of French scientists to adopt this style when attitudes in Germany were so different? Perhaps, on this point, Renan gave as good an explanation as we can ever hope to have when he stated that it was simply a question of national characteristics, a consequence of " the peculiar character of the French mind." [154]

---

[149] For a comparison of post-Restoration incomes with those of Berthollet and Laplace see footnotes 34, 49, and 107.

[150] Théodore Olivier certainly believed that a growing concern for mathematics harmed experimental science at the École polytechnique after 1816. Olivier, Théodore, *op. cit.*, pp. vi–xviii.

[151] The centralisation of French science was much criticised about 1870, not only in France but also abroad. See, for example, Playfair, L., *op. cit.*, pp. 7–11. Some of the harmful effects of centralisation are discussed in Ben-David, J., *op. cit.*, pp. 174–179, and in the same author's " Scientific Productivity and Academic Organization in Nineteenth Century Medicine ", *American Sociological Review*, XXV (1960), pp. 828–843.

[152] The harmful effects of the classical bias in secondary education were stressed in Dumas, Jean-Baptiste André, *et al.*, *Ministère de l'Instruction Publique. Rapport sur l'enseignement scientifique dans les colléges, les écoles intermédiaires et les écoles primaires* (Paris: P. Dupont, 1847).

[153] In this respect the effect was probably similar to that which the Romantic movement is sometimes thought to have had. On the latter, see Herivel, J. W., "Aspects of French Theoretical Physics in the Nineteenth Century ", *The British Journal for the History of Science*, III (1966–1967), pp. 116–118.

[154] Renan, E., *op. cit.*, p. 84.

II

The *savant* confronts his peers: scientific societies in France, 1815–1914

As it was used by the Ministry of Public Instruction in the nineteenth century, the category of *sociétés savantes* was both capacious and flexible. It included several hundred officially recognized societies throughout France, with interests that ranged widely in the pure and applied sciences, the social sciences, the arts, and the humanities. All the societies possessed a significant, if not primary commitment to the advancement and diffusion of knowledge of one form or another, which they pursued through meetings and, in the great majority of cases, publications. However, the societies were by no means all 'learned' in the sense which their collective title might imply. Though many pursued predominantly intellectual objectives, few did so exclusively, and some, like the industrial societies, were explicitly utilitarian in their aims and ideology.[1] Hence, despite the rather vague corporate sense which it helped to promote, recognition by the Ministry was entirely compatible with a variety of functions that extended far beyond the scholarly or scientific.

Moreover, the loose affiliation with the Ministry did little to lessen the determined, even aggressive independence of the societies. Throughout the nineteenth century, the *sociétés savantes* formed a distinctly uneasy appendage to the network of institutions under the control of the Ministry of Public Instruction. Though never quite divorced from the network, they were — at

It is a pleasure to thank the Royal Society of London for two grants which allowed me to make several visits to French libraries and archives between 1976 and 1978.

The later versions of this paper were written while I was a Fellow at the Shelby Cullom Davis Center, Princeton University, in the Fall Term 1978. I wish to thank the Director of the Center, Professor Lawrence Stone, for inviting me to Princeton, to work in the programme on the history of the professions, and the members of the Davis Center Seminar for their helpful response to the text I presented on 10 November 1978.

I am also grateful for the criticisms I have received from Dr E. R. B. Gibson of the University of Lancaster, and participants in seminars in the Committee on Social Thought of the University of Chicago, the University of Pennsylvania, and the Université de Montréal, where I discussed some of the material in November and December 1978.

1. Two important categories of society that were not normally classed as *sociétés savantes* for ministerial purposes were the agricultural societies (the few that were recognized had interests extending significantly beyond agricultural improvement and the organization of competitions) and the societies of *anciens élèves* of the various educational institutions.

least in the sciences — seldom at the forefront, and their indigenous, voluntarist character made them something of an oddity, hovering on the fringes of respectability. Of the exceptions to this generalization, the most obvious are the Academy of Sciences, which, with the four other national academies, made up the Institut de France, and the Academy of Medicine; there are also partial exceptions among the national disciplinary societies, some of which became important vehicles for the publication of research late in the nineteenth century. But, for all these exceptions, most societies were, at best, marginal to the national system of research and education.

It is entirely consistent with this status that the members of the *sociétés savantes* were, by and large, self-taught in their chosen interest. They rarely possessed the qualifications that were required for an appointment in the Université[2] — the familiar hierarchy of *baccalauréat, licence, agrégation*, and doctorate — and their activities were oriented very firmly towards bourgeois enthusiasms — natural history, regional topography, local history and archaeology, and, later in the century, geography — that were peripheral to the core of traditional subjects on which the curricula of faculties and *lycées* alike were founded. So competently did the best of the societies colonize these disregarded areas that, for much of the nineteenth century, a whole sector of rather pedestrian research was effectively given over to them. But from about 1860, as botany, zoology, and geology emerged as major university disciplines (along with history, the social sciences, and other 'modern' subjects), the independent *savant*'s realm of acknowledged competence shrank, until, by the early years of this century, it embraced little more than the most routine forms of description and collecting.

The diminishing status of the non-professional in French science is an underlying theme of this paper. It is especially relevant to the first two sections, in which I discuss the 'local' societies — that is, societies which existed, above all, to promote the cultural life or material well-being of a town or region, with little concern for national exposure. In section 1, I review the work of these societies, stressing the gulf that separated most of them not only from the world of the *universitaire* but also from the handful of socially exclusive provincial academies which survived as incongruous relics of the *siècle des lumières*. As the hierarchical structures of eighteenth-century culture receded into irrelevance, it was increasingly by the standards of the career academic, with his bureaucratic qualifications and his sense of professional identity, that the nineteenth-century *sociétés savantes* had to measure their attainments. In science, the new touchstone of competence had the predictable effect of detaching the local societies from the recognized mainstreams of research. As I show in section 2, although the process

2. Here and throughout the paper I use the term Université to describe the unified national system of faculties and *lycées*. Teachers in both faculties and *lycées* are described as *universitaires*.

of detachment was a protracted one, with patterns varying from discipline to discipline, the threat to the societies was perceived at an early stage, and, for some forty years, beginning in the 1830s, determined provincials strove to promote the societies as an autonomous sector of scientific research and scholarship that might claim parity with the official institutions. So vigorous was this rearguard action and so vehement its commitment to intellectual decentralization that the Ministry, already sensitive to clerical threats to the state's monopoly in education, could only regard it as provocative; as a result, the *sociétés savantes* became the focus for one of the most intense of mid-century debates about the proper extent of ministerial prerogatives.

In sections 3 and 4, I turn to the disciplinary societies which sought national standing and the right to speak for all the serious practitioners of the disciplines they represented. In these societies, the involvement of the 'savants officiels' was not a threat but a prize — though, as I argue in section 3, a prize not easily won. It was by no means obvious to, say, a professor of physics at the Sorbonne, or even to his assistants, that societies performed any useful function that was not already performed by his institution or the Ministry. But such circumscribed functions as there were did appreciate from the middle of the century, as research and publication came to be recognized as a primary, rather than a secondary concern of the professional academic. As I argue in section 4, it was through the exploitation, first, of the growth of the research ideal in a rapidly expanding academic profession, and, then, of a variety of opportunities raised by industrialization and a new commitment to international co-operation, that bodies like the Société Chimique de France and the peripatetic Association Française pour l'Avancement des Sciences eventually achieved a limited intellectual prominence and great material prosperity.

Finally, by way of introduction, I stress the obvious point that only a small minority of *sociétés savantes* were communities of professional *savants* engaged in academic careers. By the end of the nineteenth century, the minority included a handful of societies with truly distinguished membership lists, most notably perhaps the Société Française de Physique; but even this society was in no sense dominant in research and publication at the highest level, and, while it embraced the élite of French physicists, it was largely irrelevant to the fashioning of their careers. To say this is, of course, in no way to belittle the achievements of the *sociétés savantes*. For the criteria by which we should assess them are emphatically not those of modern professional science. If, as I would argue, most of the members of the great majority of societies were the recipients of what, in the British context, has been aptly called 'ornamental learning',[3] we need not to dismiss

3. The notion of 'ornamental learning' is used to particularly good effect in Arnold Thackray's suggestive paper 'Natural knowledge in its cultural context: the Manchester model'. *American historical review*, 79 (1974), 672–709 (685).

them but to analyse the values and diverse local conditions which made that sort of involvement in the world of learning attractive.

This said, however, the existence and the changing character of the academic profession did impinge crucially on the fortunes of the societies. By the time the societies began to be founded in large numbers, in the 1830s, academics were already an identifiable, if small, professional group. For the élite, there were a few posts, spanning virtually all disciplines, in the official research institutions – the Collège de France, the Muséum d'Histoire Naturelle, and the Paris Observatory; while, for the lesser *universitaires*, career structures had already emerged from the fluidity of the early years of the Napoleonic Université, assuming an inflexibility which they never lost.[4] Thereafter, but especially from the Second Empire, the professional academic became ever more obtrusive in French intellectual life. Not only did he 'take over' disciplines which had previously been pursued as avocations, he also refashioned those disciplines in ways that made them more esoteric and so less accessible to the self-taught. Late in the century, it is true, academic professionalism took a new turn, inspired by an enhanced sensitivity to the needs of industry and by the quest for a greater autonomy in the management of universities. One manifestation of this – apparent from the 1890s – was the strengthening of the links between universities and their local communities; now, the barrier between the *sociétés savantes* and the professional academic was crossed more frequently than ever before, but not – and here again I speak only of the sciences – on terms that advanced the societies as 'learned' bodies. By the early twentieth century, no amount of rhetoric extolling the virtues of the independent *savant*'s contribution to research could conceal the fact that his particular skills were no match for the assertive professionalism of the university scientist.[5]

1. The local societies: science and bourgeois values

The proliferation of the *sociétés savantes* is one of the most startling, and neglected, cultural phenomena of nineteenth-century France.[6] The bare

---

4. For studies of the changing career patterns of academic scientists, see Zwerling, 'Emergence of the École Normale', and Karady, 'Qualifications and university careers', both in *The organization of science*, pp. 31–60 and 95–124; also Maurice P. Crosland, 'The development of a professional career in science in France', *Minerva*, 13 (1975), 38–75.
5. A fine example of the rhetoric is in the presidential address for 1921, delivered to the Société Zoologique de France by Étienne Rabaud, professor of zoology at the Sorbonne; see *Bulletin de la Société Zoologique de France*, 46 (1921), 5–9. Rabaud, a professional zoologist in the fullest sense of the term, declared (p. 6) that the amateur 'symbolizes unflagging zeal, he is the essential stimulus, he is life itself. The science he does is neither less nor more valuable than that of any professional.'
6. The only attempt at a serious study of the societies extending into the nineteenth

bones of the evidence are readily accessible in lists and bibliographies that were put out at various times by the Ministry of Public Instruction.[7] The most systematic of the surveys, dating from 1886, mentions no fewer than 655 societies (with twelve in Algeria and the colonies) and an accumulation of some 15,000 volumes of *Mémoires*, *Bulletins*, and other works — a figure that was said to be growing at the rate of 500 a year.[8] Although, by the time the survey was made, the lion's share of influence and prestige was passing unmistakably to the national societies based in Paris, 'local' societies (in the sense in which I define them above) comprised well over three quarters of those mentioned. It is these societies, nearly all of them provincial in their location and aspirations, which I discuss in this section.

The provincial societies fall distinctly into two groups: the seventeenth- and eighteenth-century foundations which re-emerged after the Revolution, and the societies founded after the mid-1790s, with memberships made up chiefly of new men — 'new', that is, to corporate intellectual activity. Dominant in the pre-revolutionary group were institutions typically bearing the title Académie des Sciences, Belles-Lettres, et Arts. On the eve of the Revolution, there were some thirty-five of these academies, formally recognized by their possession of letters patent from the King.[9] All thirty-five were closed by the Convention's decree of 8 August 1793, and some, like those of Châlons-sur-Marne and Pau,[10] were never revived. But, beginning in 1796,

century is to be found in the papers prepared for the 100th Congrès des Sociétés Savantes in Paris in 1975. See the two volumes of papers, one subtitled *Colloque interdisciplinaire sur les sociétés savantes*, the other *Les sociétés savantes. Leur histoire*, which appeared as parts of *Comité des Travaux Historiques et Scientifiques. Actes du 100e Congrès National des Sociétés Savantes (Paris, 1975)* (Paris, 1976).

7. Of the lists sponsored by the Ministry, the most important are: *Annuaire des sociétés savantes de la France et de l'étranger* (Paris, 1846); Eugène Lefèvre-Pontalis, *Bibliographie des sociétés savantes de la France* (Paris, 1887); and Robert Charles, Comte de Lasteyrie de Saillant, *et al.*, *Bibliographie générale des travaux historiques et archéologiques publiés par les sociétés savantes de la France* (6 vols., Paris, 1885–1918).

     Useful unofficial lists are: Le Comte Achmet d'Héricourt, *Annuaire des sociétés savantes de la France et de l'étranger* (2 vols., Paris, 1863–5; 2nd edn in four consecutively paginated parts, Paris, 1866); and Pierre Caron and Marc Jaryc, *Répertoire des sociétés françaises de sciences philosophiques, historiques, philologiques et juridiques* (Paris, 1938). See also the list in *Annuaire des sociétés savantes de France et des Congrès Scientifiques*, 4th ser. 10 (1880), 42–66.

8. Lefèvre-Pontalis, *Bibliographie*, p. vi.

9. The incompleteness of the evidence concerning the eighteenth-century academies casts some doubt on the accuracy of the figure thirty-five. However, I adopt it, with due caution, following Daniel Roche, 'Milieux académiques provinciaux et société des lumières', in Geneviève Bollème *et al.*, *Livre et société dans la France du XVIIIe siècle* (2 vols., Paris and The Hague, 1965–70), vol. 1, pp. 93–184 (see especially pp. 96–7). Roche's valuable book *Le siècle des lumières en province. Académies et académiciens provinciaux, 1680–1789* (2 vols., Paris, 1978) appeared too late for me to be able to make proper use of it in this paper.

10. On the Châlons Academy, see Roche, 'Milieux académiques'; on Pau, see Christian

many did reappear, thinly disguised under such titles as 'Lycée', 'Athénée Libre', or 'Société Libre', which were promptly abandoned on the restoration of the Bourbons in 1814. Although not all of them prospered, by 1820 ten of the old academies had achieved recognition as 'sociétés royales', with the right to use their original titles.[11] In this way, continuity with the ideals of the Ancien Régime was asserted, and the local pride of secretaries and the traditional *notables* on whose support they depended was satisfied.

Titles, however, were a small part of the battle. Sadly for the academies, there was no way of resurrecting the conditions that had allowed at least some of them, especially in the earlier eighteenth century, to make substantial contributions to the advancement of knowledge. Now, even in their immediate localities, their claims to cultural supremacy were challenged by the newer societies, which promoted meetings and the publication of papers in ways that endorsed the less arcane styles of cultural life appropriate for a modern-minded bourgeoisie.

The case of Lyon illustrates what happened in the larger towns. Here the Académie des Sciences, Belles-Lettres, et Arts de Lyon was the main focus of cultural life from its foundation in 1724 to its suppression in 1793.[12] It is true that in the later years of the Ancien Régime there had been rivals: the Société Royale d'Agriculture (founded in 1761) was one of many such societies that were founded in France in the 1760s, and the Société de Médecine de Lyon (1789) was another typical creation of the period. But the challenge to the Academy's cultural prerogative was trifling by nineteenth-century standards. With the Revolution behind them, the people of Lyon celebrated the new century not only by restoring the Academy, as the Athénée de Lyon, but also by founding new societies with a zeal quite unknown in the eighteenth century. The Société de Pharmacie (1806), the precursor of the Société Littéraire, Historique et Archéologique de Lyon (1807), one of the earliest and most successful of the French Sociétés Linnéennes (1822), and societies for the study of architecture (1830), meteorology (1843), geography (1871), botany (1872), political economy (1876), anthropology (1881), and astronomy (1883), all reflected changing tastes and contributed to the figure of twenty-six societies that appeared in the 1886 survey (no fewer than eleven of the societies being concerned, at least partially, with the natural sciences, medicine, or technology). To all appearances, the societies were stable, and nearly all published a journal with

Desplat, *Un milieu socio-culturel provincial. L'Académie Royale de Pau au XVIIIe siècle* (Pau, 1971).

11. The ten 'restored' academies were those of Arras, Bordeaux, Caen, Cherbourg, Lyon, Marseille, Nancy, Nîmes, Rouen, and Toulouse.

12. On the history of the Lyon Academy, see Jean-Baptiste Dumas, *Histoire de l'Académie Royale des Sciences, Belles-Lettres et Arts de Lyon* (2 vols., Lyon, 1839).

which to proclaim their seriousness and (most important, if any sort of library was to be kept) to secure exchanges with other societies.

Of several respects in which the new societies differed from the academies, I shall mention just two. First, by their organization most (though by no means all) of them abandoned any commitment to a unified Republic of Letters, establishing themselves instead as the curators of specialized learning. This is illustrated for Lyon by the societies I mention above, and the point could be made equally well for any number of other towns, where specialization went far beyond the hiving off of the sciences from literature and the fine arts. In active centres like Caen, Bordeaux, Rouen, and Marseille, society after society staked out clearly defined intellectual territory and remained within it, leaving the academies to pursue their hallowed ideal of comprehensiveness.

A second distinctive characteristic of the new societies was their openness. With few exceptions, they would admit to full membership anyone who could pay the annual subscription of, typically, 20 francs. The academies, by contrast, maintained their pre-revolutionary practice of setting an upper limit of between twenty-four and sixty for the category of what were variously called *membres résidents* or *membres titulaires*. They also retained the traditional paraphernalia of classes and permanent secretaries, the hierarchy of full members, corresponding members, and associate members, and the pattern of closed sessions broken only by the rare *séances publiques*, at which the mysteries of the academic process were cautiously revealed to the profane. In abandoning these fustian structures in favour of accessibility and an undifferentiated membership, the leaders of the societies founded in the nineteenth century had the satisfaction of at once enhancing the income from subscriptions and affirming their commitment to a social order in which an aristocracy of birth had given way to the more permeable aristocracy of talent.

Faced with these challenges, the academies made few concessions. They preferred to linger as vestiges of a more ordered age, jealously maintaining their particular form of social exclusiveness, publishing increasingly puny *Mémoires*, and offering derisory rewards for competitions which often enough could not raise a single entrant. It was only late in the nineteenth century that some academies adjusted to the changing styles in provincial culture. But, as in the case of the Toulouse Academy, which was revitalized in the 1880s by the active involvement of the professoriate of the local faculties,[13] the overtures tended to come from outside; and, in any case, Toulouse was an exception. The norm was one of recurring financial difficulties (heightened by the capriciousness of municipal, departmental, and governmental support) and intellectual retrenchment.

13. See below, p. 253.

A good example of a faltering academy was the Académie des Sciences, Belles-Lettres et Arts de Dijon, which never fully recovered from the blow of closure in 1793.[14] With its handsome headquarters taken over for use by the faculties of science and letters, its funds squandered, and its fine library dispersed, it eked out a meagre existence, though one in which the outward vestiges of academic respectability were stubbornly preserved. The *séances publiques* and the prize-competitions, with prizes never exceeding a paltry 500 francs, were an empty charade. The notable contributions which had secured the scientific reputation of the Academy in the days of Guyton de Morveau were a thing of the past, and throughout the nineteenth century the once substantial *Mémoires* offered a frugal diet composed largely of unimaginative local history, travelogues, and literary trivia; science was poorly represented by occasional unco-ordinated forays into descriptive natural history.[15]

With the traditional sector of provincial academic life in decay and the new societies emerging in the ever-increasing numbers indicated in Table 1, the trends in the nineteenth century are clear. Gradually the old academies were swamped, until by 1914 (with a total membership of no more than a few hundred) they were only a small element in the burgeoning world of the *sociétés savantes* (whose subscribers, by then, must have numbered some 200,000). But if the *pattern* of growth is unmistakable, the explanations for it are far less so. Just what were the attractions of the new societies? What enabled, say, the Société Philomathique de Bordeaux to mount a series of thirteen spectacularly successful exhibitions — the later ones ambitious international affairs — between 1827 and the end of the century?[16] And what engaged the level of support that allowed Linnaean societies in towns as

---

14. For an impression of the vigour and prosperity of the Dijon Academy in the eighteenth century, see Roger Tisserand, *Au temps de l'Encyclopédie. L'Académie de Dijon de 1740 à 1793* (Paris, 1936). The declining activity in the nineteenth century has yet to be analysed, but it can be followed clearly enough in the Academy's *Mémoires*, and in the evidence accumulated in Ph. Milsand, *Notes et documents pour servir à l'histoire de l'Académie des Sciences, Arts et Belles-Lettres de Dijon* (Dijon, 1871), published as *Mémoires de l'Académie . . . de Dijon*, 2nd ser. 16 (1870).

    I give further comment on the character of the restored academies and their difficulties during the Bourbon Restoration in Robert Fox, 'Learning, politics, and polite culture in provincial France: the *sociétés savantes* in the nineteenth century', *Historical reflections/Réflexions historiques*, 7 (1980), in press.

15. My observations on the contents of the *Mémoires* are, as yet, impressionistic, but it is not an unfair commentary on their cast at the end of the century to note that the longest papers in the volume for 1899–1900 (4th ser. 7) are an account of 'Seize jours de croisière sur les côtes de Dalmatie' and a 'Notice sur la Société de charité maternelle de Dijon'.

16. On the Bordeaux exhibitions, see Charles Bénard, *Histoire des expositions de Bordeaux* (Bordeaux, 1899).

diverse as Lyon, Caen, and Amiens to publish *Mémoires* which, in girth and quality, put the publications of the local academies to shame?

In answer, one thing is clear: the success of a local society did not depend on the expertise of the teachers in the faculties or even, in most cases, in the *lycées*. This is not to say that many career academics in the provinces did not join a local society as a matter of course; the majority of them did so. But their willingness to invest energy in either the scholarly or the administrative work of the societies they patronized was by no means assured. Until the 1890s, when relations between the faculties and the educated lay public became noticeably closer, the academics who not only joined but were also active in the affairs of a society were to be found not among the mobile, visible 'stars' of the Université, but almost exclusively among those un-ambitious, entrenched professors who were content with their allotted role as educational functionaries and whose aspirations to distinction never transcended the parochial.

To illustrate the point, I descend, significantly, to the depths of academic obscurity — to Gabriel-Alcippe Mahistre, a teacher in the *lycée* of Saint-Omer who was appointed to the chair of pure and applied mathematics in the newly established Faculty of Science in Lille in 1854.[17] At the time of his appointment, at the age of forty-three, Mahistre can have harboured few hopes of national recognition as a mathematician, and he was evidently able to satisfy such ambitions as he did have by contributing a stream of papers on machines, mathematics, and steam-engines to the Société Impériale des Sciences, de l'Agriculture, et des Arts de Lille. Contrast, now, Mahistre's high degree of involvement in the Lille society with the indifference of two of his far better-known colleagues in the faculty: Louis Pasteur, who was dean of the faculty and professor of chemistry, and the zoologist Henri de Lacaze-Duthiers. While Mahistre contributed no fewer than twenty-five papers to the society's *Mémoires* in the six years between his appointment to the faculty and his death in 1860, Lacaze-Duthiers had only two contributions published in his eight years as a full member (1855—63), and Pasteur, who was a member of the society from March 1855 to November 1857, published just one paper. Admittedly this was the important 'Mémoire sur la fer-mentation appelée lactique', but it is a telling reflexion on Pasteur's priorities that he also read the paper to the Academy of Sciences in Paris and had it published both in the Academy's *Comptes rendus* and in the *Annales de chimie et de physique*.[18]

17. On Mahistre's career and work for the Lille society, see the address delivered at his funeral by J. Girardin, in *Mémoires de la Société Impériale des Sciences, de l'Agri-culture et des Arts de Lille*, 2nd ser. 7 (1860), pp. xxxvi—xxxix.
18. The paper was published in the *Mémoires* of the Lille society, 2nd ser. 5 (1858), 13—26, after an abstract had appeared in *Comptes rendus hebdomadaires des séances de l'Académie des Sciences*, 45 (1857), 913—16. A full version was published in *Annales de chimie et de physique*, 3rd ser. 52 (1858), 404—18.

Table 1. *The foundation of sociétés savantes*

Subject	Pre-1790	1790-9	1800-9	1810-19	1820-9	1830-9	1840-9	1850-9	1860-9	1870-9	1880-9	1890-9	1900-9	1910-19	Date of foundation unknown	Total
Multi-disciplinary (embracing science or technology and humanities)	23	8	10	3	3	15	10	13	25	22	16	9	3	1	5	166
General science	2	0	0	0	0	2	1	4	4	17	7	0[a]	0[a]	0[a]	2	39
Natural history (including botany, entomology, etc.)	0	1	0	1	3	4	2	3	1	3	8	0	0	1[a]	1	28
Physical sciences, mathematics	0	0	0	0	0	0	0	1	0	2	0	0[a]	0[a]	0[a]	0	3
Meteorology	0	0	0	0	0	0	1	1	1	2	0	0[a]	0[a]	0[a]	0	5
Medicine	1	4	10	3	4	6	13	8	11	6	7	0[a]	0[a]	0[a]	7	80
Pharmacy	0	0	2	0	0	2	1	1	2	5	4	0	0	0	8	25
Veterinary medicine	0	0	0	0	1	0	2	1	0	1	1	0	0	0	1	7
Geology, mineralogy	0	0	0	0	0	1	0	0	0	3	0	0	0	0	0	4
Agriculture, horticulture[b]	8	5	5	4	4	10	4	4	4	0	2	0	0	0	2	50
Industrial applications, technology	0	1	3	0	2	5	0	3	4	7	5	1[a]	0[a]	0[a]	3	38
Local history, antiquities, archaeology	0	0	1	1	0	16	21	17	16	19	12	12	21	22	6	165
History (non-local)	1	0	1	1	0	6	0	2	2	2	4	1	8	4	1	33
Anthropology, ethnology	0	0	0	0	0	0	1	2	0	0	2	1	0	0	0	6

Category															Total
Local traditions, folklore, regional studies	0	0	0	0	0	1	3	0	1	4	1	5	0	0	19
Oriental, African studies	0	0	0	1	1	0	0	2	0	0	0	1	0	0	6
Geography	0	1	0	1	1	0	2	12	3	22	3	3	2	1	37
Literature	1	0	1	4	0	0	0	3	0	3	0	0	3	3	27
Music	0	0	0	0	0	0	1	0	0	0	0	1	1	10	12
'Beaux-arts'	3	1	0	3	1	6	3	8	9	8	1	1	5	11	53
Philosophy, psychology	0	0	0	0	0	1	0	0	0	1	0	4	0	2	8
Political economy, law, economics	0	1	0	0	1	2	3	3	3	0	1	0	0	1	13
Moral welfare, philanthropy	0	0	0	0	0	0	1	0	3	0	0	0	2	0	6
Education	0	0	0	2	1	0	3	3	3	2	0	0	7	0	19
Statistics	0	0	0	2	1	0	0	1	1	0	0	0	0	0	5
Architecture (Sociétés des Architectes)	0	0	0	0	1	3	8	8	7	7	0	0	0	0	26
Bibliography (including Sociétés des Bibliophiles)	0	0	0	1	0	0	0	5	4	1	0	2	1	0	14
'Amis de l'Université'	0	0	0	0	0	0	0	0	0	1	5	2	0	0	8
Miscellaneous	0	0	0	0	1	2	3	2	7	5	1	1	5	6	31
Total	39	21	33	14	27	78	65	76	100	143	123	36	49	56	73 · 933

a. It is probable that these figures (all for societies in the sciences and technology) are significantly too low. The sources of my information for societies founded after Lefèvre-Pontalis made his survey in 1886 take virtually no account of scientific and related societies.

b. These figures give an inadequate impression of the number of agricultural and horticultural societies. As I indicate below (p. 256), no fewer than 167 agricultural societies were recognized by the Ministry of Agriculture and Commerce in 1880. I have taken account only of societies that were recognized by the Ministry of Public Instruction. Normally these societies were not devoted to agriculture or horticulture alone: a typical title would be Société d'Agriculture, Sciences et Arts.

Sources: This table, which covers both national and local societies, has been compiled from the items cited in note 7. It summarizes the dates of the foundation of new societies, and so gives only a rough impression of the total number of societies functioning in any decade.

As Pasteur knew full well, without high-level exposure of the kind that the Academy of Sciences and the major Parisian journals provided, his work would have been lost amid the miscellaneous poetry ('Les fastes de Lille et les invalides du travail', by Constant Portelette, a *lycée* professor, was a particularly stirring piece) and the articles on local industry and history which graced the relevant volume of the *Mémoires*. Pasteur and Lacaze-Duthiers, unlike Mahistre and the fourth professor in the Lille faculty, Auguste Lamy (a former *lycée* professor who was elevated to the faculty in 1855, and who published eighteen contributions in the *Mémoires* between 1855 and 1863 and served as secretary of the society), were mobile, restless, hell-bent for the top. In their quest, the high opinion of the great patrons, men like Biot, Dumas, and Henri Milne Edwards, counted for far more than local recognition, which was all that the Lille society could provide. There can be little doubt that Pasteur and Lacaze-Duthiers saw themselves as birds of passage in a town to which they were haplessly banished in the early stages of their careers.[19] Even if Pasteur took an interest in the manufacturing processes of the town during his sojourn in Lille, as we know he did,[20] his long-term ambitions lay very firmly in the capital; and the centralized procedures for the making of appointments only reinforced his indifference to the local community.

Given such career strategies, it is easy to understand the detachment of the more thrusting professors from the provincial societies, especially after the 1860s, when research-based careers became more normal. Yet, by the same token, subsequent changes in those strategies help to explain why in places the pattern of detachment was reversed late in the nineteenth century. Largely as a result of the decentralizing policies of Jules Ferry (Minister of Public Instruction, with two short breaks, from 1879 to 1883) which were epitomized in the degree of 1885 giving faculties a civil status and hence the right to accept private donations, there was a new incentive to supplement the income which a faculty received from the Ministry and from students. As it transpired, local councils and industrialists were the most lucrative sources, and it was chiefly their money which allowed the faculties to effect their remarkable and, by the standards of the earlier nineteenth

19. In this, Pasteur and Lacaze-Duthiers were not alone. About the mid-century, their sense of isolation was shared by Charles Gerhardt in Montpellier and Quatrefages de Bréau in Toulouse, to name just two of the better-known 'exiles'. However, I still feel that we may have paid undue attention to the recollections of a comparatively small body of discontented *universitaires*, while disregarding the majority of professors in provincial faculties who frequently served for thirty years or more in the same institution without any evidence of dissatisfaction; for a comment, see Fox, 'Learning, politics, and polite culture'.
20. Denise Wrotnowska, *Louis Pasteur, professeur et doyen de la faculté des sciences de Lille (1854–1857)* (Paris, 1975), pp. 55–67 and 70–5; and Paul, 'Apollo courts the Vulcans', in *The organization of science*, p. 156.

century, uncharacteristic incursion into technical education in the 1890s.[21] Suddenly, the withdrawn élitism which the academic profession had cultivated over the previous thirty years was in question. Now, as never before, in return for an unprecedented, if limited degree of autonomy, the fifteen groups of faculties (which became Liard's universities in 1896) were answerable to, and dependent upon, the communities around them.

The changed perspective of university administrators and professors was soon apparent. The tradition of giving faculty lectures to lay audiences, which had flagged since the Second Empire, was revived; popular magazines were put out; the involvement of laymen in university councils and of academics in local administration became common; and, most important for my purpose, societies too were used as a focus for joint endeavour. Several of the societies that served this purpose were new foundations: the Société Chimique du Nord in Lille (1890) and the network of Sociétés des Amis de l'Université that emerged in university towns from about 1880 are good examples. But existing societies could also provide common ground for the industrialist and the academic, as the case of Toulouse, one of the more vigorous provincial centres for science in the late nineteenth century, shows clearly. Here, from the late 1880s, Benjamin Baillaud, the dean and professor of astronomy, and Paul Sabatier, the professor of chemistry and future Nobel laureate, transformed an indifferent faculty of science into a research institution of national standing.[22] In doing so, they and other professors leant heavily on local support and, in turn, brought new life to a number of societies, including the Toulouse Academy.[23]

The efforts of the faculties to identify more closely with their regions reflects a concern, on the part of academics, to demonstrate their usefulness to an industrializing society, at a time when the euphoria of government-financed expansion showed signs of waning. It also underlines the fact that the expansion, the enhanced incomes, and the improved morale of the provincial faculties under the Third Republic at last made a serious scientific career outside Paris feasible. For indigenous provincial intellectual life, the

21. On the technical institutes that were attached to most science faculties in the 1890s, see Paul, 'Apollo courts the Vulcans', in *The organization of science*, pp. 155–81; Terry Shinn, 'The French science faculty system 1808–1914: institutional change and research potential', *Historical studies in the physical sciences*, 10 (1979), 271–332; and George Weisz, 'The French universities and education for the new professions, 1885–1914: an episode in French university reform', *Minerva*, 17 (1979), 98–128 (112–21).
22. For a study of this transformation, see Mary Jo Nye, 'The scientific periphery in France: the Faculty of Sciences at Toulouse (1880–1930)', *Minerva*, 13 (1975), 374–403.
23. It is a mark of their interest that of the ten scientific papers published in the Academy's *Mémoires* for 1890, no fewer than four were by professors from the Faculty of Science at Toulouse, with the rest (except for two papers by local doctors) coming from professional academics in other faculties and educational institutions.

integration of the professoriate with non-university élites came as a long-overdue stimulus. But we should beware of exaggerating its effect on the *sociétés savantes*. Most of the contacts between town and gown seem to have been made at a social rather than an academic level. Moreover, many societies were far from the orbit of a university, and, as the case of Dijon reminds us, not all universities were in centres that fostered the educated, industrially oriented groups most obviously congenial to a scientific professoriate.

So while in certain towns, university professors, industrialists, and businessmen discovered a common interest, in the majority of communities, the alliance was not made. Here, despite changing fads, the dominant cultural styles in 1900 were essentially those of the mid nineteenth century, with *sociétés savantes* still providing the main focus for intellectual activity, unaffected by the intrusion of academics. Of course, even to make that distinction is still to mask a great diversity in objectives and memberships. However monolithic the societies may have seemed in the disparaging eyes of many scientific professionals, they fostered no one élite, no one ideology – a fact that makes an answer to my original question, why they attracted the support they did attract, all the more elusive.

In full awareness of the complexity of the problem and the snares of the 'ideal type', I doubt whether at least one category of motives will ever be more tellingly revealed than in Flaubert's vignette of Homais, the pompous pharmacist in *Madame Bovary* (1857). It was a matter of pride for Homais that he was, to use a phrase much loved by the provincials, 'membre de plusieurs sociétés savantes'.[24] Moreover, he numbered himself not among the passive recipients of knowledge, but among its creators; he was emphatically a performer, a status he earned by his demonstrated competence. His credentials included a memoir (published at his own expense!) entitled 'Du cidre, de sa fabrication et de ses effets', and on one occasion he had even submitted his observations on the woolly aphis to the Academy, with unspecified results. Clearly, Homais saw himself as representative of a responsible modern élite, in the cautious sense in which the bourgeoisie understood modernity in the mid-century: hence his declared admiration for Voltaire, and his concern for the practical aspects of cider manufacture in Normandy.

Homais, in short, seems to have vaunted his association with his societies as a means of displaying the 'distinction' which demarcated the enlightened, forward-looking bourgeoisie from the *classes populaires* and, no less importantly, from both the Gradgrind world of business and the faded land-owning aristocracy.[25] Needless to say, there were other ways in which a particular

24. Gustave Flaubert, *Madame Bovary. Moeurs de province*, Oeuvres complètes de Gustave Flaubert (Paris, 1930), pp. 477–8.
25. I use 'distinction' in the sense in which it is used in Edmond Goblot's classic study of the French bourgeoisie, *La barrière et le niveau* (Paris, 1925).

social status could be affirmed. By the 1850s and 1860s, the progressive bourgeois family might invest in the paintings of Courbet or Millet; and, much as they abhorred the rising tide of bohemianism, many did.[26] Or they might stimulate the booming market for works of popular science by purchasing the writings of Figuier or by subscribing to one of several new journals that catered for their interests.[27] Or, like Homais, they might engage in the wholesome pursuits of a local *société savante*, enjoying the masculine clubbability of its meetings, reading in its library, and proudly ranging its unread *Mémoires* in their salon.

In the mid nineteenth century, the local *sociétés savantes* and their main specialities — natural history, and regional history and antiquities — were, quite simply, fashionable. In style and subject-matter, they were perfectly congruent with the uncomplicated tastes of the bourgeoisie — tastes enshrined in Napoleon III's work in archaeology and his commercially successful but otherwise unmemorable *Histoire de Jules César* (1865–6). Participation demanded neither lengthy training, nor intellectual gifts, nor, still less, refined taste; particularly where agricultural problems were concerned, it also had the aura of utility and public service; and, for the many members who neither gave papers nor contributed to the *Mémoires*, but whose subscriptions were indispensable, there was always the lure of an entertaining illustrated lecture on a subject devoid of controversy and replete with high moral tone.

Clearly, in insisting on the social function of the *sociétés savantes*, I run the risk of over-simplification, of extrapolating misleadingly from the single stereotype of Homais.[28] This risk is all the greater because most of the

26. Timothy J. Clark, *The absolute bourgeois. Artists and politics in France 1848– 1851* (London, 1973), and *Image of the people. Gustave Courbet and the 1848 revolution* (London, 1973).

27. Perhaps the most notable of the new periodicals was the weekly *Cosmos*, founded in 1852 by B. R. de Montfort and edited by the Abbé Moigno. An exceptionally competent publication, *Cosmos* sought both to spread a knowledge of science and, in its pungent editorial matter, to exercise informed criticism of new developments. See, on its aims, the Preface to the first volume (1852), pp. i–iv. *Cosmos* continued to be published until 1870, but in 1863 Moigno left the journal and founded a rival, *Les mondes*, which ceased publication in 1914.

   During the Second Empire and Third Republic, the number of popular scientific periodicals grew steadily, with such newcomers as the *Revue des cours scientifiques de la France et de l'étranger* (founded in 1863; retitled *Revue scientifique* in 1871), *La nature* (1873), *Le naturaliste* (1879), and *L'astronomie* (1882).

28. The social function as I describe it was similar, though by no means identical, to that identified for provincial Britain in Steven A. Shapin, 'The Pottery Philosophical Society, 1819–1835: an examination of the cultural uses of provincial science', *Science studies*, 2 (1972), 311–36; Steven A. Shapin and A. W. Thackray, 'Prosopography as a research tool in history of science: the British scientific community 1700–1900', *History of science*, 12 (1974), 1–28; and Thackray, 'Natural knowledge in its cultural context'. All of these papers have proved extremely helpful.

societies fulfilled not one but several roles. A number of the wealthier societies, for example, supplemented their programmes of meetings and publication with an active philanthropy, providing rewards for dutiful employees in local industries and serving as a last resort for the destitute. Others, like the pharmaceutical and medical societies, disseminated occupational techniques. And there was an especially important group of societies established to promote agriculture or industry.

Of the agricultural societies, I shall say only that they were ubiquitous — by 1880, 167 of them were recognized by the Ministry of Agriculture and Commerce — and that many had their roots in the eighteenth century.[29] The industrial societies, by contrast, were all founded in the nineteenth century; they were fewer in number (by 1914, there were only fourteen societies in the Union des Sociétés Industrielles de France) and they flourished particularly in the manufacturing areas of the north and east.[30] Their stated aims were to facilitate the exchange of technical information, to encourage inventiveness by means of competitions and exhibitions, and, often, to provide education for working men. The educational activities could take many forms, ranging from courses in basic literacy to the specialized instruction in spinning and weaving that was offered by the Société du Commerce et de l'Industrie Lainière in Fourmies. It is clear that in the later nineteenth century education presented an incomparable opportunity for any society to enhance its income while asserting its usefulness to the community, and it was an opportunity seized by many societies other than those formally designated as Sociétés Industrielles. Particularly successful in this respect were the Société des Sciences in Lille, with its École des Chauffeurs providing instruction in the operation of steam-engines, and the Société Philomathique de Bordeaux, with a whole range of courses in the 'three R's', hygiene, commercial geography, chemistry, and accounting.

The industrial societies are a particularly good illustration of the unduly restrictive nature of the term *société savante*. Local conditions created a great variety of opportunities for corporate action, and the most prosperous societies were clearly those which assumed as many roles as possible. Consider,

---

29. For a list of the recognized societies, see *Institut des Provinces. Annuaire des sociétés savantes de France et des Congrès Scientifiques*, 4th ser. 10 (1880), 67–77. A recent comment on the agricultural societies, with references to earlier literature, is in Yves Laissus, 'Les sociétés savantes et l'avancement des sciences naturelles. Les musées d'histoire naturelle', in *Colloque interdisciplinaire sur les sociétés savantes*, pp. 41–67 (50–3). Laissus notes peaks in the number of new foundations in the 1760s, between 1800 and 1809, and, most markedly, in the 1830s. New horticultural societies were particularly numerous between 1850 and 1880.

30. According to the list in *Union des Sociétés Industrielles de France. Compte rendu du deuxième congrès tenu à Reims les 23, 24 et 25 mai 1914* (Nancy, Paris, and Strasbourg, [1920]), p. 5, there were member societies in Paris, Amiens, Elbeuf, Nancy, Fourmies, Mulhouse, Nantes, Lille, Reims, Roubaix, Rouen, Saint-Quentin, and Tourcoing.

by way of illustration, the wealthiest of all provincial societies, the Société Industrielle de Mulhouse (1826). Here, narrow declared objectives masked an astonishing range of subsidiary interests. Aided by a tenfold increase in population during the nineteenth century, from approximately 8,000 to nearly 90,000, and by the active patronage of wealthy manufacturing families like Schlumberger, Koechlin, and Dollfus, the Mulhouse society could provide spacious headquarters, internationally famous schools of Chemistry and Design (as well as several other technical schools), and museums of art, technology, geology, natural history, ethnography, and archaeology, to say nothing of prizes of the order of 5,000 francs each in pursuit of technical innovation.[31] It is beyond question that the main purpose of the society throughout the nineteenth century was to promote the industries, in particular the textile industries, of the region; this was the declared aim of the small nucleus of founder-members in 1826, and the uncompromisingly sober contents of the *Bulletin de la Société Industrielle de Mulhouse* show that the overriding commitment to technical and commercial improvement never waned. But by the time of the annexation of Alsace to Germany after the Franco-Prussian war, the society's membership (of some 600) included a substantial number of representatives of the liberal professions with little interest in textiles. These men were clearly not lured to the society by the very active specialized committees on mechanics, industrial chemistry, trade, transport, and the like; they subscribed for the library, the museums, the diverting lectures, and the guarantee of sophisticated and congenial company. Hence, when Auguste Lamy used his presidential address in 1862 to warn the Société des Sciences in Lille of the danger of associating science too closely with utility,[32] he was issuing a timely caution. Science, as he knew, had cultural as well as industrial applications, and any society which forgot that did so at its peril.

Plainly, despite their title, the *sociétés savantes* of nineteenth-century France were not exclusively or even, in most cases, primarily vehicles for the advancement of knowledge. While some won national recognition for their work in regional history and topography, their contributions in science were decidedly patchy. In the highly professionalized disciplines of chemistry, physics, and mathematics, in fact, they were virtually non-existent, and it was only in field-work, in which the leisured naturalist with local knowledge had the advantage over his disciplinary peers in the teaching profession, that

31. On the Mulhouse society, see Achille Penot, 'La Société Industrielle de Mulhouse', forming pp. 1–136 of a separately paginated supplement to the *Bulletin de la Société Industrielle de Mulhouse*, 46 (1876), published to mark the fiftieth anniversary of the founding of the society; also *Centenaire de la Société Industrielle* (2 vols., Mulhouse, 1926), vol. 1, pp. 1–187.
32. August Lamy, Presidential address, 21 December 1862, in *Mémoires de la Société Impériale des Sciences, de l'Agriculture et des Arts de Lille*, 2nd ser. 9 (1862), xliii.

258

the local societies retained a toe-hold in research. So long as work of this kind was esteemed — and it seems to have represented the dominant style of the Muséum d'Histoire Naturelle in Paris until the mid-century[33] — so too were men like the Auxerre lawyer Gustave Cotteau and societies like the Société des Sciences Historiques et Naturelles de l'Yonne, in whose journal Cotteau published a notable series of papers on fossil echinoids about 1850.

It was not until the 1860s that the esteem for field-work began to be seriously undermined. Far from being accidental, the process was part of a calculated attempt to raise the status within the Université of the life sciences, and its chief spokesmen are readily identifiable as Claude Bernard (speaking for a new conception of physiology) and Lacaze-Duthiers (advocating a laboratory-based *zoologie expérimentale*).[34] For Bernard, the life sciences could only attain parity of esteem with the physical sciences if they were divested of the dilettantist associations of what he dismissively called the 'sciences naturelles', his term for descriptive natural history. Once this re-definition of academic respectability had been effected — and it was resplendently enshrined in the research laboratories for zoology, palaeontology, and botany that were established in the science faculties from the 1880s — even the most active of the local societies came to be regarded as backwaters of science, the bastions of an outdated and unprofessional morphological tradition.

2. Ministerial prerogatives and the challenge to autonomy

It is conceivable that the local societies might have been left to run their modest course independently of any central authority. But in nineteenth-century France, autonomy was hard to sustain, and, especially under the July Monarchy and the Second Empire, the societies were an object of considerable attention in the Ministry of Public Instruction. At times, when sensitive ministers associated the societies with a more general challenge to their authority, the attention could be malevolent.

The potentially subversive character of the *sociétés savantes* was established, in ministerial eyes, by a dispersed network of zealous provincials inspired from Caen by Arcisse de Caumont.[35] An archaeologist, antiquarian,

33. See Limoges, 'Development of the Muséum d'Histoire Naturelle' in *The organization of science*, pp. 230—3.
34. For their views, see Claude Bernard, *Rapport sur les progrès et la marche de la physiologie générale en France* (Paris, 1867), pp. 131—49 and 221–37; and Henri de Lacaze-Duthiers, 'Direction des études zoologiques', *Archives de zoologie expérimentale*, 1 (1872), 1—64.
35. Highly sympathetic but useful biographical sources on Caumont (1801—73) include Charles Richelet, *Notice sur M. de Caumont* (Paris, 1853), reprinted with supplementary material on Caumont's later years in *Annuaire de l'Institut des Provinces, des sociétés savantes et des Congrès Scientifiques* (1869), 358—400; Renault, 'Notice

musician, geologist, and natural historian of distinction, Caumont embodied
the proudly independent *savant* for whom the acceptance of an appointment
under the Ministry would have been almost an act of betrayal. It was his
dream to give the activities of the local societies the visibility and coordination they needed in order to sustain work at a high level while resisting
(to use his colourful terminology) the cancer of assimilation to Paris.

The most important of several organizations that Caumont established to
this end[36] were the Congrès Scientifiques de France, held annually for a
week or more from 1833 to 1880.[37] In so far as they took place each year in
a different provincial town, with the capital being studiously avoided, the
congresses bore an obvious resemblance to the meetings of the British
Association for the Advancement of Science, which began in 1831 in York.
But their peripatetic character and their adherence to a common model, the
Deutscher Naturforscher Versammlung, which had roamed the German
states since 1822,[38] represent the sum of their similarity. The Congrès
Scientifiques were altogether a more personal affair than B.A.A.S., depend-
ing heavily, as they did, on Caumont's boundless energies and seemingly
bottomless pocket. The two organizations also differed in the range of
subjects which they accommodated. Far from being exclusively scientific,
both the Congrès Scientifiques and Caumont's Institut des Provinces, an élite
of 200 *savants*, established in 1839 as a provincial counterpart to the Institut
de France, reflected the whole range of polite bourgeois interests, including
agriculture, local antiquities and archaeology, and the natural sciences (almost
invariably interpreted to mean natural history). Hence there is no sense in
which they expressed the nascent self-consciousness of a discrete scientific
community, in the way that the early B.A.A.S. seems to have done. If they

biographique sur M. de Caumont', *Annuaire des cinq départements de l'ancienne
Normandie* [otherwise the *Annuaire normand*], 40e année (1874), 465–99; E.
de Robillard de Beaurepaire, 'M. de Caumont. Sa vie et ses oeuvres', *Mémoires de
l'Académie des Sciences, Arts et Belles-Lettres de Caen* (1874), 324–401.

36. The other organizations for whose foundation Caumont was wholly or partly responsible were: the Société Linnéenne du Calvados (1823; renamed Société Linnéenne de Normandie in 1836), the Société des Antiquaires de Normandie (1824),
the Association Normande (1831), the Société Française pour la Conservation et la
Description des Monuments Historiques (1834), the Congrès Archéologiques (1834),
and the Institut des Provinces (1839). All were concerned with the integration of
cultural activity, either nationally or within the region of Normandy.

37. A detailed report on the work of each congress was published. The reports, frequently
extending to two volumes (five in 1861), covered every congress up to that of 1878,
which was probably the last. There was no congress in 1879, and I have no evidence
that the one planned for Brest in 1880 actually took place.

The only modern study which bears on the congresses is Marcel Baudot's valuable
paper, 'Trente ans de coordination des sociétés savantes (1831–1861)', in *Colloque
interdisciplinaire sur les sociétés savantes*, pp. 7–28.

38. On Caumont's admiration for the German congresses, see Richelet, *Notice sur
Caumont*, p. 13, and *Congrès Scientifique de France. Première session tenue à Caen
en juillet 1833* (Rouen, 1833), pp. vii–viii.

promoted any corporate identity at all, it was that of a conservative, predominantly rural élite, for whom the professional academic was not merely a rival but, as many of their members would have said, a threat to morality and good order.

But it was above all in the sustained intensity of their provincialism that the Congrès Scientifiques differed from B.A.A.S. In this, Caumont was the inspiration. Until his death in 1873, as 'the revered patriarch of intellectual decentralization',[39] he missed no opportunity of inveighing against ministerial arrogance and the professionalization of intellectual life. Everything that Caumont touched was infused with an active resentment against the capital that went far beyond the bounds of local pride, with the result that the establishment of a truly national constituency for his activities was virtually impossible. Year after year, the congresses drew audiences that were almost exclusively local in origin, with a nucleus of no more than half a dozen enthusiasts to maintain continuity from one session to the next. Of the 217 people present at the first congress, in Caen in 1833, for example, only a handful came from outside Normandy, and no more than six were from Paris; it was a pattern of recruitment that remained unchanged to the end.[40] There was a similar consistency in the social composition of the support. It is one indicator of this that throughout the history of the congresses, the honoured guests were generally not *savants* of national standing, but mayors, landowners, and clergy. Even local *universitaires* were welcomed only to the extent that they subscribed to the overwhelmingly provincial ethos of the gatherings. For them, the test of provincial integrity was a rigorous one, passed, it would seem, by very few candidates. A notable scientist who did pass — with obvious distinction — was Jacques-Amand Eudes-Deslongchamps; the 'Cuvier normand', he was a close friend of Caumont who served for many years as secretary of the Institut des Provinces.[41] Though he held chairs of natural history and, later, zoology in the Faculty of Science at Caen, Eudes-Deslongchamps's overriding loyalty

39. Geslin de Bourgogne, report as general secretary to the 38th Congrès Scientifique (1872), in *Congrès Scientifique de France 38e session . . . Saint-Brieuc . . . 1872* (2 vols., Saint-Brieuc, 1873–4), vol. 1, p. ii.
40. Of the 133 registrants for the Nice Congress in 1878, for example, all but twenty-six gave addresses in Nice; and of these twenty-six, only thirteen were from outside Provence. More than a quarter of the registrations came from doctors.
41. Eudes-Deslongchamps was also the central figure in the scientific work of the Congrès Scientifiques from 1833 until his death in 1867. For a biographical sketch, see A. de Saint-Germain, 'Recherches sur l'histoire de la Faculté des sciences de Caen de 1809 à 1850', *Mémoires de l'Académie Nationale des Sciences, Arts et Belles-Lettres de Caen* (1891), 42–104 (97–9).

    The only scientist of unquestionably national standing who allowed his name to be associated, over a long period, with the Congrès Scientifiques and the Institut des Provinces was Quatrefages de Bréau, an inveterate joiner of societies. However, his signs of active interest were slight.

was not to the Ministry but to his region, a preference he demonstrated by his deep involvement in local societies and his work for Caen's remarkable Museum of Natural History.

Caumont's relations with the Ministry were, understandably, turbulent. His campaign for intellectual decentralization was readily identified with the clerical threat to the state's cherished monopoly in education, and Caumont, an ultramontane in religion and a legitimist in politics, did nothing to conceal his sympathies. Indeed, by giving such well-known enemies of the state's monopoly as Mgr Dupanloup, Bishop of Orléans, and Montalembert, a prominent place in his enterprises, he was being downright provocative; and the provocation was heightened by Caumont's prickly intransigence in his dealings with the Ministry. Even the benign attentions of the Comte de Salvandy were resented, and although relations between the two men were good during Salvandy's first period as minister (1837–9), they deteriorated badly during the second (1845–8).[42] With this deterioration there began a period of some twenty years in which Caumont believed, with some justice, that the Ministry victimized both him and his organizations.

The main prize for which Caumont contended with the Ministry was the allegiance of the *sociétés savantes*. As the first minister to take an interest in the societies, Guizot in the 1830s saw them merely as a useful adjunct in his plans for the publication and study of documents relating to the history of France. But from 1837, under Salvandy, the interest began to take a more tangible form. In circulars to the secretaries of societies and departmental prefects in 1838, Salvandy offered the services of his Ministry as a clearing house for information and the exchange of publications.[43] As it happened, secretaries and prefects alike proved recalcitrant correspondents, and the project foundered. In 1845, with Salvandy back at the Ministry, another initiative was tried, but again it elicited little response: only the first issue of a projected *Annuaire des sociétés savantes* was published,[44] and, despite a regular allocation for the societies in the budget from 1847,[45] other projects,

---

42. Baudot, 'Trente ans de coordination', pp. 16–20. In fact, Salvandy's second term of office began favourably for Caumont. In August 1846, Salvandy appointed Caumont as his unpaid 'delegate' to the *sociétés savantes*. However, Salvandy's aim, which was to strengthen the links between the Ministry and the societies, was at odds with Caumont's ideal of independence, and by the summer of 1847 the association had become soured; see Fox, 'Learning, politics, and polite culture'.

43. Circulars dated 5 July 1838 and 7 August 1838, in *Bulletin universitaire*, 7 (1838), 284 and 333–4.

44. *Annuaire des sociétés savantes de la France et de l'étranger* (1846).

45. The sum allocated in the education budget for what were normally called 'subventions et encouragements aux sociétés savantes' varied interestingly during the nineteenth century. They rose from an average of about 50,000 francs a year during the Second Empire (70,000 francs in the late 1860s), to between 120,000 and 130,000 francs in the late 1870s. They reached a peak (as did most items in the budget) in the early 1880s (176,000 francs in 1883, for example) and thereafter fell appreciably

such as a central library in which all societies would deposit their publications, were never completely realized.

During the Second Empire, as first Fortoul (1851–6) and then the gallican Rouland (1856–63) strove to impose order in their dispersed province, ministerial determination to associate the *sociétés savantes* with the national educational system gained new strength. In 1854, just two years after Caumont launched a bulletin summarizing the publications of the provincial societies, Fortoul killed it by inaugurating an identical official publication, which was soon transformed, in 1856, into the *Revue des sociétés savantes*.[46] This huge periodical, containing annually well over a thousand closely printed pages of news, announcements, and a selection of papers from the societies' publications,[47] was an incomparable vehicle for publicizing the activities of the societies, in Paris as well as in the provinces: its scale and comprehensiveness pointed all too clearly to the extent of ministerial resources and the hopelessness of Caumont's cause. It reduced the somewhat comparable but far smaller *Annuaire de l'Institut des Provinces* to insignificance, though it is a mark of the tenacity of the editors of the *Annuaire* that their volume continued to appear without interruption from its inception in 1851 until 1880.

The flagrant duplication of effort was a matter of deliberate, if covert policy. With a determination that accorded fully with the authoritarianism of the earlier years of Napoleon III's Empire, Fortoul and Rouland were striving to protect state monopolies in all sectors of education and intellectual life. Even if this led to conflict with certain of the clerical and monarchist interests whose sympathy the Empire would manifestly have welcomed, there could be no holding back; the ambitions of the ultramontane clergy in particular had to be resisted. Hence it was in the context of a general resolve to enforce central control that Fortoul and Rouland faced Caumont's challenge. Their response, characteristically, was to give high priority to the establishment of their own rational structure for research that would at once add lustre to the Ministry and weaken potential rivals. Clearly the Congrès Scientifiques and the Institut des Provinces were intended

(to a mere 78,000 francs in 1896). However, these figures do not represent the whole of government support for provincial *savants*: sums of between 180,000 and 200,000 francs were allocated to 'encouragements aux savants et aux gens de lettres', and between 140,000 and 200,000 francs were available each year for 'souscriptions scientifiques et littéraires'.

46. The publication of the Institut des Provinces was entitled *Bulletin bibliographique des sociétés savantes des départements*. The last issue, no. 12, which was dated December 1853 but published in the spring of 1854, records the sense of indignation at the intervention of the Ministry; see the pained editorial note by A. Du Chatelier on pp. 321–2.

47. Even this figure gives an inadequate impression of the *Revue*, for in 1862 a separate series of the publication, devoted entirely to science, was launched. It was published weekly, the issues for a year totalling some 800 pages.

to be the main victims of this politically inspired priority, and there is much evidence that through the 1850s and early 1860s both organizations were harassed. In 1852, for example, three professors from the local faculty of science who had agreed to serve as officers for the Toulouse Congress withdrew after receiving a ministerial warning;[48] and conflict was never more bitter than in 1861, when, to Caumont's chagrin, a representative of the Ministry secured election to the organizing committee of the Bordeaux Congress.[49]

Caumont's organizations, however, were not the only targets. Both Fortoul and Rouland were equally concerned to diminish the status of the Institut de France, which under the Empire was not only aggressively independent but politically suspect. Elections of known critics of the Université and the imperial régime like Dupanloup and the legitimist Berryer were one obvious irritant; another was the veiled criticism of the Empire that repeatedly found its way into the much-publicized *discours de réception* of new academicians.[50]

The chief weapon that Fortoul and Rouland used in response to these various signs of waywardness was the committee of scientists and scholars which, after numerous changes of name and function, is now known as the Comité des Travaux Historiques et Scientifiques.[51] Founded by Guizot in 1834 to supervise the publication of historical documents, the committee engaged for some twenty years in an unobtrusive but effective programme that united independent *savants* and the trained products of the École des Chartes; by 1852, ninety-four substantial volumes in the series 'Collection de documents inédits de l'histoire de France' had been published. But under Fortoul and, more conspicuously, under Rouland, the committee assumed a political as well as a scholarly significance. Major reorganizations between 1852 and 1858 gave it responsibility for co-ordinating the *sociétés savantes* throughout France (hence the new title of Comité des Travaux Historiques et des Sociétés Savantes, adopted in 1858). Since the members of the committee were appointed and closely supervised by the Ministry, this innovation bore an unmistakable implication. Control in the world of learning was re-affirmed as the prerogative of the Minister of Public Instruction, and neither Caumont nor the widely favoured alternative of the Institut de France were to be allowed to fulfil the role.[52] More specifically, Fortoul and Rouland

---

48. Renault, 'Notice sur Caumont', pp. 491–2.
49. Baudot, 'Trente ans de coordination', p. 24.
50. On the recurring conflicts between Fortoul and the Institut de France, see Paul Raphael and Maurice Gontard, *Un ministre de l'instruction publique sous l'Empire autoritaire. Hippolyte Fortoul 1851–1856* (Paris, 1975), pp. 283–96.
51. On the history of the Comité, see Xavier Charmes, *Le Comité des Travaux Historiques et Scientifiques (Histoire et documents)* (3 vols., Paris, 1886), especially vol. 1, pp. i–ccxxvi. The Comité's present title was adopted in 1881.
52. The case for attaching the *sociétés savantes* to the Institut de France was put most

were agreed that the *sociétés savantes* should be attached to the Université, even though they differed over the precise mechanism of control — Fortoul favouring supervision through his regional rectors, Rouland preferring to use the professors of the faculties.

In science, which had been one of its responsibilities since 1835, the committee was palpably ill-conceived. Though its Section des Sciences included most of the great names of imperial science, including the special favourites of the régime — Dumas, Le Verrier, and Henri Milne Edwards — its programme of publishing series of *Oeuvres complètes* (Lavoisier, Fresnel, and Cauchy were among those favoured) amounted, scientifically, to very little.[53] In historical and archaeological research and topography, however, the activity of the first twenty years was intensified, and most notably in the multi-volume *Dictionnaire topographique de la France* (begun in 1861), there is ample evidence that the committee's support, far from stifling the work of the societies, gave it vigour and prominence.[54]

But, whatever its scholarly achievements, there can be no doubt that between 1851 and 1863 the committee was judged in the Ministry by its political effectiveness. Its success, in this respect, is nowhere more evident than in the annual meetings in Paris to which the committee invited delegates from all the societies of the Empire, provincial and national, in science and the humanities. The first of these Congrès des Sociétés Savantes was held in November 1861, in a self-congratulatory atmosphere of patriotism and ministerial benevolence.[55] In Rouland's opening address, there was only a brief reference to the Institut de France, and Caumont's work was pointedly ignored.[56] It was an omission which the obdurate critics of the attempt to subject provincial academic life to the rule of the *universitaires* could not fail to observe, and the affront was only aggravated by the patronizing way in which prizes and encouragement were meted out.[57]

fully and forcibly by the philosopher Francisque Bouillier in his *L'Institut et les académies de province*. The first version of this text was published in Lyon in 1857; a greatly enlarged edition appeared in Paris in 1879.

53. The sense of inactivity in the science section is vividly conveyed in the *Extraits des procès-verbaux des séances du Comité Historique des Monuments Écrits* (Paris, 1850). Time and again, the section had no business to report.

54. Several volumes were published in collaboration with a local society, and the co-operation of societies seems almost invariably to have been sought. On the *Dictionnaire topographique* and its less successful companions, the *Répertoire archéologique* and the *Description scientifique*, see Fox, 'Learning, politics, and polite culture'.

55. For reports on the meeting, see *Revue des sociétés savantes*, 2nd ser. 5 (1861), 393—487, and *Revue des sociétés savantes. Sciences mathématiques, physiques et naturelles*, 1 (1862), 1—143.

56. *Revue des sociétés savantes. Sciences mathématiques . . .* 1 (1862), 3—8.

57. Charmes, *Comité des Travaux Historiques*, vol. 1, pp. clxix—clxxi. The Second Empire's system of regional and national competitions, which gave the Congrès des Sociétés Savantes the air of a *lycée* prize-giving, was abandoned in 1872; see *ibid.*, vol. 1, p. clxxii, and vol. 2, pp. 241—2 and 248.

However, the indignant reaction of some of the leaders of the provincial societies was not that of the rank and file. For the majority of members, the excursion to Paris and the heady atmosphere of the *séances solennelles* were irresistibly alluring; and so it was that five hundred or more delegates crowded each year into the Sorbonne to hear their efforts lauded by the Minister and by Parisian *savants* who would have shuddered at the thought of a sojourn in the provinces. By comparison, Caumont's very similar annual congresses (for the delegates of provincial societies only, of course) were austere affairs; held in the headquarters of the Société d'Encouragement pour l'Industrie Nationale in the Rue de Rennes, they had no prizes to offer, no ministerial financing, and only the occasional celebrity on display. However, like much that Caumont initiated, they preceded the official gatherings (Caumont's first congress met in 1848, and regular meetings began in 1851) and they continued valiantly until 1880.[58] But by 1880 — indeed, ever since Caumont became incapacitated in 1871, two years before his death — the fire had gone out of his enterprises, and only the Congrès Archéologiques (peripatetic gatherings which Caumont inaugurated in 1834) survived into the twentieth century as a last vestige of the attempt to establish a truly independent sector in scientific and historical research.

3. The national societies: constraints and opportunities

The organizations I have considered so far were rooted in a world distinct from that of the professional academic in which livelihood and advancement depended upon intellectual competence formally demonstrated in examinations, *concours*, and, increasingly, in nationally visible publications. The divergence between the two worlds became especially marked in the second half of the nineteenth century as more disciplines passed into the realm of university study and as traditional, *lycée*-based academic career-patterns gave way to the research-based patterns that have persisted ever since.

As I indicated in section 1, the effect of this creeping professionalization was less significant in the physical sciences and mathematics, for which the 'lay' audience had been negligible ever since the Revolution, than it was in the life sciences and geology, which still retained an honourable place for the researching non-academic quite late in the nineteenth century. In the latter subjects, before their total assimilation into the Université, allegiances were more commonly determined by competence in the discipline than by commitment to an institution (the norm in the highly professionalized disciplines). For this purpose, active local societies could provide a useful context, but disciplinary allegiances at the highest level were more commonly

---

58. The proceedings of Caumont's Paris congresses were published each year in the *Annuaire de l'Institut des Provinces*.

displayed in a new, 'first generation' of disciplinary societies which laid claim to national status.

It was the common aim of national societies to represent all the serious practitioners of a particular discipline. They made a virtue of the comprehensiveness of their membership, though comprehensiveness had to be on terms that would preserve a sufficiently high level of competence to attract such professional academics as there were in the field. Hence a substantial publication in which no contribution would be too advanced or too specialized, and from which no tradition of competent research would be excluded, was indispensable. But this requirement immediately raised possible tensions. What intellectual compromises were to be made in pursuit of the minimum of about 300 subscribers that was necessary if a society was to put out a journal of any consequence?

Outstandingly successful among the first-generation societies which managed to achieve disciplinary purity and the requisite support were the Société Géologique (1830), the Société Entomologique (1832), the Société Météorologique (1852), and the Société Botanique (1854). They show a recognizable pattern. Each had a nucleus of Parisian *savants* in the senior honorific posts (presidents, vice-presidents, honorary members), with non-academics doing the donkey-work as secretary and treasurer. In all cases, an overwhelming proportion of their members were not engaged professionally in research or higher education; they were drawn by a variety of motives that ranged from, in a minority of cases, the quest for national exposure for their observations, to the more nebulous gratification of being associated, however distantly, with the celebrated masters of the discipline.

To illustrate the constraints and opportunities of the first generation of national societies, I shall restrict my comments to the Société Géologique, a society that drew its 'lay' audience not only from the independent provincial *savants*, many of whom were turning to the exciting new discipline of geology by 1830, but also from the large state corps of mining engineers.[59] Regional meetings were used to particularly good effect to drum up a broadly based support in the provinces, and an emphasis on observation was cultivated as a way of suppressing the theoretical conflict which bedevilled the early years of geology.[60] The result was a rare tradition of lively debate and ambitious publications of high quality. The six members of the geological section of the Academy of Sciences were almost invariably members, and, at least in the 1830s and 1840s, the small body of academic geologists, led by Constant Prévost (at the Sorbonne) and Élie de Beaumont (at the Collège de

59. On the early history of the Société Géologique and its membership, see Albert de Lapparent, 'Rapport d'ensemble sur les travaux de la Société Géologique de France depuis sa fondation', *Bulletin de la Société Géologique de France*, 3rd ser. 8 (1879–80), xix–lvii.
60. *Ibid.*, p. xxvi.

France and the École des Mines), mixed freely with the competent non-academics, among whom the Vicomte d'Archiac, a retired cavalry officer, was outstanding.[61] All groups were happy to publish in the society's *Bulletin* and *Mémoires*, though it is a reminder of the limits beyond which the Société Géologique could not go that Élie de Beaumont, long the most powerful of French geologists, freely exercised the various options that were open to him in the matter of publication; in particular, he used the *Annales des mines* and, increasingly, the Academy's weekly *Comptes rendus*, which offered a speed of publication for short papers that no other journal could match. There was no sense, therefore, in which the Société Géologique monopolized the provision of scholarly aids in geology, but it resembled the other societies of the first generation in winning immediate acceptance as an adjunct to the traditional networks of patronage and power.[62]

The national societies in chemistry, physics, and mathematics — the disciplines at the heart of the Université's curriculum in science — form a separate, second generation, with constraints and opportunities very different from those I have just described. Unlike their predecessors, these societies drew heavily on recruits formally trained and often professionally engaged in the teaching or application of their science. Hence they benefited far more than the first-generation societies had done from the enhanced status of research in academic life, and from the proliferation of science-based careers, academic and non-academic, during the second half of the nineteenth century.

There is abundant evidence that a new interest in research, allied with the first assaults on the monumental insularity of French science, made its presence felt in higher education in France early in the Second Empire.[63] By the 1850s, many French academics, especially those teaching in the faculties of science and medicine, wanted to abandon the traditional preoccupation of their profession with the refinement and diffusion of knowledge and with a variety of bureaucratic functions that had no bearing on innovation. The movement for reform can be interpreted most convincingly as a quest for

61. Effectively d'Archiac made his reputation through the society, publishing nearly all his work in the *Bulletin* and *Mémoires*.
62. The Société Entomologique, in particular, prided itself on its closeness to the highest reaches of academic life, though an examination of the society's *Annales* does not suggest that the degree of active involvement on the part of the élite of French entomology quite matched the effusive declarations of support by Cuvier, Geoffroy Saint-Hilaire, Duméril, and Blainville. For these declarations, see the first volume (1832) of the *Annales de la Société Entomologique de France*, pp. 17–22.
63. For comments on this movement, and the mounting discontent with provision for research, see Robert Fox, 'Scientific enterprise and the patronage of research in France 1800–70', *Minerva*, 11 (1973), 442–73 (442–3 and 459–62); and George Weisz, 'Le corps professoral de l'enseignement supérieur et l'idéologie de la réforme universitaire en France, 1860–1885', *Revue française de sociologie*, 18 (1977), 201–32 (227–9).

enhanced professional status, in particular for a means of challenging the long-established conflation of the professoriate in higher education and the professoriate of the *lycées* as one *corps enseignant*. But it is the effect rather than the motivation of the research-oriented ideology that concerns me here. For my purpose, it is enough to make two points. First, the ideology won many adherents through the 1850s and 1860s — despite total indifference in the École Polytechnique and the other technical *grandes écoles*, and despite a fiercely chauvinist rearguard action led from the École Normale Supérieure by Désiré Nisard.[64] Secondly, and even more importantly, under the Third Republic the ideology became an institutional reality in the research facilities with which most faculties were endowed and in career patterns that reinforced the heightened appreciation of the professional academic's innovative function.[65]

At the same time as the unprecedented emphasis on research, there came educational expansion. Here the doubling of the *lycée* population between 1850 and 1876 and the proliferation of clerical secondary schools following the Falloux Law of 1850 conspired to increase the teaching body at the secondary level; and this in turn helped to stimulate, or at least to justify, the remarkable growth in higher education which took place from the mid-1870s, especially in the faculties. Even in the comparatively unfavourable years between 1873 and 1877, nineteen chairs were created in the provincial science faculties alone, and numerous research assistantships and junior teaching posts were introduced. Thereafter, backed by a booming educational budget,[66] the establishment of new posts proceeded apace. By the early 1890s, with the expansion by no means over, Louis Liard could look back smugly on two decades without precedent in the history of French education. Now, he noted, there were more than one thousand teaching positions in the faculties, compared with 650 in 1870, and the case of the Paris Faculty of Science, in which the number of positions had almost doubled since 1870 (from eighteen to thirty-five), was given as just one indicator of what had occurred throughout France.[67]

64. Claude Digeon, *La crise allemande de la pensée française (1870–1914)* (Paris, 1959), pp. 115–16.
65. On the improvement in research facilities, see Shinn, 'The French science faculty system', *passim*. On career patterns, see the items cited in note 4.
66. The budget for education had risen from 51 million francs in 1876 (representing under 1.7 per cent of the national budget) to 180 million francs in 1894 (representing more than 5 per cent). See *Annuaire statistique de la France*, 15 (1892–4), 564.
67. Louis Liard, *L'enseignement supérieur en France, 1789–1889 [1893]* (2 vols., Paris, 1888–94), vol. 1, pp. 373–4. It should be noted that Liard's statistics refer to 'enseignements'. They include not only chairs but also *maîtrises de conférences* and auxiliary positions associated with a *cours complémentaire*; and they take no account of *cumul*. For more systematic data on the expansion in the faculties of science and medicine, see Lundgreen, 'German perspective', Table 7, and (for the medical

An important consequence of these changes was the growth in the number of academics who were engaged professionally in research but for whom the traditional rewards were far less easily attainable than they had been a generation earlier. A good illustration is the highest of all rewards for a scientist, election to the Academy of Sciences. Here, between 1815 and 1914, the number of resident members rose by only three, from seventy-five to seventy-eight (the additions being, in any case, in the marginal area of geography and navigation); and, predictably, the result was steady ageing. The average age on election rose from forty-four during the First Empire to fifty-four in the period 1900—14, and the number of members under forty years of age declined from eight in 1815, to four in 1850, until by 1867 there were no members aged less than forty-five. Thereafter, as the number of potential aspirants pursuing research-based careers increased, membership of the Academy became an ever more remote ambition for a young scientist. It might, at best, crown his career; it would be unlikely to form an integral part of it, in the way it had done for the generation of Biot, Arago, and Gay-Lussac early in the century.

The implication of this discussion is not that the various changes about the mid-century — the growth in academic research, educational expansion, and the diminishing accessibility of traditional rewards — created the *necessity* for more disciplinary societies of national standing. But, without the changes, it is difficult to see how the national societies, at least in the 'hard' sciences that dominated the curriculum of the Université, could have prospered to the extent that they did under the Third Republic. In the 1850s and 1860s, for the first time, there began to emerge the pool of otherwise unattached recruits which the societies needed; more specifically, they were recruits with both a formal training in their discipline and a material interest in participation in scientific debate and publication.

4. The national societies: new audiences, new roles

Founded in 1857, the Société Chimique de Paris (the Société Chimique de France, as it was called from 1906) was the first society of national standing to benefit from the new availability of informed audiences for physical science at the research level. It was established, modestly enough, by three young assistants in Parisian laboratories, who appear to have conceived it chiefly as an organizing body for series of lectures on chemistry directed at candidates for the *licence*.[68] At first, the group was small and undistinguished,

faculties) Weisz, 'Reform and conflict', Tables 3 and 4, both in *The organization of science*, pp. 72—3 and 328—9.
68. On the founding of the society, see the introductory note in *Société Chimique de Paris. Bulletin des séances de 1858—1860* (Paris, 1861), pp. 1—2, and Charles Paquot (ed.), *Mémorial de la Société Chimique de France 1857—1949. Histoire et développe-*

and its *Bulletin* was little more than a news sheet. There were only ten members at the first meeting,[69] and eighteen months after its foundation the membership had only risen to forty-nine.[70] Quite suddenly, however, in December 1858 its character was changed. At a meeting on 28 December, it was decided, after debate and a far from unanimous vote, that the society should consciously extend the range of its activities.[71] The resolve was evidently aimed at the involvement of more senior academics, and the implications became apparent immediately after the vote, when it was announced that no less a figure than Dumas had agreed to become president, though he was not even a member at the time. Significantly, the key figure in the transformation of the society, Adolphe Wurtz, became secretary.

At the time of the transformation of the society, Wurtz was just forty years of age, already an organic chemist of considerable stature.[72] He had been joint-editor of the *Annales de chimie et de physique* since 1852, had succeeded Dumas in the chair of organic chemistry at the Faculty of Medicine in Paris in 1853, and was a member of the Academy of Medicine from 1856. Yet, although he was close to the recognized 'establishment' in his discipline, he was not quite of it. It was, for example, another ten years before he was elected to the Academy of Sciences, and only in 1874 did he leave the somewhat incongruous context of the medical faculty for the more appropriate chair of organic chemistry at the Sorbonne. Still more importantly, his conception of chemistry as a bond between the academic world and the expanding world of industry, though not without precedent, was radical by the standards of the 1850s.

By 1857, in fact, Wurtz was recognized as one of the leading spokesmen for the younger generation of chemists striving to fashion a new disciplinary professionalism in which both esoteric research and service to the community, in particular to industry, were stressed. Academically, Wurtz's ideal was collaborative research under a master of the kind he had learned at Liebig's laboratory in the 1840s, when he was one of the strikingly few French visitors to work at Giessen.[73] As part of this ideal, he promoted a

ment de la Société Chimique depuis sa fondation (Paris, 1949), pp. 3—7. A useful general account of the first fifty years of the society is Armand Gautier, 'Le cinquantenaire de la Société Chimique de France', *Revue scientifique*, 5th ser. 7 (1907), 641—59 and 680—9; also in *Centenaire de la Société Chimique de France (1857—1957)* (Paris, 1957), pp. 3—89.

69. *Société Chimique. Bulletin des séances 1858—1860*, p. 1. The first meeting was held on 4 June 1857 in a café in the Cour du Commerce.
70. *Ibid.*, p. 5 (in a report of the meeting on 3 November 1858).
71. *Ibid.*, p. 6 (easily located, but note the errors of pagination at this point; it is about here that the new title, *Bulletin de la Société Chimique de Paris*, is adopted).
72. For biographical details, see Charles Friedel, 'Notice sur la vie et les travaux de Charles-Adolphe Wurtz', *Bulletin de la Société Chimique de Paris*, 43 (1885), l—lxxx.
73. Armin Wankmüller, 'Ausländische Studierende der Pharmazie und Chemie bei

determined internationalism, making good use of his rich linguistic background in Alsace; and, in the absence of research facilities in the medical faculty, he even financed his own private laboratory for the training of foreign as well as French students.

With ideals such as these, it is not hard to imagine the lure which the Société Chimique de Paris had in Wurtz's eyes. Above all, it offered him the prospect of control over an independent but securely financed publication in which disinterested research in the German fashion and industrial applications would have their place. It also gave him a context in which he, his pupils, and his immediate circle, among whom fellow-Alsatians like Charles Friedel were especially prominent, might appear before the leaders of the discipline. The only problem was to involve these leaders in a society that had little, if anything, to offer them. To this, personal friendships provided one possible solution, and it was almost certainly as an act of friendship towards Wurtz that not only Dumas but also the other great patron in academic chemistry, A.J. Balard, agreed to take their turn in the annually rotating presidency, along with Pasteur, Sainte-Claire Deville, and others of the younger generation. Not surprisingly, although both Dumas and Balard had interests in applied chemistry that made them sympathetic to Wurtz's ideals, there is no evidence that they sacrificed their multifarious activities in the Academy and other prestigious institutions, or in governmental committees, to become seriously involved in the society; their priorities were simply not Wurtz's.

The presence of the likes of Dumas and Balard, access to a journal, and Wurtz's mixture of amiability and high-minded commitment to reform provided sufficient bait for the young academic chemists who formed the active nucleus of the Société Chimique through the 1850s and 1860s. But in those early years, academic support was not enough, even though the society remained deliberately accessible to all schools and traditions.[74] Hence there were considerations of both ideology and expediency which led Wurtz to court the industrial interest. In the struggle to win this support, the *Bulletin* was the main weapon. Especially after 1864, when it incorporated the *Répertoire de chimie appliquée*, edited by Wurtz's friend Charles-Louis Barreswil, it was without a serious rival as the national journal for industrial chemists.[75]

Liebig in Giessen', *Tübinger Apothekengeschichtliche Abhandlungen*, Heft 15 (Stuttgart, 1966), p. 3–15 (13).

74. The accessibility was demonstrated, in the early years of the society, by the attempt to secure an alternation of followers of Berthelot and Sainte-Claire Deville in the presidency.

75. Barreswil's journal, like Wurtz's own *Répertoire de chimie pure* (which was incorporated in the *Bulletin* in 1863), had appeared, since its foundation in 1858, under the auspices of the Société Chimique. However, it was not an official publication of the society. Charles-Louis Barreswil, a former pupil of Pelouze, was adviser on chemical matters to the Ministry of Agriculture and Commerce.

Within a few years, therefore, Wurtz and his circle had established a context in which all of several degrees and styles of commitment to chemistry could be pursued without any sense of conflict. But, despite their judicious policy of recruitment, the rise in membership during the Second Empire followed much the same slow course as expansion in the career-openings for chemists. By 1870, there were still only 283 members, and concern was expressed at both the negligible rate of growth and the poor attendance at meetings[76] (a chronic problem that has afflicted nearly all national societies throughout their history). Success, however, was imminent. By 1880, membership had reached 400, and thereafter it rose even more rapidly, topping 1,000 in the early years of the twentieth century. The growth owed something to the influx of academic chemists, but even more to the industrial constituency of technically trained personnel and their corporate and individual employers. The contribution of the employers was particularly important in effecting the 'take off' of the society, for it went far beyond the mere act of subscribing, to acts of conspicuous generosity. In the 1880s and 1890s alone, some 180,000 francs were received in the form of donations from firms and individual industrialists.[77]

Another organization which benefited at about the same time from the convergence of academic and industrial interests was the Association Française pour l'Avancement des Sciences, set up in 1872, amid an aura of emotional patriotism, in the aftermath of the Franco-Prussian war.[78] In that it was peripatetic and drew heavily on provincial support, A.F.A.S. resembled the Association Scientifique de France, another national society with interests slanted towards popular astronomy and meteorology which Le Verrier founded in 1864 and which eventually became part of A.F.A.S. in 1886.[79] A.F.A.S., however, was very much a creature of the Third Republic, in a way that the Congrès Scientifiques and the Association Scientifique were not. As such, it was divided into sections organized around the main

76. Thiercelin *et al.*, 'Rapport sur les comptes du trésorier pour l'exercice 1869', pp. 9–10; separately paginated but bound in *Bulletin de la Société Chimique de Paris*, 2nd ser. 13 (1870).

77. For many years, the names of major benefactors headed the list of members in the *Bulletin de la Société Chimique*. The largest single sum, 10,000 francs, was donated by the Solvay company in 1894.

78. The work and administration of A.F.A.S. are best studied in the annual *Comptes rendus* of its meetings, but see also *Association Française pour l'Avancement des Sciences. Notice historique publiée à l'occasion du centenaire de la création de l'Association (1872–1972)* (Paris, 1972).
   The need to restore France's fortunes after the defeat by Prussia was a recurring theme in the early years, and it is no coincidence that several of the leading founders of A.F.A.S., including Wurtz, were from Alsace.

79. In the absence of any account of the Association Scientifique, its activities have to be traced through its *Bulletin hebdomadaire*, which appeared from 1865 to 1887. Throughout the Association's history, the Parisian element in the administration was dominant, and, after the early years, provincial meetings became infrequent.

disciplines of the faculties of science and medicine, together with a miscel-
lany of other pursuits, like anthropology, geography, political economy,
statistics, education, and even archaeology, which began to lay serious claim
to scientific status in France in the 1870s. A.F.A.S. also conveyed its acceptance of the new corporate capitalism through its leaders, who were divorced
at once from the conservative, landed interests that dominated the Congrès
Scientifiques and from the imperial overtones that lingered with any body
headed by Le Verrier. The preparatory meetings were organized by Charles
Combes, the director of the École des Mines and a noted writer on steam
locomotives; its first officers included at least the more liberally inclined
leaders of the academic reform movement, such as Bernard, Paul Broca, and
Wurtz;[80] and its main benefactors were drawn from industrial, commercial,
and, more particularly, railway interests.[81]

For the mass of its support, however, A.F.A.S. looked, perforce, to a less
committed public; in fact, it looked to the very same public which had
turned out for earlier peripatetic gatherings. As it wandered from provincial
town to provincial town, A.F.A.S. enlisted audiences, the administrative
competence of local societies, and municipal subsidies by a programme that
combined patriotism ('Par la science pour la patrie' was an irresistible motto
in the 1870s), an industrially oriented conception of utility, and support for
indigenous provincial culture.

In also espousing the cause of decentralization, A.F.A.S. blatantly usurped
the central ideology of the Congrès Scientifiques, and it was this above all
that piqued the remnants of Caumont's circle in the 1870s. For the hard-line
provincials, A.F.A.S. was yet another symbol of Parisian arrogance, an
attempt to bleed the provinces of their vitality and independence. As the
Breton antiquarian and military writer J.H. Geslin de Bourgogne observed,
in his capacity as secretary for the Congrès Scientifique at Saint-Brieuc in
July 1872, the forthcoming inaugural meeting of A.F.A.S. at Bordeaux was
tantamount to a rejection of forty years of accumulated experience, one
more in a long line of studied insults emanating from the capital:

if, here and there, serious efforts [at decentralization] are being made, would it not be
better to unite rather than fragment them? Why not bring together *all* men of good will in
great scientific gatherings? If Paris thinks it necessary to place its gloved hand in the rude
hand of the provinces, let it not be done half-heartedly. The future of France lies in the
union of . . . all the energies of the country.[82]

80. Bernard was the first president (as a replacement for Combes, who died just before
    the inaugural meeting at Bordeaux). Wurtz was vice-president in 1872, president in
    1873.
81. A particularly generous and active benefactor was the Alsatian banker Adolphe
    Seligman d'Eichthal: he was one of only two founder-members who subscribed
    5,000 francs. The four leading railway companies of France subscribed 2,500 francs
    each — a sum only rivalled, among the industrial benefactors, by the 2,000 francs
    donated by the Compagnie du Gaz Parisien.
82. *Congrès Scientifique de France 38e session . . . Saint-Brieuc . . . 1872*, vol. 1, pp.
    iv—v.

However, two months later in Bordeaux, the provincial *savants* whom Geslin claimed to represent proved less fastidious. Receptions in the glittering Hôtel de Ville and free concerts in the Grand Théâtre were not everyday events, and they were enjoyed with obvious relish by the *congressistes* of A.F.A.S. More than a dozen local societies, from the opulent Société Philomathique to the more sedate Bordeaux Academy, provided publicity, organizational skills, and, above all, a public that had imbibed the customs of scholarly discourse for two generations or more. In fact, it is difficult to imagine a more favourable venue, for A.F.A.S. came to Bordeaux at the end of thirty years in which there had been a Congrès Archéologique (1842), the best-attended of all the Congrès Scientifiques (1861), a national medical congress (1865), and a meeting of Le Verrier's Association Scientifique (1866).

The social and economic conditions of the early Third Republic were such that it was virtually inevitable that A.F.A.S. would prosper while the Congrès Scientifiques were doomed to fade. In the 1870s, with war and defeat an all too vivid memory, A.F.A.S.'s audience was ready to see, in its multifarious objectives, the coherent programme of controlled reform which France needed. Inexorably, membership of A.F.A.S. soared as that of the Congrès Scientifiques fell to unprecedentedly low levels.[83] By 1880, at Reims, A.F.A.S. drew an attendance of 712 from a total membership of 3,156, and in the 1890s subscribers regularly numbered more than 4,000.[84] Financially, too, A.F.A.S. was an unqualified success (see Table 2). Already, in the late 1870s, its capital exceeded 300,000 francs, and, aided by a massive donation of 200,000 francs from the director of the state tobacco factory at Lyon, Girard, it grew to almost 2,000,000 francs by 1914.

A.F.A.S.'s intellectual contribution is less easy to assess, but it was certainly not negligible. Wherever A.F.A.S. went, it stimulated local endeavours: the Société de Géographie Commerciale de Bordeaux was founded as a direct result of the meeting of 1872. And it was a significant source of support for both private and faculty-based research. By 1885, it had distributed 136,781 francs for the purchase of equipment and in aid of publication and travel, and thereafter it disbursed sums of the order of 40,000 francs each year until 1914.[85] Needless to say, such largesse won the warm approval of professional academics, but it was not common for their approval to be translated into sustained, active involvement in the work of A.F.A.S. Members of the Academy of Sciences attending meetings seldom numbered more than four or five, and even at the 1900 meeting, held exceptionally in Paris to

83. For the Congrès Scientifique at Nice in 1878, there were only 133 registrants; see note 40.
84. Note that subscribers were far more numerous than those who actually attended the meetings. The Association Scientifique boasted a membership of about 6,000 in the early 1870s.
85. Roughly half of this sum came from the Association's general fund, the rest from the Fonds Girard.

Table 2. *National societies: finances and membership*

	Date of foundation	1860		1885			1910		
		Members	Income (francs)	Members	Income (francs)	Capital (francs)	Members	Income (francs)	Capital (francs)
Société Botanique de France	1854	420	15,000[a]	422	15,790	32,112	371	15,339	88,693
Société Chimique de France	1857	238[b]	?	579	16,839	110,000[a]	1,124	61,822	400,000[a]
Association Française pour l'Avancement des Sciences	1872	—	—	3,800	79,002	493,808	3,050[a]	87,103	1,800,000[a]
Société Française de Physique	1873	—	—	629	10,998	27,540	1,558	23,203	243,343
Société Zoologique de France	1876	—	—	258	7,784[c]	2,832[c]	343	9,000[d]	2,723[e]

a. Estimated figure
b. Figure for January 1862
c. Figures for 1884
d. Approximate figure for 1911
e. Figure for 1911
Source: This table has been compiled from the membership lists and accounts that appear in the publications of the societies concerned.

276

mark the Exposition Universelle in that year, only seven of the seventy-eight resident members bothered to attend, though forty-two of them subscribed. However compelling its ideology, A.F.A.S. in practice was simply inappropriate to the needs of the élite of French scientists. The overwhelmingly lay audience made the presentation of a paper, at best, an exercise in *haute vulgarisation*, and the delay in the publication of the annual *Compte rendu*, which often did not appear before the following year's session, was totally at odds with the accelerating pace of academic publishing in the Third Republic.

It is an obvious but not insignificant inference of this discussion of the Société Chimique and A.F.A.S. that a flourishing society did not emerge by chance or necessity, even in the relatively propitious conditions that prevailed by the 1870s. Success in the primary task of securing a buoyant membership while maintaining high standards depended on the deliberate policies of organizers who were ready to exploit or create opportunities. A good example of this is the Société Zoologique de France (1876), which was guided to a respected place in zoology after a decade of precarious existence in which it was almost totally ignored by the 'establishment' of the discipline.[86] Since the episode illustrates not only the importance of personal enterprise but also some of the new circumstances that could be turned to the advantage of a society in the Third Republic, I propose to recount it briefly.

What I see as a decisive entrepreneurial role was played, in the case of the Société Zoologique, by a young parasitologist, Raphaël Blanchard, who became secretary of the society in 1879.[87] When he was appointed, at the age of twenty-one, Blanchard was a rather weakly placed but intensely ambitious academic novice with a strong commitment to research and an envious respect for German universities (fostered in a tour of Germany in 1878). An assistant in the physiological laboratory at the Sorbonne, with indifferent patrons and his doctorate in medicine still to complete, he assumed, almost single-handed, the running of the society. His administration was masterly. By the time he retired from the secretaryship, twenty-two years later, the early history of the society, when it was beset by internal dissent, inanition, and imminent financial ruin, was a distant memory; he left the society, if not dominant in French zoology, then at least the only institution that embraced all those with a serious interest in zoology, from academicians to *lycée* professors, and even including the beetle-hunters, travellers, and dealers in specimens who had loomed large among the founder-members in 1876. As Blanchard was at pains to demonstrate, there

86. Robert Fox, 'La Société Zoologique de France. Ses origines et ses premières années', *Bulletin de la Société Zoologique de France*, 101 (1976), 799–812; translated as article IV in this volume.
87. For biographical information on Blanchard, see *Deuxième supplément à la notice sur les titres et travaux scientifiques de M. le Dr. Raphaël Blanchard* (Paris, 1893–1908).

was no necessary incompatibility between, on the one hand, popular lectures and publications, merry excursions, and boisterous annual dinners – all of which helped to maintain a strong non-academic element, chiefly in the provinces – and, on the other, the serious business of publication which most interested him and his research-minded colleagues.

But essential and effective though it was as a means of attracting a lay audience, the mounting of a popular programme left one objective unfulfilled: it did nothing to lure the great patrons. This, as every secretary of a national society knew, was the most difficult task of all, and it was not until 1889 that Blanchard had his first true success.[88] The success arose from an unprecedented combination of events associated with the Exposition Universelle, held in that year to mark the centenary of the Revolution. The part of the programme that lent itself to Blanchard's purposes was the series of some seventy international congresses – in many disciplines, as in zoology, the first of their kind. Money and encouragement flowed freely: in an attempt to outshine the Germans in an activity – research – in which they were coming to be regarded as pre-eminent, the various ministries involved in the Exhibition were only too glad to sponsor the congresses and the subsequent publications, entrusting the task of organization to whatever competent body presented itself. For zoology, the obvious choice as organizer was the Muséum d'Histoire Naturelle. But, with Blanchard's friend Alphonse Milne-Edwards (an influential professor of zoology and future director of the Muséum) deliberately withdrawing, Blanchard was left to mount an exceptionally successful Zoological Congress, to publish the proceedings, and thereby to give his society a hitherto inconceivable status internationally. At such a prestigious gathering, the leaders of the discipline in France could not afford to stand aloof; and since the organizing committee was *de facto* a committee of the Société Zoologique, they could no longer stand aloof from the society either.[89] Quite suddenly, the society found itself being courted by academicians and professors of the Sorbonne and the Muséum who had virtually ignored it for thirteen years.

Over the next quarter of a century, Blanchard consolidated the role he had secured for his society as the representative of France's zoologists

---

88. The low degree of involvement on the part of major zoologists before this date is indicated in the first membership list of the society, published in the *Bulletin de la Société Zoologique* for 1877. The list contains the names of only four chair-holders in major Parisian institutions. Of these, Quatrefages de Bréau, Henri de Lacaze-Duthiers, and Charles Robin were totally inactive, and Edmond Perrier (the only representative of the five professors at the Muséum with zoological interests) resigned in 1879. Of the eleven holders of chairs of zoology in provincial faculties, one was a member.

89. Twenty-four of the thirty members of the organizing committee for the congress, including all eight of the executive officers, were members of the Société Zoologique. The six non-members included three professors at the Muséum (Émile Blanchard, Gaudry, and Georges Pouchet) and two at the Collège de France (Balbiani and Ranvier).

abroad. In this, he was helped by an expansion of international collaboration, notably in the committee on zoological nomenclature that continued to meet far into the twentieth century, and at the triennial congresses which followed the first gathering in Paris. At home, he used the society's international status to promote it as an independent national forum for the presentation and publication of papers. But, paradoxically, this task proved more difficult. For while the new international functions were not hard to exploit, if only because they in no way infringed the established roles of other institutions, it was far less easy to intrude on domestic academic life. At best, the society could hope to complement existing structures; and that is precisely what it did, with its *Bulletin* and *Mémoires* providing an outlet for research which, for a variety of reasons, did not appear in the more sought-after publications of the Academy of Sciences and other major institutions. The duplication of functions was judiciously avoided.

The Société Zoologique is a telling exemplar of the problems of the national societies and the stratagems to which they resorted in their quest for intellectual respectability and financial security. The support of the leaders of a discipline was never easily won, and Blanchard was but one of many conscientious secretaries taxed by the need to cater both for these leaders and for a wider audience. In pursuit of its audience, the Société Botanique (like the Société Géologique) paid special attention to its 'polite' provincial membership, organizing regular meetings in the provinces as a means of sustaining interest. Other societies, most notably the Société Astronomique, sponsored popular publications which made use of new techniques for illustration.[90] And, as I have shown, the Société Chimique cultivated its constituency of industrial and academic chemists (using a network of 'sociétés filiales' to good effect), while A.F.A.S. built on the non-professional traditions of the provincial societies. Clearly, the conditions for success varied with the disciplines concerned — a point that is amply substantiated by the last of the disciplinary societies I shall consider: the Société Française de Physique and the Société Mathématique de France, both founded in 1873.

The main constraint on both of these societies was the exceptional and growing abstruseness of their subject-matter. In the 1870s, it was virtually inconceivable to be self-taught in physics or mathematics, and it was almost as difficult to maintain even the competence of one's student days without being professionally engaged in the discipline. It is a clear indication of this limitation that, in the 1880s, at a time when the Société Zoologique drew only 20 per cent of its membership from the academic profession, more than

90. By 1911, the *Bulletin de la Société Astronomique de France* appeared as a subsidiary part of the popular journal, *L'astronomie*. A similar change took place in 1925, when the severe *Annuaire de la Société Météorologique de France* was transformed into the far more attractive journal, *La météorologie*.

70 per cent of the members of the Société Française de Physique were employed in higher and secondary education. The comparable figure for the Société Mathématique was lower, only 43 per cent, but this figure was supplemented by another 28 per cent of members who held technical posts in the state engineering corps or the army and who therefore had received a thorough mathematical training, often at the École Polytechnique.[91]

Constraint though it was in the early years, the exclusiveness of both societies eventually proved their strength. The Société Mathématique, though always small (218 members in 1885; 285 in 1910), was an important disciplinary refuge in a subject which by 1900 was becoming too esoteric even for presentation at the Academy of Sciences. Academicians remained consistently loyal to it, and there is no doubt that a number of them had reason to be grateful to the society for help in the fashioning of their careers; in a very obvious sense, while it welcomed competent members from the state corps, the Société Mathématique also served academics as a waiting-room for the Academy, in the way that the multi-disciplinary Société Philomathique is often said to have done early in the nineteenth century. The success of the Société Française de Physique was of a different, more spectacular kind, with the growth in industrial engineering and an allied growth in the teaching of physics in *lycées* and technical institutions providing an audience of a size which, even in 1873, would have been unthinkable. As a result, the society became a giant, bigger even than the Société Chimique (see Table 2). However, it was a giant that used its following to foster work at the highest level; it moved increasingly into the more recondite branches of mathematical physics which had been poorly represented in the early years of the society (when conceptually straightforward experimental physics was dominant both in the meetings and in publications). By 1900, while retaining strong links with the electrical and other physics-based industries, it provided facilities, notably its highly competent *Journal de physique théorique et appliquée*, which physicists of the stature of Lippmann, Perrin, and Poincaré valued and used for the presentation of their own work.

Conclusion

It is no coincidence that this study has been suffused by the presence of the Ministry of Public Instruction and the official institutions that catered for French science and scholarship. This is not to say that centralization was complete: the Ministry's protracted and sometimes frenetic attempts to control the local societies call for an obvious qualification to that stereotype. Still less was ministerial control necessarily harmful. Hence I could not

91. By 1910, the proportion of members holding posts in higher or secondary education had risen to 58 per cent, with 14 per cent in the state corps and the army.

endorse the criticism on this point, which was widespread in the nineteenth century and which still pervades historical writing: indeed, the patronage of research in science and the humanities seems to have been effective enough once academics began to express their need for funds, as they did from the 1850s. Yet this very effectiveness set clear limits to the activities of the *sociétés savantes*. Time and again, ministerial action, much of it in response to the agitation of men like Wurtz and Pasteur, reduced the range of appropriate objectives for voluntarist activity.

The point is most easily demonstrated by the systematic duplication of the efforts of Caumont and his circle. But it also obtrudes in the struggles of the national societies. As illustration, we need only compare the French societies (Table 3) with their counterparts in Britain, which were founded earlier, by periods extending randomly to as much as sixty-seven years,[92] and which were nearly always both larger and more prominent in academic life. To return once again to zoology, while the Société Zoologique was struggling to attain a membership of the order of 300 between the 1870s and 1914, the Zoological Society of London, founded fifty years earlier in 1826, had a fellowship of more than ten times the size and was a byword for prosperity, even opulence.[93]

Why the contrast? One answer lies in the range of unfulfilled roles that were available to the London society. It had the only specialist zoological library in London; it offered the most coveted of social assets for any English gentleman — an elegant London club in its West End headquarters; and, most importantly of all for its income, it possessed the leading zoological garden in the country, in the immensely lucrative Regent's Park. (By comparison, the national museum of natural history remained an underfinanced department of the British Museum, competing for space with antiquities and works of art, until it was rehoused in South Kensington in 1880;[94] even then, the desiccated delights of stuffed animals and skeletons in South Kensington were a poor substitute for the fashionable outdoor excursion, with real living creatures, which private enterprise afforded in Regent's Park.) And when the London society engaged in its central scholarly function, it had no sense of being an accessory to a higher, official

92. Of the metropolitan disciplinary societies, only the Entomological Society of London and the Physical Society were founded after their French equivalents, by one year in both cases (1833 and 1874). The sixty-seven-year gap applies to the astronomical societies, founded in London in 1820, in Paris in 1887.

93. See Henry Scherren, *The Zoological Society of London. A sketch of its foundations and development* (London, 1905). In the early years of the twentieth century, the London society had an annual income of well over £30,000 (approximately 750,000 francs), many times that of even B.A.A.S. or A.F.A.S. On the income of A.F.A.S. see Table 2. On the income of B.A.A.S., which averaged about £4,000 in the late nineteenth century, see the financial statements in the annual reports on the meetings of B.A.A.S.

94. A. E. Gunther, *A century of zoology at the British Museum* (Folkestone, 1975).

Table 3. *Dates of foundation of some national disciplinary societies*

Académie d'Agriculture de France	1761 (reconstituted 1804)
Académie de Pharmacie de Paris	1803
Société de Géographie	1821
Société d'Histoire Naturelle de Paris	1821 (reconstituted 1833)
Société Asiatique	1822
Société d'Horticulture de France	1827
Société Géologique de France	1830
Société Entomologique de France	1832
Société de Chirurgie	1843
Société des Ingénieurs Civils de France	1848
Société de Biologie	1848
Société Météorologique de France	1852
Société Zoologique d'Acclimatation	1854
Société Botanique de France	1854
Société Chimique de France ('de Paris' until 1906)	1857
Société d'Anthropologie de Paris	1859
Société de Statistique de Paris	1860
Société Française de Physique	1873
Société Mathématique de France	1873
Société Zoologique de France	1876
Société Française de Minéralogie et de Cristallographie	1878
Société Internationale (*later* Française) des Électriciens	1883
Société Mycologique de France	1884
Société Astronomique de France	1887
Société d'Océanographie de France	1897
Société Française de Psychologie	1901
Société de Chimie Physique	1908
Société Ornithologique de France	1909
Société de Chimie Biologique	1914

network of institutions. In England, there was no equivalent to the powerful professoriate of the Muséum or the zoological section of the Academy. Of course, the Royal Society of London harboured zoologists, but even after the Society's revitalization in the mid-century, it offered few of the disciplinary aids which active researchers like Richard Owen and A.R. Wallace required. Unlike the Academy's heavily subsidized and exceedingly prompt *Comptes rendus*, the Royal Society's publications were slender, certainly no superior to the *Transactions of the Zoological Society* or a number of other private publications.

Another severe constraint on the disciplinary societies in France was the faltering but still powerful system of personal patronage, which tended to bind young men more closely to a master than to peers within the discipline. In return for loyalty, a powerful patron could give decisive help in the advancement of a career, often by providing access to what was effectively his own publication. This pattern of allegiances was reinforced throughout the nineteenth century, as journals devoted to the work of one school or one

282

laboratory proliferated. Again the availability of support from the Ministry, or from the institution, was the key. House-journals never wanted for a subsidy, with the result that what in the early nineteenth century had been a handful of such publications became, by 1900, a comprehensive array of generously financed journals in which masters and pupils alike were glad to place their work. Scarcely a university or any other institution of research or higher education ignored the practice.[95]

By 1914, with so many possible functions pre-empted by official bodies, the *sociétés savantes* occupied positions in French science that ranged from the completely inconsequential to, at best, the supportive. A glance at the 3,000 or more pages that appeared each year in the *Bulletin de la Société Chimique de France* suggests that for societies with a large and informed clientele the role was not inconsiderable. But, important though its functions were, most obviously perhaps in the abstracting of foreign papers (in which it had no official rival), the centres of real power in chemistry remained much as they had been in the 1850s. What we witness in the second half of the nineteenth century, therefore, is the emergence of the major disciplinary societies as a new sector in scientific research, but not a sector that significantly weakened the authority invested in the traditional structures.

95. Especially striking is the proliferation of journals put out by the provincial faculties from the mid-1880s. Often all the faculties in a town would collaborate to produce a joint publication — a first step towards the establishment of Liard's universities in 1896.

III

Learning, Politics and Polite Culture in Provincial France: The Sociétés Savantes in the Nineteenth Century

A ministerial survey conducted in 1886 revealed the existence of no fewer than 655 sociétés savantes in France.[1] They were distributed fairly evenly throughout the country (though with some obvious concentrations not only in Paris and the larger towns, but also in certain regions, notably Normandy), and they had intellectual interests that embraced most branches of pure and applied science, the arts, and the humanities (see Table 1). Even when the 142 societies with their headquarters in Paris are discounted, as they will be in this paper, the yards of society journals alone make the subject a daunting one. For the savants of provincial France were zealous publishers: it was through the printed word that they proclaimed their seriousness, and through *Mémoires* and *Bulletins,* however meagre and irregular, that they secured the all-important exchanges with other societies which allowed them to maintain a library. With the thousands of volumes put out by the societies in the nineteenth century still largely unread by historians, it may appear premature to attempt a general survey, but in this short paper I shall try to do three things: to outline what I see as the main patterns of development in the period from 1815-1914, with special reference to the shifting balance between the restored academies of the Ancien Régime and the new, nineteenth-century foundations; to examine the difficult relations that existed between the societies, with their strong tradition of independence and voluntarism, and the official ministerial world of learning; and, in the context of a review of the diverse and changing functions of the

This text originally appeared in a special issue of *Historical Reflections/Réflexions historiques*, (volume 7, numbers 2–3, 1980), edited by Donald Baker and Patrick Harrigan and entitled, *The Making of Frenchmen. Current Directions in the History of Education in France, 1679–1979.*

1. Lefèvre-Pontalis. *Bibliographie.* There were another twelve societies in Algeria and the colonies.

544

societies in education and polite culture, to suggest some tentative explanations for the following which they enjoyed in provincial France between 1815 and 1914.

First, the pattern of growth. The Convention's decree of 8 August 1793 abolishing the national academies and other learned societies effected a decisive break between the old and the new orders at many levels. This was nowhere truer than in the provinces. All of some thirty-five provincial academies that were functioning on the eve of the Revolution were closed,[2] and a number of them in smaller towns, such as Châlons-sur-Marne and Pau, never revived. But under the Directory, the main academies began to reemerge, thinly disguised as Sociétés libres, Lycées, or Athénées, until by 1820, with the Bourbons firmly restored, ten of them had achieved recognition by the Ministry of the Interior as sociétés royales.[3] With that recognition, there went the right to use an original title (usually of the form Académie Royale des Sciences, Arts, et Belles-Lettres) and at least the prospect of being able to reassert old cultural privileges.

In some respects, it appeared and was meant to appear as if nothing had happened. In towns as different as Bordeaux, Caen, and Nancy (see Table 2) groups of local notables re-formed and, through their activities in the academies, declared their allegiance to the restored Bourbons and to the aristocratic style of learned culture and the political ideology which bound the Restoration to the Ancien Régime. The mere passage of time meant that the notables of the 1820s were seldom those of the 1780s, but notables they certainly were. Throughout the Bourbon Restoration, academy after academy pressed honorary membership on Prefects, rectors of the Université Royale, and mayors, and all of them welcomed leading officers in the local appeal courts or (where appropriate) chairholders in the faculties as respected full members. The presence of this ubiquitous core of prominent administrative and legal officials was clearly not fortuitous. Even if, as in some cases, the core was inactive, it set an aristocratic tone (normally reinforced by a smattering of the culturally inclined nobility[4]) and it served as an unmistakable gauge of political conformity.

What was afoot was a process of cautious social integration, an attempt to assimilate a restricted circle of the educated higher bourgeoisie—predominantly "new men" who might be faculty professors at Caen or doctors at Bordeaux—to a core of centrally appointed administrators and, where it existed, the rump of an old cultural aristocracy. The goal was a unified local élite, with a mission for leadership legitimated by adherence to a national, even international Republic of Letters, and, still more strongly, to

2. On the history of the provincial academies of the Ancien Régime, see Roche, *Siècle*.

3. These were the academies at Arras, Bordeaux, Caen, Cherbourg, Lyon, Marseille, Nancy, Nîmes, Rouen, and Toulouse.

4. Nobles were usually to be found among the honourary members of an academy; frequently they were the senior administrators to whom I have just referred. To take just one example, in the Bordeaux Academy throughout the Restoration and the July Monarchy roughly half of the dozen or so honourary members were titled; very few ordinary members had titles.

an ideology of improvement and public service. At the annual séances publiques, presidents and secretaries might concede the modest scope of the purely intellectual pursuits of their academies: most of their excursions into natural history or the study of antiquities were acknowledged to be parochial in character and, to that extent, they were offered as mere complements to the more synthetic work of the "savants officiels" in Paris. But on the practical utility of the academies, especially to local agriculture, their leaders brooked no reservations. For Raymond Vignes, an innovative landowner who was president of the Bordeaux Academy in 1825, the glory of the institution lay in its contribution to the control of the dunes, the planting of the barren Landes, the wine industry, and the navigation of the Garonne; science for him, as for virtually all his fellow academicians in the Restoration period, was justified by its usefulness.[5] An earlier president, Jean-Claude Leupold, had put his own very similar position as follows in 1820:

> Ce que les sciences ont fait pour la prospérité publique a indiqué leur véritable but; on attend d'elles des lumières positives, des moyens de perfectionnement, des résultats utiles à la société; le bien qu'on leur doit est la mesure de l'estime qu'on leur porte.[6]

The determination of members of the provincial academies that they should at once lead and serve their communities dictated an organizational structure adapted to the clear definition of an élite. Hence the exclusiveness of the academies (most of them had no more than forty full members), the hierarchical structure of membres honoraires, membres titulaires, and membres correspondants, and their attachment to the formal manifestations of Bourbon power were not incidental fripperies; they were part and parcel of an assertion of their quasi-official status. Assertion, however, was one thing; acceptance by the local community quite another. Even in the Restoration period, let alone in the mid-century, it was all too easy to dismiss the academies as irrelevant vestiges of a bygone age. For one thing, the buildings and fine libraries which most academies had possessed on the eve of the Revolution had invariably been sequestered. The Dijon Academy's headquarters had been allocated to the faculties of science and letters, and its library was now part of the Bibliothèque Municipale; the Bordeaux Academy's 40,000 volumes likewise had passed to the state and then to the civic authorities, along with the sumptuous hôtel bequeathed to it by Jean-Jacques Bel in the eighteenth century.

But the most insidious and demeaning problems for the restored academies were financial. They now relied, to an extent they had not done in the Ancien Régime, on municipal and departmental support. This support was not only notoriously fickle and therefore a recurring source of insecurity, it also made the academies dependent on the very communities which they sought to lead. The consequences are obvious in the tales of chronic penury and of academies being shunted from one unwanted civic

5. See Vignes's presidential address for 1825 in *Académie Royale des Sciences, Belles-Lettres et Arts de Bordeaux. Séance publique du 10 mai 1825* (Bordeaux, 1825), 7-12.

6. Leupold, Presidential address for 1820, *Académie...de Bordeaux. Séance publique du 26 août 1820* (Bordeaux, 1820), 3-17.

building to another.[7] Academicians might continue to act out the role of cultural arbiters, but now they did so with few of their old accoutrements—the buildings, libraries, and so forth—and in a society which became less and less interested in winning a derisory sum in a prize competition for eloquence or agricultural innovation or in attending preposterously formal séances publiques.

The irrelevance of the academies became all the more palpable after 1830. Not only were the faculties (including the hitherto almost lifeless faculties of science and letters) beginning to emerge as a more visible force in intellectual life—a point to which I shall return later—but also an unprecedented burgeoning of societies was transforming the independent sector of learning out of all recognition. In the 1830s alone, 78 new sociétés savantes were recognized by the Ministry of Public Instruction, and that level of innovation was maintained at least until the 1880s.[8] These new societies were very different from the academies. In the first place, most of them abandoned any commitment to an undifferentiated Republic of Letters. In their organization and publications, though not necessarily in the work of their individual members,[9] they rejected the cult of the generalist which decreed that the typical academician should be equally au fait with, say, the geology and the antiquities of his region. There were Linnaean societies for the study of botany and descriptive zoology, antiquarian societies for the study of archaeology and local history, societies of bibliophiles, and so on. Even more strikingly, the societies founded after 1830 were almost invariably open: they would admit anyone who could pay the annual subscription of 20 or 30 francs.

The openness of the societies, allied to their receptiveness to the main learned enthusiasms of the nineteenth century—local history, natural history, topography, and, from the 1870s, geography—did more than just attract members; it attracted money, and money meant activity on a scale

7. Even when a municipal council undertook to house a restored academy, the changed conditions of the nineteenth century were reflected in the common practice of creating a "Hôtel des Sociétés Savantes" shared by all the main societies of the town; Bordeaux and Rouen provide good examples of this. For a revealing account of the protracted (but successful) attempts of the Lyon academy to regain its confiscated possessions, in particular its library and prize funds, see Dumas, *Histoire*, 1:428-453.

8. See Table 1 in Fox, *"Savant,"* 250-251, and Table 1 of this paper.

9. In fact, many of the individuals who were prominent in the newer, more specialized societies had far from narrow interests. Arcisse de Caumont, whom I discuss below, belonged to a circle of savants and érudits who were active in more than one of the specialized societies in Caen. Although he was a member of the Caen Academy, he clearly regarded the other societies as more effective vehicles for his diverse intellectual interests.

Among the leading antiquarians of the first half of the nineteenth century, the cult of the polymath with both historical and scientific interests seems to have been particularly pronounced. Charles-Adrien Duhérissier de Gerville, a geologist and botanist as well as an archaeologist and student of church architecture, illustrates the point no less well than Caumont; see Delisle, "Notice," iii-xl (especially p. xvi concerning Gerville's determination to "faire marcher de front l'étude des sciences naturelles, de l'archéologie et de l'histoire").

that few of the faltering academies could match. In Caen, for example, the Academy in 1846 was relying on a patently inadequate income of no more than 700 francs a year, of which 600 francs came as a "grace and favour" payment by the departmental council of Calvados.[10] In a period of 25 years between 1825 and 1850, only seven rather slender volumes of *Mémoires* were published, and in 1851 the Academy had to take the humiliating step of asking its members to pay a subscription.[11] Thereafter, a volume of some 400 or 500 pages appeared every year until the first world war. Quantitatively, however, this still appears a modest effort when seen in the wider context of the combined publications of the societies of the town; and intellectually the Academy's *Mémoires* were not a serious rival to the specialized journals of the Société Linnéenne, the Société des Antiquaires de Normandie (both founded in 1823), and the Association Normande (founded in 1831 to promote agriculture and industry in the old province of Normandy)—to name just three of the more vigorous younger societies with their headquarters in Caen.

By the mid-century, therefore, the world of the provincial societies was becoming more closely synonymous with that of the nineteenth-century foundations. To make the transition to the second part of the paper, it should be added that the embourgeoisement which this trend clearly displays was condoned not only by most conseils généraux and municipal councils but also by the Ministry of Public Instruction. As Minister in the 1830s, Guizot saw the historical societies as an invaluable tool in his quest for a systematic study of provincial archives and antiquities, and it was to this end that, in 1834, he set up what has come down to us as the Comité des Travaux Historiques et Scientifiques to mediate between the Ministry and the voluntarist efforts of the local societies. Salvandy, in the later 1830s and the mid-1840s, was another Minister who warmly encouraged the societies, at least to the extent that they were prepared to work and produce serious publications rather than merely pose as the ornaments of local communities. It was Salvandy who first set aside an item in the educational budget for "subventions et encouragements aux sociétés savantes";[12] and he repeatedly

10. *Annuaire des sociétés savantes de la France et de l'étranger* (Paris, 1846), 430. The information in this volume suggests that the Caen Academy may have suffered from financial hardship rather more than most academies. In Toulouse, for example, it was reported that the Academy received 3,000 francs p.a. from the municipal council, though even this comparative prosperity left the Academy without the botanical garden, observatory, library, and significant private income which it had had in the eighteenth century; ibid., 566-567 and supra.

11. *Mémoires de l'Académie des Sciences, Arts et Belles-Lettres de Caen* (1855): 527.

12. The subsidies were paid either as a general contribution to the funds of societies (sums of between 300 and 500 francs were normal, though they could rise as high as 2,000 francs for a Parisian society) or to support a specific project, usually a publication. A sum of between 50,000 and 70,000 francs was made available each year during the Second Empire; this figure increased under the Third Republic and reached a peak of 176,000 francs in 1883, before falling back to less than 80,000 francs in the mid-1890s.

offered the services of the Ministry as a clearinghouse for information about the societies and as a centre for the distribution of their journals.

Tellingly, Salvandy had only limited success. Just one volume of an ambitious *Annuaire des sociétés savantes* appeared in 1846, and his overtures to the societies frequently fell on deaf ears.[13] Salvandy's successors—Fortoul, Rouland, and Duruy in the Second Empire, Simon and Ferry in the Third Republic—had precisely the same problems when they too tried to yolk the voluntarism of the societies to the network of official institutions. Some of the more ambitious societies, it is true, were ready to collaborate with the Ministry and the Comité des Travaux Historiques in the production of the multi-volume *Dictionnaire topographique de la France* and the *Répertoire archéologique,* both of which began to appear during the Second Empire.[14] But the response was patchy, and even the Ministry's desire to maintain a record of the activities of the societies was constantly thwarted by responses that varied from indifference to active non-cooperation. One reason for this is clear. It lies in the widespread mistrust between, on the one hand, the Minister's own representatives in the provinces—the universitaires employed in the faculties and lycées—and, on the other, the leaders of many of the societies, who feared domination by Paris and by the local purveyors of the Ministry's vision of "official" culture.

In order to elaborate this point about the tension between official, Parisian culture and indigenous, provincial culture, the attitudes of the faculty professors would have to be analyzed in more detail than is possible here; there were local variations to which a brief discussion cannot do justice. But a broad distinction can be drawn between those chairholders who sought assimilation in the world of the local notables—something they might achieve through an academy or a flourishing société savante—and those whose aspirations were oriented firmly towards the longed-for call to Paris and the acquisition of a national reputation. Until the mid-century the great majority of chairholders in the provincial faculties of science and letters fell into the former category. They were often men promoted from local lycées, and they either possessed or were ready to establish roots in their localities,

13. In *"Savant"* I refer to the frustrations which Salvandy experienced in his attempts to establish a central Bibliothèque des Sociétés Savantes. Other marks of the indifference of many societies to ministerial initiatives, most notably their patchy response to requests for information, can be found throughout the Second Empire and the early Third Republic. From about 1890, the signs of indifference are less numerous, chiefly because of the ministry's own loss of interest in the societies; by the 1890s far more attention was being paid to the development of the faculties.

14. The *Dictionnaire topographique de la France,* which began to appear in 1861, was the most successful of three publications intended to give a complete description of France, department by department. The publication, conceived under Rouland as one of the main responsibilities of the newly reorganized Comité des Travaux Historiques et des Sociétés Savantes (the title of the Comité from 1858 to 1881), continued well into the Third Republic; by the first world war, most departments were covered. Only a few volumes of the *Répertoire archéologique* were published, and the plans for a *Description scientifique,* treating the geology, botany, zoology, and meteorology of each department, were abandoned in the early 1860s, after the appearance of only one volume (for the Bas-Rhin).

with the result that they often held their appointments for periods of thirty years or more, with every appearance of contentment.[15] But as the pattern of academic career-making changed in the later Second Empire, with success coming to depend increasingly on originality displayed in nationally circulated publications, so the separation of the intellectual activity of the faculties from that of the academies and other societies became more apparent.[16] The new generation of ambitious, mobile, researching academics, epitomized by Louis Pasteur (professor of chemistry and dean of the science faculty at Lille from 1854 to 1857) and Henri Lacaze-Duthiers (professor of zoology in the same faculty from 1855 to 1863), saw little point in a close association with men outside their profession, however distinguished they might be locally. Predictably, as I have pointed out elsewhere,[17] the contributions of Pasteur and Lacaze-Duthiers to the *Mémoires* of the Société Impériale des Sciences, de l'Agriculture, et des Arts de Lille, were trifling—in sharp contrast with their professorial colleagues in the faculty, Gabriel Mahistre and Auguste Lamy, who published heavily in the *Mémoires* and were evidently satisfied with their traditional role as educational functionaries. The prospect of eventual promotion to a major Parisian chair, which did so much to colour the career strategies of the likes of Pasteur and Lacaze-Duthiers, was not, for Mahistre and Lamy, a realistic or, so far as I can judge, alluring one.

So it was that the provincial sociétés savantes tended to be ignored by the thrusting, research-oriented professionals who began to transform the faculties from about 1860. But the attitudes that prevailed in many of the societies also contributed to their detachment from at least the intellectual élite of the Université. In an organization that had evolved to serve a variety of local needs—and the societies were nothing if not indigenous in this sense—nationally directed career objectives of the kind harboured by Pasteur and Lacaze-Duthiers were likely to be a disruptive intrusion.

No one expressed this last point more vehemently, even obsessively, than Arcisse de Caumont of Caen.[18] An archaeologist, local historian, organist,

15. The loyalty of the majority of nineteenth-century university professors to their institutions, and their apparent contentment with long service in a provincial town, have all too often been overlooked, in favour of the protests of a vociferous minority. To take the case of the Bordeaux faculty of science, we may have paid undue attention to the alacrity with which Auguste Laurent left the Bordeaux faculty after six years as professor of chemistry (1838-1844). Contrast Laurent's career with that of J.J.B. Abria, who was professor of physics in the faculty from 1839 to 1886 (and dean from 1845 to 1886). A similar contrast could be made with Alexandre Baudrimont, Laurent's successor in the chair of chemistry, who, despite some initial unhappiness, was professor from 1849 until his death in 1880, or with Victor Raulin, professor of natural history for over forty years at about the same time. Abria, Baudrimont, and Raulin all published extensively and were deeply involved in the intellectual life of Bordeaux and the town's societies.

16. On the changing nature of academic careers, see Zwerling, "Emergence"; Karady, "Educational Qualifications."

17. Fox, *"Savant,"* 249-252.

18. For biographical information on Caumont (1801-1873), see Charles Richelet, "Notice sur M. de Caumont," *Annuaire de l'Institut des Provinces, des Sociétés*

naturalist, geologist, and internationally celebrated partisan of Gothic architecture, Caumont dedicated his life to giving the disparate activities of the local societies the coordination and encouragement they needed in order to resist (to use his emotional rhetoric) the "cancer of assimilation to Paris." With the aid of the private fortune that came to him with a startlingly biddable wife, Caumont devoted some fifty years, from the 1820s until his death in 1873, to his ideal of intellectual decentralization.

At first, Caumont had not ruled out the possibility of some sort of association with the administrators of higher education and research, who from 1832 were to be found in the newly founded Ministry of Public Instruction. His relations with the first Minister, Guizot, were amicable, without ever being close, and in August 1846 Salvandy went so far as to appoint Caumont as an unpaid "delegate" to the sociétés savantes, with a roving mission to strengthen the links between the Ministry and the societies throughout France.[19] Fleetingly, Caumont was delighted by this mark of approval, but his tendency to present himself as an autonomous leader of provincial intellectual life rather than as the Minister's ally in an attempt to extend ministerial control, quickly, and inevitably, led to a break. By the summer of 1847 Caumont could not conceal his frustration at Salvandy's failure to associate any real powers or clear responsibilities with the appointment, and Salvandy in turn was irritated by what he saw as Caumont's self-seeking waywardness.[20]

Thereafter, with his appointment effectively terminated, Caumont's view of the Université as an insidious threat to provincial vitality hardened, and in the 1850s and 1860s what had once passed for his sturdy independence became open hostility. Even the ministerial allocations which the unthinking secretaries of certain societies had welcomed were seen to carry the threat of subservience, and repeated assertions by Ministers that they only wanted to help the societies while leaving them free to run their own affairs were

Savantes et des Congrès Scientifiques (1869): 358-400; Renault, "Notice biographique sur M. de Caumont," Annuaire Normand, 40e année (1874): 465-499; E. de Robillard de Beaurepaire, "M. de Caumont. Sa vie et ses oeuvres," Mémoires de l'Académie Nationale des Sciences, Arts et Belles-Lettres de Caen (1874): 324-401; Baudot, "Trente ans."

19. Caumont's appointment as "délégue général du Ministère de l'Instruction Publique auprès des sociétés savantes du Royaume" was announced in a letter of 28 August 1846 from Salvandy to Caumont; AN F[17] 3026.

20. In a letter of 9 October 1846 Caumont asked Salvandy for guidance about his role and for information concerning the funds that would be made available for the work of the societies. A helpful reply from the Minister on 23 November 1846 was followed by correspondence showing a steady worsening of relations. Caumont's proposal for a 120-man Conseil Général des Académies du Royaume was summarily rejected, and by 6 July 1847 Caumont wrote to the Minister reviewing his appointment with obvious signs of disenchantment: "Je ne tardai pas à reconnaître que ce titre était vain, que les employés du Ministère ne voulaient rien faire...." Salvandy's reaction to Caumont's hectoring tone in this and other letters was to strengthen his resolve to oppose the Institut des Provinces (see below). Prefects, for example, were instructed to forbid societies in their Department from associating with the Institut. The relevant documents on these matters are in AN F[17] 3026.

dismissed by Caumont as pernicious cant.

Of several institutions which Caumont established to resist the ministerial takeover of learned culture, the most important were the annual Congrès Scientifiques (which began in Caen in 1833) and his provincial counterpart to the Institut de France, the Institut des Provinces (f.1839). Like the Institut des Provinces, the Congrès Scientifiques—held each year for a week or so in a different provincial town—were not scientific in the modern restricted sense. They embraced all the enthusiasms of the sociétés savantes: local history and archaeology, of course, natural history, and a strong utilitarianism, with the emphasis firmly on agricultural rather than industrial improvement. And they did so with an assertive ideology of independence. As Caumont repeatedly reminded his audiences, there was to be no truckling to Paris.[21]

Everything that Caumont touched was infused with an active resentment against the capital, and, for this reason alone, his relations with the Ministry would have been turbulent enough. But by involving in his cause such men as Montalembert and the ultramontane Bishops of Orléans and Arras, Dupanloup and Parisis, he was being downright provocative. By an unmistakable implication, he was wedding the campaign for intellectual decentralization to the clerical challenge to the state's monopoly in education. So, inevitably, in the eyes of two ministers who were particularly dedicated to enforcing central authority—Fortoul (1851-56) and Rouland (1856-63)—he became a prime target.

Repeatedly, as they tried to impose order in their dispersed and growing province, and to restrain the ambitions of the ultramontane clergy, Fortoul and Rouland duplicated initiatives already taken by Caumont, with the allegiance of the sociétés savantes very plainly as the prize. Fortoul's monumental *Revue des sociétés savantes,* launched in 1854, immediately killed an existing bibliography of the work of the societies put out by Caumont;[22] the Congrès des Sociétés Savantes, begun in 1861 and still held each year in Paris, was the Ministry's response to very similar meetings which Caumont had been organizing in Paris for ten years.[23] Moreover, as well as the mere duplication of effort, there were instances of blatant harassment,

21. Caumont's preferred forum seems to have been the Congresses of the delegates of the provincial societies, held each year in Paris from 1851 (see note 23). His aggressive speeches, which he delivered in his capacity as president of the Congresses from 1851 to 1868, are reproduced in the *Annuaire de l'Institut des Provinces.*

22. Caumont's bibliography, published under the auspices of the Institut des Provinces, was entitled *Bulletin bibliographique des sociétés savantes des départements.* The last issue (no. 12), published in the spring of 1854, contains an indignant editorial note on the *Revue des sociétés savantes* by Caumont's collaborator, A. Du Chatelier.

23. The Congresses which Caumont held each year in Paris (for delegates from the *provincial* societies only) are not to be confused with his peripatetic Congrès Scientifiques. The aim of the Paris Congresses, like that of the Institut des Provinces, was to strengthen the sense of continuity in Caumont's various decentralizing enterprises. A first congress was held in 1848, but regular annual meetings only began in 1851; the last Paris Congress took place in 1880. Each year the proceedings were fully reported in the *Annuaire de l'Institut des Provinces.*

with faculty professors who showed an interest in the Congrès Scientifiques receiving official ministerial warnings.[24] Most wounding of all was the Ministry's systematic refusal to grant Caumont and his organizations any form of official recognition; indeed, even though its existence was condoned, the Institut des Provinces was declared illegal.[25]

The episodes of the *Revue des sociétés savantes* and the Congrès des Sociétés Savantes—which were held in the heady atmosphere of the Sorbonne, while Caumont organized his meetings at the Société d'Encouragement pour l'Industrie Nationale—point clearly to the inevitability of the outcome. In the end, the Ministry just had to prevail. Gradually, though they never won over Caumont and his immediate circle, Fortoul and Rouland used prizes and subsidies to wean many a secretary to the idea of an association with the Ministry, and so to emasculate the opposition. By the early 1870s, the struggle was all but over. Now, in the aftermath of Sedan, intellectual decentralization was realized in the unprecedented level of financing for provincial faculties and in the Association Française pour l'Avancement des Sciences, which began in 1872 and quickly replaced the Congrès Scientifiques as the main peripatetic body of scientists. But this, of course, was decentralization by outward diffusion from Paris, the decentralization of a quest for national unity. In so far as it stressed the oneness of Parisians and provincials, of professional academics and devotees, it was a far cry from Caumont's notion of an autonomous provincial culture. Happily, Caumont did not live to see the collapse of his campaign. In 1871 he was incapacitated by a stroke, in 1873 he died, and by 1880, despite the efforts of his closest associates, both the Congrès Scientifiques and the Institut des Provinces had been wound up.[26] The episode was virtually erased from the official record.

24. For example, in 1852 three professors in the local faculty of science who had agreed to serve as officers at the forthcoming Congrès Scientifique in Toulouse withdrew after receiving an official warning. See Renault, "Notice biographique sur M. de Caumont," 491-492.

25. It was the name of the Institut des Provinces which gave the Ministry its pretext for declaring the organization illegal. By article 41 of a law of 11 floréal an X, the title "Institut" was reserved exclusively for the Institut de France. However, as the Minister of Justice advised Fortoul in a letter of 3 March 1852, the law prescribed no sanction against an offending institution (AN F[17] 3021). Correspondence (AN F[17] 3021 and 3026) shows that the Ministry's wish to make the Institut des Provinces change its title remained as strong as ever in the 1870s, but the legal obstacles to taking effective action were never removed.

26. I use the word "despite," though I am tempted to say "because of," for in the 1870s the secretary, and effectively the leader, of the Institut des Provinces, J.-E. Druilhet-Lafargue, was quite as provocative in his relations with the Ministry of Public Instruction as Caumont had been. Correspondence and drafts of letters in AN F[17] 3026 suggest that Druilhet-Lafargue was immovable in his determination to retain the title (and, more importantly, the complete independence) of the Institut des Provinces, despite the assurances of the Minister that if the change of title were made, there would be no objection to the granting of official ministerial recognition. Druilhet-Lafargue was a naturalist from Bordeaux, secretary of the Société Linnéenne de Bordeaux, and an active and prominent Catholic in the city.

And so to the third part of my paper, in which I ask, why? Why did the provincial societies attract the interest they did attract? Why, by 1886, do we have to deal with some 500 of them, with perhaps more than 100,000 members and several thousand volumes of accumulated publications to their credit?[27] An answer is not easy, if only because the sociétés savantes in the nineteenth century were adapted to a wide variety of roles on which I still find it impossible to impose a simple pattern. Some of them, like the Société des Sciences, de l'Agriculture, et des Arts in Lille, served as important channels for philanthropy, sponsoring housing projects, caring for the destitute, and rewarding faithful service to employers. Another very important function, albeit one pursued in a comparatively small number of societies, chiefly in the industrial areas, was the encouragement of technical innovation and the dissemination of intellectual and practical skills. This might be done in the traditional way, through the organization of prize-competitions or through exhibitions of trade and industry which, as in the case of the Bordeaux Exhibition of 1895 mounted by the Société Philomathique de Bordeaux,[28] could occasionally rival the better-known enterprises of the capital.

But the most obvious way of advancing local manufactures, trade, and agriculture was through education. In this field, at a time in the mid-century when state and municipal provision was sparse, there were ample opportunities for private initiatives; and societies in the economically more active areas responded impressively, especially in the second half of the century. The courses they offered varied greatly. In Lille, for example, the Société des Sciences ran a successful Ecole des Chauffeurs, founded in 1857 to give highly specialized practical instruction to the operatives of steam engines. The range of courses offered by the Société Philomathique in Bordeaux was at once greater and less specific. What had begun as a modest programme in 1839 had grown, by the end of the century, into an array of some 80 courses ranging from the most elementary (essentially instruction in basic literacy) to advanced studies in the chemistry of wine manufacture, economics and foreign languages.[29] Like its exhibitions, the society's educational activities were hugely successful. In the last two decades before the first world war, membership stood at over 700; some 2,000 students enrolled each year for courses and (with the aid of a municipal subvention of 10,000 francs) the society's annual turnover regularly exceeded 80,000 francs.[30]

27. In the ministerial survey of 1886 it was noted that by then the societies of Paris and the provinces together had published some 15,000 volumes and that the figure was increasing at a rate of about 500 a year. See Lefèvre-Pontalis, *Bibliographie,* iii.

28. On the Bordeaux exhibitions in the nineteenth century, all of them organized by the Société Philomathique de Bordeaux, see Bénard, *Histoire,.* 371-456.

29. For a general account of the Société Philomathique and its educational activities, see the anonymous publication of the *Revue philomathique de Bordeaux et du Sud-Ouest,* simply entitled *Centenaire de la Société Philomathique 1808-1909* (Bordeaux, n.d. [1909]).

30. The finances of the Société Philomathique are most easily studied from the summary of the accounts and the proposed budget for the following year, a

I leave until last what was clearly the commonest, if not the most lucrative, of the functions of the provincial sociétés savantes. Nearly all societies, even many of those with predominantly practical objectives, existed, at least in part, to gratify cultural needs. The needs in question would most obviously be those of members, for improvement or just straightforward intellectual amusement. But vigorous societies would look far beyond the confines of their membership. A flourishing Linnaean Society (at Lyons, for example) would probably have a museum of minerals, botanical specimens, and stuffed animals. Others, like the Lille society, might have an art collection to display; the Lille society's headquarters, in fact was effectively the town's art gallery. And Bordeaux and Cherbourg are just two of several towns in which the voluntarist expertise of a society and municipal finance combined to support a popular botanic garden.

The diversity of these functions leads me to two fairly obvious comments. First, the roles of the sociétés savantes in their local settings were far richer than their title might lead us to believe. Although the publication of scholarly papers was an activity in which they almost invariably engaged, they were not merely "learned" bodies; indeed, they prospered to the extent that they multiplied their functions. Witness, as evidence, the fabulously wealthy Société Industrielle de Mulhouse, with its internationally celebrated Schools of Applied Chemistry and Design serving the needs of the region's textile industry, and with its Cité ouvrière at once solving a scandal in housing conditions and drawing a much-needed work force into the town.[31] It also had museums of archaeology, natural history, art, ethnography, geology, and technology to provide instruction and polite amusement; it had a substantial *Bulletin* and numerous occasional publications devoted not only to studies of industrial and commercial matters but also to work in local history and literature; and its fine headquarters met what was evidently a considerable demand for cultivated, masculine clubbability (allied, after 1870, to a stubborn French nationalism).

My second comment concerns the social context in which the societies should be set. Briefly, it seems helpful to consider more than just the immediate circle of leaders and members. To use Timothy Clark's very useful model, there was an ever-present "public" beyond the "audience" of those who paid their subscriptions and attended meetings.[32] In the first place,

publication which the society put out each year in the later nineteenth and early twentieth centuries. I am grateful to the secretary of the society for giving me access to these and other papers during my visit to Bordeaux in July 1979.

31. On the varied activities of the society, see Achille Penot, "La Société Industrielle de Mulhouse," forming pp. 1-136 of a separately paginated supplement to the *Bulletin de la Société Industrielle de Mulhouse* 46 (1876), published to mark the 50th anniversary of the founding of the society in 1826; also *Centenaire de la Société Industrielle* (2 vols.; Mulhouse, 1926), 1:1-187. The cité ouvrière was not formally controlled by the Société Industrielle, but the organization which raised the funds and administered the cité from its foundation in the early 1850s was effectively a committee of the society, headed by Jean Dollfus.

32. Clark, *Image*, 12. My thoughts on this model have been greatly helped by discussions with the Davis Center Fellows in 1978-1979, in particular with Professor Donald Scott.

this public provided commercial opportunities, which the exclusiveness and penury of the academies prevented them from exploiting; lecture-courses, art galleries, and botanical or zoological gardens could be profitable, and a society which administered such facilities was certainly more likely to win municipal approval and financial support. Also there was a quite distinct social function which depended entirely on the existence of a public of non-members. With the eighteenth-century academies and the nineteenth-century foundations alike, we are a world away from the secluded cercles littéraires, dedicated, as Maurice Agulhon has argued,[33] to sophisticated sociability. Unlike the cercles and their precursors, the salons, the sociétés savantes sought visibility, whether through their publications, the ceremonial of séances publiques, public service, or the prize-competitions. All of which points in a now familiar direction, to the quest for status and authority as an important goal.[34]

Of course, simply to speak of status and authority is not to say very much. For the status which came with the public espousal of a scientific or humane culture could be sought for many different reasons. In the restored academies of the 1820s, as I have already suggested, the assertion of an aristocratic cultural superiority was part of an attempt by newly formed groups of local notables to establish continuity with the eighteenth century, and so to lay claim to a position of leadership in the community which the turmoil of the Revolution and the Empire had left unoccupied. This was a position from which the notables hoped not simply to dominate culture but, more importantly, to strengthen their control of local affairs, in particular economic affairs.[35] Even in the mid-nineteenth century the objectives of the academicians in this respect seem to have changed rather little, even if by then they had become largely illusory. With the passing of régimes, the political cast of the Prefects and other government officials who made up the leading notables might change, but high administrative office in the provinces continued to carry with it the assurance of an invitation to join an academy. However great their financial problems, and however realistic their assessment of their diminished intellectual importance, the academies never completely lost their nostalgia for the days when they, and not the

33. Agulhon, *Cercle.*

34. For suggestive studies of this quest in very different contexts, see Thackray, "Natural Knowledge;" Haskell, *Emergence.*

35. The activities of the Bordeaux Academy in the Bourbon Restoration illustrate the point very clearly. The Academy was one of the most forceful advocates of the commercial advantages to be gained from the construction of a bridge across the Garonne. It also used its prize competitions to encourage initiatives designed to stimulate the economy of Bordeaux. The case of Mulhouse is an interesting variant on this theme. Since the independent Calvinist Republic of Mulhouse had only joined France in 1798 there were none of the traditional administrative or aristocratic notables of the Ancien Régime. Instead, the notables came from a closely knit circle of old industrial and business families. Founders of the Société Industrielle, they would appear to have been motivated primarily by a quest for immediate commercial advantages rather than by a desire for status (which they already possessed not only through their commercial activities but also by their effective control of the municipal council).

faculties, had been recognized as the main regional representatives of high culture.

The societies founded in the nineteenth century also had a public declaration to make, though with intentions that had less obvious political or economic overtones. These were the societies of the successful, self-consciously cultivated bourgeoisie; by their openness, they endorsed the new social order, and yet they bestowed an exclusiveness of sorts as unmistakably as the academies. Although it could be bought, membership implied wilful commitment to a transcendental world of disinterested science or scholarship. In Edmond Goblot's sense of the word, a 20-franc subscription brought with it "distinction."[36] More specifically, it brought a distinction that had nothing to do with birth or the position a man held in the state bureaucracy, and everything to do with cultural attainment. So, while the function of the academies was primarily to reinforce, unify, and legitimate local élites, the newer societies could actually *confer* social status on anyone who possessed the necessary superior taste, regardless of his office or profession. Hence, I believe, the attractiveness of these societies to self-made men as diverse as the conservative, legitimist Caumont, and the modern-minded, scientistic Homais in *Madame Bovary*.

So long as they either confirmed or bestowed status and authority, both the academies and the newer societies had a very real social function. But it is clear that, as the century passed, so that social function diminished in importance. One obvious reason for this is that from the mid-century, several of the intellectual pursuits which the provincial savants had long considered their own preserve passed irretrievably into the hands of the professional academic. By the 1860s and 1870s, Claude Bernard and Henri de Lacaze-Duthiers were asserting that the life-sciences should emulate physics and chemistry by becoming experimental and laboratory-based; and what they claimed as their ideal became a reality from the 1880s in the refurbished faculties of science throughout France.[37] In botany and zoology, of course, there was still taxonomy and fieldwork to be done, and Linnaean societies and natural history societies continued to perform and publish work in this tradition. But as academic professionalism gained ground under the Third Republic, it became clear that the scientific universitaires regarded the research of the collector and the classifier as of distinctly inferior standing.

Generally, the local humanistic érudits managed to preserve a more substantial toe-hold in research at the highest level: a number of the wonderful volumes of documents put out in the later nineteenth century by the Comité des Travaux Historiques et Scientifiques were the result of a collaboration in which "chartistes" at the head of provincial archives and libraries directed the efforts of a society.[38] But even in historical studies, the

36. For Goblot's use of the term "distinction," see his classic study, *Barrière*.

37. On the views of Bernard and Lacaze-Duthiers, see Bernard, *Rapport*, pp. 131-149 and 221-237, and Henri de Lacaze-Duthiers, "Direction des études zoologiques," *Archives des zoologie expérimentale* 1 (1872): 1-64.

38. In this enterprise it was not uncommon for a trained archivist to work in the context of a local society and for a volume to appear jointly under the auspices of the society and the Comité. For further evidence see works cited in note 14.

self-taught antiquarian had at best a supportive role. There was no sense in which, as a mere gatherer of facts, he could compete with the growing body of professional historians in the faculties of the Third Republic pursuing their scientific Rankean history in the style of Monod and Lavisse.

To the extent that the professionalization of academic science and scholarship made it harder to win even local status through intellectual activity, the societies suffered. But academic professionalism was not the only threat. The position of the societies was further undermined by the fact that the various public roles which had once been their preserve—the administration of philanthropy, the provision of education, the purveying of intellectual gratification for their members and polite amusement for the "grand public"—were passing from the realm of local voluntarist organizations to that of national bodies. By the end of the nineteenth century, for example, the Ministries of Public Instruction and Commerce both had networks of technical schools and university institutes which lessened the demand for the societies' offerings.[39] And even purely cultural needs could be amply satisfied in other ways. Now national disciplinary societies were providing an effective alternative focus for the more serious devotees.[40] An interest in mushrooms, for example, might now be more effectively indulged in the context of the Société Botanique de France or the Société Mycologique de France, both of which were administered from Paris and harboured the leading experts in the field. Moreover, ever since the middle of the century, popular books and journals of science and the arts had blossomed. With their large circulations, handsome illustrations, and authoritative tone, publications like the Gazette des beaux-arts (f. 1859) or La nature (1873) contributed to a noticeable nationalization, and hence a privatization, of culture in the later nineteenth century—to the inevitable detriment of the societies. There was an unmistakable tendency for the wonders of science, the beauties of art and literature, and travellers' tales to be consumed at home through the writings of nationally celebrated authorities.

Of course, the solitariness of a personal subscription to a national journal or a Parisian society did not satisfy everyone. Even at the end of the century, the provincial sociétés savantes could still satisfy a residual need. No longer did they bestow status in the way they had done fifty, seventy years earlier; still less could they strengthen political or economic power; but they, and they virtually alone in the provinces, could give culture a corporate quality. In analyses that stress hidden motives and social aspirations, it is all too easy to forget that, for a professional man who had gone through the enseignement classique, it was quite simply fun to maintain a foot, with his intellectual peers, in the universal, non-utilitarian culture he had imbibed in his lycée. In busy professional lives, the involvement would almost invariably be that of a passive recipient, implying no more than a vicarious presence at the frontiers of knowledge. But, at least from the 1890s and in the major towns, that somewhat second-hand experience of university-level

39. On these developments in technical education, see Day, "Education for the Industrial World"; Paul, "Apollo."
40. On the development of the national societies, see Fox, "Savant," 265-282.

research was possible. It was possible because then academics in the provinces tried to bridge the gulf between themselves and their local communities wHich, in the wake of a self-conscious professionalism, had widened since the mid-century.

The reason for this sudden reversal of attitudes among universitaires is clear: with the euphoria of the expansionist 80s behind them, the universities (as the various corps de facultés became in 1896) needed the financial and political support of their region.[41] As a result of this need, the professoriate became unprecedentedly and (by the standards of the previous thirty or forty years) uncharacteristically sensitive to the local audiences for polite culture. Needless to say, academic professionalization had gone too far for there to be any sense of intellectual parity between purveyors and recipients; the discourse was very much that of the certified expert talking down to the interested layman. But now even the most eminent of the provincial universitaires—men like the future Nobel laureate Paul Sabatier at Toulouse, for example—were willing to appear before local audiences and to involve themselves in the affairs of at least the more workmanlike societies. The contrast between Sabatier's service in the Académie des Sciences of Toulouse at the end of the nineteenth century and Pasteur's indifference to the Société des Sciences de Lille in the 1850s is striking.

On the eve of the First World War, times were hard for most sociétés savantes. Governmental interest in them had manifestly declined since the 1890s, and municipal and departmental support was more likely than ever to be directed to the network of universities, which provided a new focus for civic pride. In a way that few cared to lament in the secular, commercially active *République des Professeurs,* leadership in local cultural life lay unmistakably with the faculties; and national societies and attractive, nationally circulated literature posed a debilitating threat at all levels of membership. The inflation of the post-war period only aggravated the problems of the provincial societies and forced them towards retrenchment in their activities and an ever greater parochialism. The emergence of new societies for local studies, notably the study of regional folklore, in this later period would suggest that the trend was not in every way a damaging one. But the consolations for most societies were slight. By 1920, even the most zealous champions of provincial intellectual life could not mistake the trend. The sociétés savantes remained bastions of bourgeois gentility, and those with industrial interests in particular retained a certain economic role.[42] But in most other respects they had suffered, inexorably and irretrievably, from the professionalization of culture and the systematic assumption of their diverse functions by bodies better suited to the tasks they had once fulfilled almost as a monopoly.

41. The strengthening of the bonds between university academics and their local communities, in particular the industrial and commercial sectors of the communities, is discussed (with referrence to the sociétés savantes) in Fox, *"Savant,"* 252-254; more generally, in Weisz, "French Universities."

42. A mark of the sustained activity of the industrial societies is the triennial peripatetic congress of the Union des Sociétés Industrielles, which began in 1911 and continued to be held throughout the interwar period.

Table 1

The foundation of sociétés savantes in Paris and the provinces by date of formation and main area of interest(a)

	Before 1790	1790-1805	1806-1815	1816-1825	1826-1835	1836-1845	1846-1855	1856-1865	1866-1875	1876-1885	formation not known	Total (b)
Multi-Disciplinary (embracing science and humanities)	22(1)	14	6	10	18	17(1)	10	21(1)	14	23(1)	0	159
Local History, Antiquities, Archaeology	0	1	1	2	9	19	6	11	20(1)	13(2)	0	85
History (non-regional)	0	0(1)	0	0	1(2)	0(1)	0(2)	0(1)	0(1)	0(2)	0	11
Foreign/Classical Cultures and Languages	0	0	0	0(1)	0	0(1)	1	0(3)	1(2)	1(3)	0	13
Literature	1(1)	0	0	0	1	0	0(1)	1	0	5(1)	0	11
Bibliography, Societies of Bibliophiles, etc.	0	0	0	0(1)	0	0	0	1	5(2)	2	1	12
"Beaux-arts"	1(1)	0	0	1	1(1)	3	3	4	3(1)	12(1)	0	32

											Total
Geography	0	0	0(1)	0	0	1	1	2(1)	27	1	34
Anthropology	0	0	0	0	0(1)	0	0(2)	0	2	0	5
Science (General)	0(2)	0	0	1	0	3(1)	2	3(1)	6(3)	0	22
Natural History (including Botany, Entomology, etc.)	0	0	3	1(1)	2(1)	2(2)	1	8	10(4)	0	35
Physical Sciences, Mathematics	0	0	0	0	0	0	0(1)	0(2)	0(1)	0	4
Medicine	1(1)	7(3)	1	4(1)	9(2)	3(3)	4(1)	1(3)	5(5)	0(1)	57
Pharmacy	0	1	0	1	1(1)	0	3	3	5	7(1)	24
Veterinary Surgery	0	0	0	0	1(1)	1	0	0	2	0	5
Agriculture and Horticulture	6(1)	4	2	2(2)	2	0(1)	2(1)	1(1)	3	0	29
Industrial Societies	0	0	0	3	0(2)	2(2)	3	7(1)	3	0	24
Societies of Architects	0	0	0	1	1(1)	2	3	7(1)	10	0	26
Education	0	0	0	2(1)	0	0	1	1	1(4)	0	10

Micellaneous	0	0	0	0	0	2	2	2(1)	5	5	1	18
Total(b)	38	31	12	22	53	69	48	71	98	162	12	616

(a) The figures in parentheses indicate the number of societies founded in each period in Paris; the figures not in parentheses apply to the provincial societies only. Societies in Algeria and the colonies are ignored, as are societies which, although they appear in the survey, were founded after 1885.

(b) "Total" in both cases refers to both Parisian and provincial societies combined.

The data are drawn from Eugène Lefevre-Pontalis, *Ministère de l'Instruction Publique et des Beaux-Arts. Comité des Travaux Historiques et Scientifiques. Bibliographie des sociétés savantes de la France* (Paris, 1887). They only concern societies that responded in 1886, when Lefèvre-Pontalis made his survey, and they take no account of societies that had gone out of existence by then. As a result, there is quite clearly some numerical weighting in favour of societies founded in the more recent decades.

Table 2

The membership of three Academies by occupation
Bordeaux (B), Caen (C), and Nancy (N)

Occupation	1825			1855			1885			1914		
	B	C	N	B	C	N	B	C	N	B	C	N
Education												
Chairholders in												
Fac. Science	0	3	0	3	6	3	4	5	1	3	0	2
Fac. Letters	0	2	0	1	5	4	3	8	5	6	3	6
Fac. Law	0	1	0	0	5	1	2	4	2	2	1	2
Fac. Medicine	0	7	1	2	1	4	4	4	4	7	6	2
Fac. Theology	0	0	0	2	0	0	1	0	0	0	0	0
Ecole Sup. de Pharmacie	0	0	0	0	0	0	0	0	1	0	0	0
Rectors, inspectors	2	2	3	0	1	1	0	1	3	1	0	1
Chairholders in lycées	0	0	1	0	0	0	0	1	2	0	4	2
Teachers in other instns of higher specialized education	2	0	0	1	0	4	0	0	2	1	0	1
"Professeurs de..."	3	2	0	3	0	1	1	0	0	1	0	0
Law												
Officials in Appeal court, Cour royale (imp.)	2	1	8				2	6	2	5	3	2
Magistrates	0	0	0				1	1	1	0	2	3
'Avocats,' 'Notaires'	2	1	1				1	1	0	0	1	0

Administration												
Prefects	1	1	0	5	2	0	1	1	0	1	0	0
Other officials in prefectures	0	2	0	0	0	0	0	2	0	0	0	1
Municipal officials, Councillors, Mayors, etc.	1	3	0	4	0	0	1	0	0	1	2	0
Lower Govt. Admin. (e.g. taxation, customs)	0	0	0	1	0	1	0	0	3	1	1	1
Technical government services												
Ponts et Chaussées, Mines, etc.	4	2	2	2	2	1	0	1	0	3	0	0
Medicine												
Doctors, surgeons	11	3	3	4	0	0	1	0	0	1	0	0
Veterinary surgeons	1	0	0	0	0	0	0	0	0	1	1	0
Pharmacists	2	0	0	1	0	0	0	0	0	0	1	0
Archivists, librarians, keepers of museums	0	1	2	0	1	1	0	2	1	0	4	2
'Hommes de lettres' 'naturalistes,' etc. (i.e. defined by avocation or membership of another society)	3	1	1	8	8	4	4	2	7	2	1	2

Nobility, with no stated occupation	4	1	0	0	0	0	2	0	0	0	0	0
"Propriétaires"	2	0	0	0	0	0	0	0	0	0	0	0
Architects	2	0	0	1	0	2	1	0	1	1	0	0
Painters, sculptors, artists	1	0	0	2	0	0	3	1	0	1	1	1
Skilled Craftsmen (e.g. 'Horloger')	1	0	2	0	0	0	0	0	0	0	0	0
Military, Naval Personnel	0	0	1	0	2	0	0	0	1	3	1	3
Clergy	0	0	0	1	1	0	2	1	2	0	1	2
Business Usually bankers, "négociants"	2	0	0	1	0	0	1	0	0	1	1	0
Miscellaneous	0	0	0	0	2	0	1	1	0	0	0	0
Not known	0	1	0	1	0	0	2	0	0	0	0	0
Total	46	34	25	48	38	34	38	42	38	42	35	33

Only full members of the Academies are included. Honourary, associate, and corresponding members are omitted.

IV

The early history of the
Société Zoologique de France

In France, as elsewhere, specialized scientific societies were a creation of the nineteenth century. Interpreted in the simplest possible terms, they appear as a straightforward response to the growing complexity of science, meeting needs that could no longer be satisfied by such general all-embracing bodies as the Académie des Sciences or the Royal Society in London. This comfortingly straightforward interpretation has a strong element of truth: the fragmentation of the seamless web of science into an array of separate disciplines proceeded apace during the nineteenth century, creating an essential condition for the establishment of the new societies. However, the growing preoccupation of men of science with their own speciality, rather than with the broad spectrum of the sciences, does not entirely account for what occurred. It does not explain, for example, why most specialized societies in France were created later than their counterparts in Britain but earlier than those in Germany, Italy, or the United States.[1]

The fact that, despite the advancing internationalization of science, societies for the same discipline emerged at different times in different countries, in accordance with recognizable national patterns, points to the decisive role of local conditions, and it is with this role that my paper will be chiefly concerned. Quite deliberately, therefore, I shall be setting the history of the Société Zoologique de France in the context of French and, more particularly, Parisian zoology.

The origins of the society

By 1914, the Société Zoologique de France was firmly established as a forum for the nation's leading zoologists. At the time of its foundation in 1876, however, it had been a far more modest affair, made up of a disparate group of enthusiasts with no significant standing in the scientific community. To understand the transition, reference must constantly be made to the institutional structure of French science in the later nineteenth century, since it was that structure which fashioned both the constraints and the opportunities for action. Individual initiative too played an important

---

[1.] For a brief comment on the difference in the timing of the foundation of national disciplinary societies in France and Britain, see Robert Fox, "The *savant* confronts his peers", in this volume, article II, p. 280.

2   The Société Zoologique de France

role. For the Société Zoologique resembled virtually all scientific societies
in the nineteenth century in being created and then sustained in its early
years by the zeal of an active core of enthusiasts rather than by a broad swell
of spontaneous support. Often the activity would be, in effect, the work of
one man. The Société Chimique de France, for example, was fired chiefly
by the zeal of Adolphe Wurtz, just as the Société d'Anthropologie de Paris
owed much of its early vigour to Paul Broca. In the case of the Société
Zoologique de France, the corresponding role was played by a man of rather
less distinction, an obscure, not to say shady dealer in specimens by the
name of Aimé Bouvier. For the history of the society, however, Bouvier's
contribution was no less decisive than those of Wurtz and Broca.

The zoological world in which Bouvier moved was a varied one,
characterized by competing conceptions of the methods and even the extent
of the discipline. In the distinctly dismissive view of Claude Bernard,
zoology, like botany, was a mere *science naturelle*. It was a science of
observation and classification, inferior in its methods and conceptual
framework to physiology, a *science expérimentale* of which Bernard was
recognized as the great spokesman and theorist.[2] A very different position
was that taken by Henri de Lacaze-Duthiers, who maintained that zoology
could be a true experimental science, provided it went resolutely beyond the
limitations of description and external anatomy.[3] In 1876, Lacaze-Duthiers
at the Sorbonne (like Bernard at the Collège de France) was at the height
of his influence, advancing his views in the journal he edited, the *Archives
de zoologie expérimentale*, and putting them into practice in his research at
the marine laboratory of Roscoff.

In the history of the discipline of zoology, this controversy between two
of the leading professors of the capital was of outstanding importance. But,
for Bouvier and his colleagues who founded the Société Zoologique, it was
a marginal affair pursued in a world of great *savants* far removed from their
own activities of collecting, description, and elementary classification.
Nevertheless, even those intellectually modest activities contributed to the
dynamism that fired the life sciences in France throughout the nineteenth
century. The founders of the Société Zoologique, in fact, were the inheritors
of a tradition with roots in the golden age of the provincial academies of
the Ancien Régime, when devotees pursued the study of *flora* and *fauna* with
the same zeal they bestowed on local history and archaeology. Far from

[2]   Claude Bernard, *Rapport sur les progrès et la marche de la physiologie en France* (Paris, 1867),
       pp. 141-2 and 231-4.
[3]   Henri de Lacaze-Duthiers, "Direction des études zoologiques", *Archives de zoologie
       expérimentale et générale*, 1 (1872), 1-64.

decaying, this tradition of natural history had been greatly invigorated in the mid-nineteenth century, in particular during the Second Empire, when (much as in England at about the same time[4]) there had been an explosion of interest in the beauties and wonders of nature.

The pace had been set by polite middle-class society, intent on celebrating its growing prosperity in ways that embraced not only the study of the natural world but also the buying of pictures, visits to the Opera, and travel on the hugely expanded railway network. It was just one manifestation of this new departure in cultural consumption that scientific writers began to enjoy an unprecedented vogue. By the 1860s, no cultivated drawing room would be complete without its handsomely bound volumes of such works as Louis Figuier's *La terre avant le déluge* or *Les merveilles de la science*, while a more serious commitment to natural history would entail the purchase of specialized manuals as aids to the management of the proudly displayed private collections of zoological and botanical specimens of the kind that Bouvier and other *marchands-naturalistes* were only too ready to supply.

Once the immediate disruption of the Franco-Prussian war was past, the vogue for natural history resurfaced. Books, magazines, lectures, excursions, and microscopes and other equipment for use in the home and in the field were as popular as ever, boding well for any enterprise that sought to cater for the growing numbers of bourgeois enthusiasts. The tide of interest did not pass unnoticed by the founders of the Société Zoologique. The circular announcing their plan for a society, which they put out in the spring of 1876, was directed to the widest possible public (see Appendix).[5] In it, a clear distinction was drawn between the prosperity of the "anatomical and physiological sciences", pursued by eminent *savants* and fostered by a healthy range of publications, and the institutional weakness of descriptive zoology. In this field, it was said, only entomology was properly provided for; work in other zoological specialities often foundered or failed to see the light of day for want of encouragement and means of publication.

In stating as categorically as they did that their intention was to advance descriptive zoology, Bouvier and the other forty signatories of the circular set the tone of the society. One inevitable consequence was that the national leaders in zoology showed little interest in the announcement. Of the five

---

[4] David Elliston Allen, *The naturalist in Britain. A social history* (London: Allen Lane, 1976), *passim* but especially chapter 8, on British field clubs.

[5] The circular, in addition to being widely distributed, was published in *L'explorateur. Journal géographique et commercial et hebdomadaire*, 3 (1876), 596. The text I have translated in the Appendix is taken from that source.

4   The Société Zoologique de France

professors at the Muséum d'Histoire Naturelle whose work was more or less related to zoology, only one became a member. This was Edmond Perrier, the malacologist, later to become known for his defence of the doctrines of Lamarck. The two professors who held formally designated chairs of zoology at the Sorbonne, Henri Milne-Edwards and Lacaze-Duthiers, were also conspicuous absentees, as were ten of the eleven holders of chairs of zoology in the provincial faculties of science.[6] (The exception among the provincial chair holders was Eugène Eudes-Deslongchamps, professor of zoology and animal physiology at Caen.)

The explanation for this indifference was, quite simply, that the Société Zoologique had nothing to offer the élite of French zoologists. The ordinary member of the society was too insignificant and the official *Bulletin* too modest to win the attention of someone of the stature of Milne-Edwards. As a professor at the Sorbonne, dean of the Faculty of Science, and a member of the Académie des Sciences, Milne-Edwards enjoyed abundant contacts at the highest level in the scientific community. His pupils were able, and the journal that he edited, the zoological series of the *Annales des sciences naturelles,* served as a prominent and prestigious vehicle for publishing the work of his school. Plainly, for Milne-Edwards to venture outside this world of power and influence would have been to turn his back on the fruits of a brilliant career.

Young, aspiring zoologists, on the other hand, could not afford to take such an olympian view, and it is not surprising that *aides-naturalistes* from the Muséum d'Histoire Naturelle and assistants from various Parisian laboratories were among the society's early members. But they were few in number. Even professional zoologists at the beginning of their careers, it seems, took rather little interest in the society. What advantage could there be for them, either in giving papers at meetings or in publishing in the *Bulletin,* when, as they knew, the public they would address contained so few men who might help them to rise in their discipline? It was far more important for them to foster a reputation in the orbit of one of the great patrons than to give time to a society whose members, even those who were most active, sought little more than an engaging, wholesome diversion.

There was something symbolic, therefore, about the inaugural meeting

6.   According to Blanchard, Milne-Edwards declined the invitation to become an honorary member in 1878 ("on the grounds that any such society ought to have originated in the Muséum and nowhere else"). Lacaze-Duthiers did accept honorary membership, but, as Blanchard later observed, without displaying "the slightest interest" in the society. For Blanchard's recollections, see his presidential address for 1914, in *Bulletin de la Société Zoologique de France,* 39 (1914), 10.

of the society, held on 8 June 1876 amid a jumble of stuffed animals and skeletons in Bouvier's gloomy apartment at 55 quai des Grands-Augustins on the Left Bank.[7] Bouvier himself was the first general secretary, and Jules Vian, a lawyer and amateur ornithologist, was an entirely characteristic choice as president. Around them, there soon formed a small circle of enthusiasts, all eminently competent amateurs, in the mould of Vian. They included the traveller, entomologist, and arachnologist Eugène Simon, the doctor and malacologist Félix Jousseaume, and another doctor, Louis Bureau, all three of them future presidents of the society.

The handful of twenty members who gathered for the first meeting grew rapidly. Within two years, the membership stood at 160, including a quarter from the provinces and as many again from abroad. Bouvier must have been overjoyed. There is no evidence that he regretted the indifference of the leaders of the discipline; indeed, he probably welcomed their absence. If this was his attitude, he would certainly not have been alone, as Jousseaume's intemperate criticism of the collections of the Muséum and, by implication, of the world of professional zoology indicates.[8] Neverthe-less, there were members who took a different view, aspiring to closer links with the academic zoologists and to the reflected intellectual eminence which it was supposed those links would bring.

Crisis and resurrection

Those who sought to build bridges between the society and the far loftier reaches of "official" zoology had an early success, when Perrier, the only active member among the professoriate of the Muséum, accepted the presidency for 1879. But the success was short-lived. In October 1878, when Perrier had still not assumed the presidency, the society was thrown into turmoil. Publication of the *Bulletin* had ceased, the minutes of meetings were non-existent, and articles and donations were arriving at the society without any attempt at acknowledgement. Even more disquieting was the state of the society's finances. The records of subscriptions were in such turmoil that a sum of almost 5,000 francs (corresponding to roughly half the annual salary of a professor at the Muséum) could not be accounted for.

Inquiries over several months pointed inexorably to Bouvier as the

---

[7.] The meeting is described in Blanchard's presidential address of 1914, cited in note 6, pp. 5-6.

[8.] Jousseaume's attack caused predictable offence. It appears all the more inappropriate in retrospect since the main butt of the criticism was Edmond Perrier. On the affair, see Max Vachon, "Méditations d'un Président", *Bulletin de la Société Zoologique de France*, 93 (1968), 13-15.

6   The Société Zoologique de France

culprit.[9] His protestations of innocence convinced no one, and in June 1880, after repeated attempts to shift the blame (notably onto one of his children, who, as he said, had torn up one of the society's record books), he finally accepted the inevitable and resigned. The affair sparked off a spate of resignations, including that of the president, Perrier, and urgent action became necessary if the society was to be saved. With Bouvier removed from the council for 1880, Jules Vian returned, to serve once again as president. The choice had the calming effect that was intended. Vian was not only universally respected and hence a natural conciliator; he also possessed a private fortune that helped to ease the financial crisis.

In his efforts to restore the society, Vian was not alone. His closest allies were Fernand Lataste and Raphaël Blanchard, two young histologists in the early stages of their academic careers, whose training and professional aspirations could hardly have been further removed from his own or from those of most of the members of the society. Lataste was an assistant in Charles Robin's laboratory at the Collège de France, while Blanchard (who was a *protégé* of Robin's younger collaborator Georges Pouchet) worked as a *préparateur*, under Paul Bert, in the physiology laboratory at the Sorbonne.

Predictably, with Lataste as vice-president and Blanchard as general secretary, the society's affairs improved. At last, the Société Zoologique de France had some prospect of living up to its rather grand name by attracting both amateur and professional zoologists across the whole spectrum of the discipline. In Lataste's ambitious words, written in 1880:

> Our concerns must embrace not just a few branches but the
> whole of zoology in all its aspects, descriptive and
> geographical, systematic and anatomical or physiological.[10]

The next ten years saw Lataste's hopes largely realized. No branch of zoology was either excluded or particularly favoured, although in practice the members' main interest remained descriptive zoology. Over the same period, the membership grew steadily, from 161 in 1878 to over 270 in 1889. Even more significant were the rise in the proportion of active members pursuing careers in higher education and the launching of the society's *Mémoires,* a journal that could accommodate contributions of a length inappropriate for the already hard-pressed *Bulletin.*

[9.]   On the Bouvier affair, see: Blanchard's recollections in his presidential address of 1914, cited in note 6, pp. 6-7; the manuscript minutes of the Council of the society for 1879-80, in the volume entitled "Procès-verbaux des séances de la Société Zoologique de France. Commencé le 1er Janvier 1880" (Archives of the Société Zoologique de France); and the minutes of the meetings of the society published in *Bulletin de la Société Zoologique de France,* 4 (1879), pp. VI-VIII, and 5 (1880), pp. I-V, X-XII, XIV-XV, and XXV-XXVI.

[10.]   *Bulletin de la Société Zoologique de France,* 5 (1880), p. IV.

It was Blanchard, even more than Lataste, who must take the main credit for this increased level of activity.   In 1880, he was still only 22.   Yet he fulfilled not only the functions of general secretary but also those of archivist and librarian. Many years later, he recalled, without exaggeration, that he had "organized everything".[11]   With the Bouvier affair at its height, it was Blanchard who settled the society in new premises at 7 quai des Grands-Augustins, in a building that it shared with the Société Géologique. And it was he who managed its increasingly weighty publications and achieved a crucial reconciliation with the Muséum d'Histoire Naturelle in 1886 when Alphonse Milne-Edwards accepted honorary membership of the society.   As the professor of the zoology of mammals and birds at the Muséum, Milne-Edwards was universally acknowledged, after the death of his father, Henri, in 1885, as France's leading zoologist, and the coup for the society was a correspondingly important one.

The 1889 Congress

Nothing did more to enhance the society's standing, however, than its initiative in organizing the first International Congress of Zoology on the occasion of the Exposition Universelle of 1889. Like earlier congresses in other disciplines, the Congress of Zoology benefited from generous governmental funding and the publicity associated with an international event calculated to enhance the reputation of France. Already, in the 1860s and 1870s, as French science had moved to break with the streak of chauvinism that had characterized it in the first half of the nineteenth century, several international congresses had taken place in Paris.[12]   The novelty of these gatherings, the attractions of the capital, and the determination of governments to put on the best possible show had all helped to make them successful occasions, popular with French and foreign scientists alike.

It is not hard to imagine the excitement that a visitor used to the drabness of, say, Victorian England would have felt, in 1889, at the sight of Haussmann's elegant boulevards and Garnier's resplendent Opéra, to say nothing of the Exposition itself.   And there were few *savants* of any nationality who could resist the attraction of being recognized on such a prominent stage by their disciplinary peers.   Moreover, a zoologist attending

---

[11.] *Bulletin de la Société Zoologique de France*, 39 (1914), 8.

[12.] For studies of the history of international scientific congresses, see the special issue of *Relations internationales*, no. 62 (1990), in particular the article by Anne Rasmussen, "Jalons pour une histoire des congrès internationaux au XIXe siècle: régulation scientifique et propagande intellectuelle", pp. 115-33.

8   The Société Zoologique de France

the congress could achieve that personal gratification while advancing his discipline and simultaneously displaying both discrete national sentiment and an ennobling commitment to international cooperation. It was an occasion not to be missed, and even the élite of French zoologists gave its enthusiastic and virtually unanimous support. Any anxieties they may have had were wholly allayed by the fact that, even though the Société Zoologique was to be the main beneficiary of the congress, the early plans were laid in the Muséum, in particular by Alphonse Milne-Edwards.

It was Milne-Edwards who responded to the announcement of the Minister of Trade and Industry offering financial support for a series of congresses that he wished to see organized on the occasion of the Exposition Universelle.[13]   It was not to be expected, however, that Milne-Edwards could mount an international congress single-handedly, and, once the principle of a zoological congress was accepted, he followed the customary practice of formally approaching the relevant specialized society, in this case the Société Zoologique de France. More specifically, he approached Blanchard, his friend since he had joined the society in 1886 and a man who possessed the necessary but rare combination of administrative skill, social graces, and standing in the discipline. Already, by the mid-1880s, Blanchard enjoyed a reputation among zoologists that was remarkable for someone of his age. His research in parasitology and, even more so, his teaching in the Faculty of Medicine in Paris (where he had been *professeur agrégé* since 1883) had been greatly admired; within the Société Zoologique, he had shown himself a gifted and energetic secretary; and he was respected for his combination of patriotic zeal and openness to the wider international community of zoologists and to the ideals of cooperation that grew naturally from his love of travel and command of several languages.

The alliance between the best-known zoologist at the Muséum and the efficient, ambitious general secretary of the Société Zoologique was a powerful one. On formal public occasions, in the middle of group photographs and when making welcoming addresses, Milne-Edwards would be firmly at centre-stage. But Blanchard was never far away, never lost in the crowd. As the congress's general secretary, he took a leading part in all the activities, ranging from the grand formal sessions to the far jollier receptions and excursions. But the lion's share of his energy was expended

13.   The ministerial announcement was made in 1887. On Milne-Edwards's role in the organization of the congress, see Blanchard's address as president, cited in note 6, pp. 10-11, and his obituary of Milne-Edwards in *Bulletin de la Société Zoologique de France*, 25 (1900), 88. No fewer than 69 congresses were held in Paris in 1889, roughly half of them concerned more or less closely with the life sciences.

on the intensely serious matter of zoological nomenclature, which emerged as one of the most keenly debated subjects of the whole congress.[14]

For some years, several circumstances had conspired to make the need for an international agreement on nomenclature particularly urgent. The growth of academic research, the unprecedented interest in exploration and oceanography (reflected in the *Travailleur* and *Talisman* expeditions of the 1880s), and the increasingly international character of science had made it all the more necessary to secure uniformity in the presentation of results. In botany and geology, some progress had already been made, but in 1889 the task in zoology had scarcely begun.

Guided by Blanchard, the congress made progress. It formally approved the binomial system, endorsed the use of a terminology based on Latin (as opposed to Greek or a modern language), and established the principles that should govern the common but often contentious practice of associating a discovery with its discoverer through the use of a Latinized name, such as *Linnei* or *Cuvieri*. But, by the end of the congress, there were still many loose ends, and over the next quarter of a century the prospects of achieving general agreement diminished. As they did so, Blanchard's disenchantment grew. The responsibility, in his view, lay squarely with the zoologists of Germany.[15] Only two Germans were present at the Paris congress and again at the second congress, in Moscow, three years later, when the debates on nomenclature were resumed. But once German representatives began to attend in large numbers, as they did from the time of the third congress, in Leyden, in 1895, they made no attempt to conceal their disregard for the rules that had been adopted in Paris and Moscow. Initially, the German zoologists simply carried on using their own conventions, agreed at a meeting of the Deutsche Zoologische Gesellschaft in 1893. But by 1901, when the International Congress of Zoology was held in Berlin, their attitude had hardened from one of contempt to overt hostility.

With the Germans and the French implacably at odds, Blanchard was tempted to resign the presidency of the Commission Permanente Internationale de la Nomenclature Zoologique, which had been created at the Leyden congress in 1895. But, in an attempt to protect French interests, he stayed on as president and, for twenty years, endured what he regarded

[14.]   The minutes of the sessions on nomenclature are published in the proceedings of the Congress: *Congrès International de Zoologie; Paris 1889. Compte rendu des séances du Congrès International de Zoologie*, ed. Raphaël Blanchard (Paris, 1889), pp. 333-508. The Société Zoologique derived considerable prestige from this volume, which was published under its imprint.

[15.]   For evidence of Blanchard's bitterness, see his "Souvenirs d'Allemagne", *Bulletin de la Société Zoologique de France*, 40 (1915), 19-21.

as the unmerited insults and provocations of his German peers. Despite mounting frustration and irritation, it was not until the dark days of January 1915 that he decided to break his silence in a venomous account of his "Souvenirs d'Allemagne".[16] Blanchard's bitterness was directed at a country which had unleashed its "barbarous hoards" on a France that only sought peace. It is true that his outburst has to be viewed in the context of the many attacks that were being hurled back and forth across the Rhine by scientists and scholars patriotically committed to their country's cause.[17] But, even in that context, Blanchard's statement is remarkable for its violence, and it remains one of the most extraordinary texts to appear in the society's *Bulletin*.

Roles and publics

    In 1889, of course, the years of wrangling over nomenclature could not have been foreseen, and Blanchard had every reason to be satisfied with the outcome of the congress. The number of academicians and leading professors who had remained aloof had been small, and, for the most part, the Académie des Sciences, Muséum, and faculties of science had given their enthusiastic support to the congress and hence also to Société Zoologique which had organized it. Feelings towards the society were warm, the more so as there was no suggestion that it would take the place of the other institutions; indeed, its prosperity depended on its having a function that complemented those of the other, more established institutions. Even when the society began to grow significantly in size and importance, as it did in the wake of the congress, the six members of the zoological section of the Académie des Sciences remained the undisputed élite of French zoology, with their power as patrons and as the official advisors to the government undimmed.   Likewise, the Muséum retained its pre-eminence as France's leading centre for zoological research, challenged (though in no case eclipsed) by only a few of the best laboratories of the faculties of science, which at last had been adequately equipped by the governments of the Third Republic after decades of complaints.[18]

    So by studiously fashioning a role that never threatened the established seats of power, the Société Zoologique was able to emerge as a new and

---

[16.]   See note 15.
[17.]   On this battle of words, see Harry W. Paul, *The sorcerer's apprentice. The French scientist's image of German science 1840-1919* (Gainesville, Fa.: University of Florida Press, 1972), pp. 29-76.
[18.]   Harry W. Paul, *From knowledge to power. The rise of the science empire in France, 1860-1939* (Cambridge, 1985), chapter 3.

independent force in French zoology. As a body that could represent the discipline internationally and as a forum offering facilities for debate and publication to zoologists of every persuasion, both amateur and professional, it met needs that only became apparent as zoology developed new sub-specialities, attracted more practitioners, and eased away from its old rather parochial structures to embrace the spirit of internationalism to which scientists at least paid lip-service in the last quarter of the nineteenth century. The obvious losers in this process, which had its counterparts in virtually all the sciences, were those traditional societies, such as the Société Philomathique de Paris, that sought to embrace all disciplines.

The anxiety felt in the multi-disciplinary societies emerges clearly from two speeches at the centenary dinner of the Société Philomathique in 1888.[19] One speaker, Léon Vaillant, who held the chair of the zoology of reptiles and fish at the Muséum d'Histoire Naturelle from 1875 to 1910, spoke nostalgically of the affection which the intimacy of the society inspired in its members. Another gave a lyrical account of the informality and friendliness of the society's meetings. He looked forward to the day when the Société Philomathique would resume a leading intellectual role as a meeting place to which *savants* would come, as he put it, "in a spirit of friendship and freedom that has disappeared from the Académie, to exchange information, elaborate ideas, and establish the rich contacts from which discoveries will flow in ever greater abundance". Behind that facade of affection and optimism, however, there lay the hard realities of falling attendances at meetings, financial difficulties, and slim, irregular publications.

It was beyond question that the age of the universal *savant* was over. By the last twenty years of the nineteenth century, scientists found it far more attractive to present their best work before a specialized society than to the Société Philomathique, for example. This trend, which was equally marked among those with an established reputation and among the young in search of advancement, reflected a sea-change in the intellectual allegiances of French scientists. The consequences for the Société Zoologique, as for other similar societies, were profound. With its early troubles behind it and the successes of 1889 a vivid memory, a number of leading figures in the discipline began to look to the society as a useful context for the communication of their work. As men of the stature of Vaillant, Perrier, and Alphonse Milne-Edwards became involved, the quality of the papers

---

[19.] An account of the dinner appears in *Bulletin de la Société Philomathique de Paris*, 8th ser. 1 (1888-9), 1-12.

12   The Société Zoologique de France

soared, and the emphasis shifted increasingly to the more sophisticated realms of experimental zoology of the kind pursued in advanced research laboratories.

It was clearly satisfying for Blanchard and his circle to see the intellectual prestige of the society rise, but, as it did so, an old problem became more acute. How was the balance to be struck between the relatively coherent nucleus of professional zoologists who earned their living in research or higher education and the great majority of members whose competence and commitment were far more varied but without whose subscriptions the Société Zoologique could not hope to survive? The solution entailed compromises, in particular measures aimed at catering for the non-specialist element of the membership. One such measure was the introduction, in 1894, of annual "assemblées générales" lasting several days.[20] The aim of the assemblies, which were always held in Paris, was the serious one of keeping members, in particular provincial members, in touch with the latest developments in zoology. But the proceedings also had their more frivolous side. The annual banquet at the Marguery restaurant was the highlight of every programme, with the gaiety of the occasion expressed in menus that celebrated the zoological interests of the *président d'honneur* for the year. Louis Bureau's presidency, for example, was marked in 1898 by a menu adorned with puffins, the chubby representatives of the class of sea-birds in which he specialized (see Plate I).[21] In 1903, Charles Schlumberger's work was honoured by a design reflecting his interest in foraminifera, tiny marine animals of the type that had been collected during the voyages of the *Travailleur* and the *Talisman* twenty years earlier (Plate II). In celebrating Schlumberger's year of office in this way, the society was acknowledging not only some fine research but also a representative of the many members who, as unpaid devotees, made scientific contributions that would have been worthy of their professional peers in academic life. Schlumberger, in fact, was a former *polytechnicien* whose virtually full-time engagement in zoology and the work of the Société Zoologique (which he served as Treasurer from 1890 to 1905) only began after his retirement as a naval officer.

[20.]   Another innovation with a similar purpose was the series of *causeries scientifiques*, inaugurated in 1900 to appeal to a broad lay public. The *causeries* took the form of lectures of general interest which were then reworked as modest popular publications.

[21.]   A founder-member of the society, Bureau was director of the Musée d'Histoire Naturelle in Nantes and a professor in the Faculty of Medicine there. His dedication to zoology was legendary. When asked by his friend Henri Filhol why he had never married, he is said to have replied: "But what would I do with a wife? Am I not already married to science?" See Filhol's speech to the banquet, reported in *Bulletin de la Société Zoologique de France*, 23 (1898), 32.

The same high spirits that surfaced in the banquets were also encouraged by excursions to the countryside around Paris. These outings, at which ladies would be present, combined field work with relaxation and even flirtation. At one of them, in 1890, Blanchard had the greatest difficulty in controlling certain members of his "flock" who were more taken with the charms of an attractive female colleague than with the collecting of specimens.[22] But his protests were light-hearted. As he knew, the fostering of the less sombre side of zoology was essential to the success, or even the survival, of the society. The size of the membership, which stood at about three hundred from 1890 to 1914, could not have been maintained but for Blanchard's skill in reconciling the divergent interests of the academic scientists, active devotees, and interested amateurs on whose allegiance he depended.

The quarter of a century or so before the first world war was something of a golden age for the society, which was now stable, well run, and comparable in size with other national scientific societies. By 1896, its success was formally acknowledged by the granting of the status of a *utilité publique*. This not only gave the society a number of practical privileges, including the unfettered right to benefit from legacies, but also, and even more importantly, set it among the learned bodies officially recognized by the government. It was a rich reward for Blanchard's careful stewardship, and it opened the way to his withdrawal from an active engagement in the management of the society. Four years later, in 1900, his long reign ended when he was succeeded by his pupil Jules Guiart in the position of general secretary.

After Blanchard

Blanchard's resignation after two decades as the effective leader of the society might conceivably have reopened a period of instability. But the passage of the secretaryship to Guiart went smoothly, and even a potential financial difficulty, when the society had to face the cost of installing itself in the Hôtel des Sociétés Savantes in the rue Serpente in 1900, was averted by the Council's successful appeal for contributions.[23] Once established in the Hôtel, which it shared with a number of other societies, the Société

[22.] On this excursion, to Mortefontaine, see the presidential address by Blanchard, cited in note 6, pp. 14-15.

[23.] The details of the subscription appear in the council minutes, cited in note 9, pp. 121-2. The Hôtel des Sociétés Savantes had been inaugurated under the Third Republic to house societies unable to meet the cost of maintaining their own premises.

Plate 1   Menu for the annual banquet of the society in 1898, honouring Lois Bureau, *Président d'honneur*. Courtesy of the Société Zoologique de France.

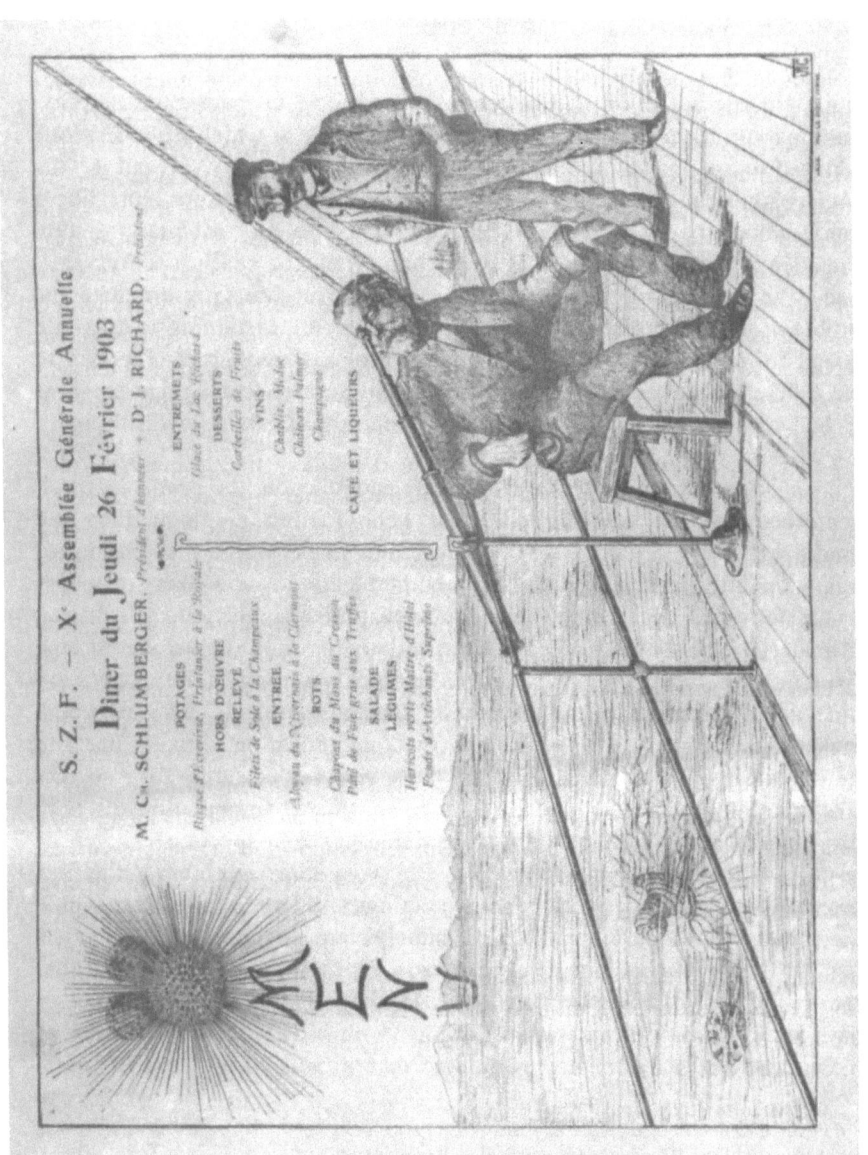

Plate II Menu for the annual banquet of the society in 1903, depicting Charles Schlumberger (*Président d'honneur*) and Jules Richard (President). Courtesy of the Société Zoologique de France.

14   The Société Zoologique de France

Zoologique entered on half a century of solid if unspectacular activity, culminating in the unmistakable signs of faltering to which Etienne Wolff referred in his presidential address in 1958. By then, as Wolff put it, "the society's meetings, despite a core of faithful members, were more like a confessional than a forum".[24] Ten years later, in his "Méditations d'un Président", the arachnologist Max Vachon (who, for nearly twenty years, had been another of the society's dedicated Treasurers) returned to the problem of poor attendances and the difficulty of sustaining worthwhile debate in meetings spanning the whole spectrum of zoology.[25] The drift to specialization, it seems, was threatening the Société Zoologique much as it had threatened and, in due course, marginalized multi-disciplinary societies in the later nineteenth century.

Since the gloomy reflexions of Wolff and Vachon, the problems of all societies, however specialized, have been further aggravated by the emergence of alternatives to meetings and even journals as means of communication. The paper read at a regular meeting is now likely to be less a vehicle for conveying new ideas or observations than the polished version of material already distributed and assimilated by more immediate means. Even the "pre-print" circulates now more quickly and effectively than it did a decade ago, thanks to the ubiquitous fax and the electronic mail which makes the composing of a paper and its communication virtually one and the same act. There will certainly be those who see no future for a society like the Société Zoologique de France in this changing climate: both professional zoologists and the once numerous non-specialist audiences seem to be catered for in other ways. But workshops and other gatherings more precisely focussed than general meetings still have a function to which the society could contribute.[26] Refreshingly, too, the notion that scientists benefit from contacts outside their narrow speciality, again of a kind that the society could facilitate, dies hard. My hope is that it dies sufficiently hard to ensure future celebrations, of the 150th and 200th anniversaries, as memorable and encouraging as the one that stimulated this text in 1976.

[24.] *Bulletin de la Société Zoologique de France*, 83 (1958), 12.

[25.] See page 17 of the "Méditations d'un Président", cited in note 8.

[26.] The success of the society's conference on "Electropheresis and taxonomy" in 1983 illustrates the continuing interest in thematic meetings. For a comment on this example and on the problems discussed here, see Jean-Loup d'Hondt, "Histoire de la Société Zoologique de France. Son évolution et son rôle dans le développement de la zoologie", *Revue française d'aquariologie herpétologie*, 16e année, no. 3 (1989), 65-100 (93-4).

Appendix
SOCIETE ZOOLOGIQUE DE FRANCE

Every year, zoological studies become more prominent and attract an ever increasing number of followers. In France, the anatomical and physiological sciences in particular have had a central role in this movement and have made rapid progress. Taught by eminent men, they have contributed significantly to the improvement of methods of classification and they have yielded important discoveries. In most of our great scientific institutions, these sciences form schools and they have access to a large number of publications. But DESCRIPTIVE ZOOLOGY, through which we learn to distinguish animals from one another and which teaches us about their form, characteristics, habits, and conditions of life, as well as their habitats and geographical distribution, appears for some time to have been neglected by us. In almost all other European and American countries, however, there are numerous societies concerned with the whole range of the zoological sciences. Each year, these societies publish important studies of both local and foreign *faunas*.

In France, only entomology has managed to create a thriving society, with a journal that is universally respected. But the various other branches of zoology, which have many practitioners in our country, especially in Paris, still lack a meeting-point for their activities. Many studies and observations remain unpublished; others are not completed or are neglected, for want of a society able to bring them together, facilitate the publication of their work, and keep them informed of what is published in France and abroad.

Concerned to fill this gap and convinced that we shall receive the willing support of everyone who is interested in general zoology or the still little known *fauna* of France, we publish the programme of the association we have just founded under the title of the

SOCIETE ZOOLOGIQUE DE FRANCE

MM.   Dr E. Alix, B. du Boccage, Bémer, Berard, A. Bouvier, H. Bouvier, Cte Constantin Branicki, Louis Bureau, P. Carbonnier, Albert Cretté de Palluel, Delessalle, Eudes Deslongchamps, Adrien Dollfus, D.-G. Elliot, Ed. Fairmaire, Jules de Gaulle, Z. Gerbe, de Giveuchy, Dr Goyard, Jules Gros, S. Gugenheimer, Bon d'Hamonville, Charles Hertz, Dr Félix Jousseaume, Prince Ladislas Lubomirski, Dr A. Lucan, G. Lunel, A. Maingonnat, Alfred Marche, Dr Marmottan, E. Mulsant, G. Pouchet, Alexandre Quinet, Cte de Riocourt, Dr Ch. Robin, R.-B. Sharpe, Maurice Sedillot, D.-N. Severtzow, Ladislas Taczanowski, Eug. Simon, Jules Vian.

Acknowledgements

I am grateful to the Royal Society for a grant under its European Science Exchange Programme, which allowed me to work in Paris in preparation for the centenary conference of the society, held in Paris in September 1976. In writing this text, which is based on the French version published in the *Bulletin de la Société Zoologique*, 101 (1976), 799-812, I have retained some of its character as an evening address read at the 1976 conference.

16   The Société Zoologique de France

It is a pleasure to acknowledge the friendly help I received, while preparing the original paper, from the late Professor Max Vachon, of the Muséum National d'Histoire Naturelle. I am also grateful to the current President (Professor A. Beaumont) and the Council of the society for permission to publish this English version. Finally, I express my special thanks to the society's Vice-President, Professor B. Condé, and General Secretary, Monsieur Jean-Loup d'Hondt, who helped me to secure the photographs reproduced as Plates I and II.

Monsieur d'Hondt is the author of an important recent study of the history of the society: "Histoire de la Société Zoologique de France. Son évolution et son rôle dans le développement de la zoologie", *Revue française d'aquariologie herpétologie*, 16e année, no. 3 (1989), 65-100.

V

Presidential address: Science, industry, and the social order in Mulhouse, 1798–1871

THERE is a story, which historians of modern France often tell, of the ministerial official in Paris who had only to glance at his clock in order to know the exact passage of Vergil being construed and the law of physics being expounded in every school throughout the country. Invariably, the story is told for a purpose. It is used to demonstrate the high degree of centralization and the attendant rigidity of the French educational system, usually with special reference to the nineteenth century. The story, which has its roots in the rich corpus of Napoleonic legend, serves this purpose very well, but unfortunately it is both apocryphal and misleading. For while it is true that most nineteenth-century ministers with responsibility for education aspired to the ideal of total control, not one of them came close to it in reality.

This disparity between ideal and reality has become very obvious to me in my own recent research. I am now convinced that provided we move out from Paris and look at France as a whole, then it is diversity rather than uniformity which emerges as the dominant characteristic not only in education but in intellectual life generally. It would be very odd if this were not so. To the eve of the Revolution, the old provinces of France—Picardy, Provence, Normandy, and so on—jealously preserved their various identities. Through their local administrators, they fought vehemently for their diverse economic interests and cultivated distinctive traditions in dress, folklore, and even language. In the nineteenth century, these traditions were still tenacious survivors. Regional patois continued to be used, as they are to this day, and Breton, Provençal, and Flemish were only

* Department of History, University of Lancaster, Lancaster LA1 4YG.

This is a revised version of the Presidential Address delivered at the Annual General Meeting of the British Society for the History of Science in Manchester 15 May 1982.

I am grateful to the Royal Society of London for a grant towards the cost of research in France and Britain. I have also drawn on work, financed by the Joint SERC/SSRC Committee, which forms part of a more general study of the relations between scientific education and research and industrial performance in Europe since c. 1850.

In preparing the text for publication, I have been greatly helped by my recent appointment to a British Academy Readership in the Humanities and by a discussion of some of the material in the paper at the Parex seminar on 'Science, medicine, and technology in Restoration France, 1814–30', held at the Maison des Sciences de l'Homme, Paris, 31 August–2 September 1983.

three of several distinct languages which as late as the 1860s, continued to infuriate Parisian officials intent on demonstrating the oneness of France.[1]

I believe that the existence of these deeply ingrained local traditions, in the face of governmental attitudes that were usually hostile and never truly encouraging, is as important for historians of science as it is for other historians of French culture. As I have argued elsewhere, one important focus for cultural provincialism in the nineteenth century was the network of learned academies;[2] here, nostalgic but by no means incompetent gerontocracies advanced their claims to be regarded as the local arbiters of culture and the champions of economic improvement. Especially after the 1830s, the academies were joined by a flood of less select societies, with interests, notably in antiquities and natural history, that were usually even more parochial. Municipal and departmental authorities also emerged as the patrons of initiatives which served as a bulwark against the take-over of French intellectual life by Paris. As a result, in towns of even modest consequence, learned societies, municipal lecture-courses, museums, and botanical gardens abounded, and, through them, such determinedly provincial *savants* as Boucher de Perthes (in Abbeville), Félix Pouchet (in Rouen), and Henri Lecoq (in Clermont-Ferrand) were able to fashion national reputations.

Of course, my assertion of the neglected vigour of the provincial traditions in nineteenth-century French science is not offered as an argument for disregarding the 'savants officiels' of the great national institutions. But I do believe that a tendency to view French science exclusively through the distorting prism of Paris has left us with an unbalanced secondary literature. Signs that the view from the periphery might at last be attracting more serious attention are, in this respect, encouraging;[3] but, for the time being, the way ahead seems to lie in the case-study rather than in synthesis. Hence, despite my programmatic generalizations, I shall devote the rest of my paper to an examination of

[1] François Furet, Jacques Ozouf, *et al., Lire et écrire. L'alphabétisation des Français de Calvin à Jules Ferry* (2 vols., Paris, 1977), vol 2, pp. 324–48, and Eugen Weber, *Peasants into Frenchmen. The modernization of rural France* (London, 1977), pp. 67–94, 310–16, and 498–501.

[2] Robert Fox, 'The *savant* confronts his peers: scientific societies in France, 1815–1914', in Robert Fox and George Weisz (eds.), *The organization of science and technology in France, 1808–1914* (Cambridge and Paris, 1980), pp. 240–82 (244–58), and 'Learning, politics, and polite culture in provincial France: the *sociétés savantes* in the nineteenth century', *Historical reflections/Réflexions historiques*, 7 (1980), 543–64.

[3] See, for example, Terry Shinn, 'The French science faculty system, 1808–1914: institutional change and research potential in mathematics and the physical sciences', *Historical studies in the physical sciences*, 10 (1979), 271–332: Harry W. Paul, 'Apollo courts the Vulcans: the applied science institutes in nineteenth-century French science faculties', in Fox and Weisz, *The organization of science*, op. cit. (note 2), pp. 155–81; Mary Jo Nye, 'The scientific periphery in France: the Faculty of Sciences at Toulouse (1880–1930)', *Minerva*, 13 (1975), 374–403; and George Weisz, 'The French universities and education for the new professions, 1885–1914: an episode in French university reform', *Minerva*, 17 (1979), 98–128, and *The emergence of modern universities in France, 1863–1914* (Princeton, New Jersey, 1983), pp. 134–95.

Illuminating though they are, these studies are all concerned with 'official' science in the provinces, chiefly in the faculties of science. They throw little light on the more indigenous traditions of provincial science.

just one type of science—industrial science—in one Alsatian cotton-town. In doing so, I hope to demonstrate the locally conceived nature of the functions and problems of science in Mulhouse and to justify the misgivings that I feel about the supposed existence of a single entity called 'French science'.

Culture and Authority in the Traditions of Mulhouse

Mulhouse was one of the wonders of nineteenth-century France.[4] Its population grew from about 6,000 at the time it became part of France in 1798 to 30,000 in 1848, and finally to almost 60,000 on the eve of the war of 1870, as a result of which it passed, with the rest of Alsace, to the German Empire as part of the annexed eastern territory or *Reichsland* (see Table 2). Expansion on this scale was rivalled only by that of Roubaix, among the major industrial towns. In a mere seven decades, it brought a community which at the beginning of the century had not even appeared among the hundred most populous towns of France to an incomparably more prominent position as the seventeenth town of Napoleon III's Empire.[5]

The leading industrialists of Mulhouse always insisted that they and their town were truly French. But Alsace, tucked away behind the Vosges and with lines of communication that led more naturally northwards and eastwards than to the west, was always an oddity when viewed from Paris. The language of all but the highest classes was one of the German dialects, and the society was dominated by an economically and politically powerful protestant minority. And if Alsace as a whole seemed odd, often menacingly so, to Parisian administrators, Mulhouse in the southern Alsatian department of the Haut-Rhin seemed odder still and even more suspect. As a local sub-prefect observed ruefully in 1821, the people of Mulhouse were 'a race apart'—'apart' that is, from other Alsatians.[6] And a

---

[4] The point is reflected in a vast secondary literature. In this paper, I draw in particular on: *Histoire documentaire de l'industrie de Mulhouse et de ses environs au XIXe siècle (Enquête centennale)* (2 vols., Mulhouse, 1902), and Paul Leuillot, *L'Alsace au début du XIXe siècle. Essais d'histoire politique, économique et religieuse (1815–1830)* (3 vols., Paris, 1959). For a convenient economic history of the region, based on the *Histoire documentaire* and other standard sources, see Henry Laufenburger and Pierre Pflimlin, *Cours d'économie alsacienne* (2 vols., Paris, 1930–2), vol. 2 ('L'industrie de Mulhouse'). An older but still valuable study cast in Durkheimian terms is Robert Lévy, *Histoire économique de l'industrie cotonnière en Alsace. Étude de sociologie descriptive* (Paris, 1912). Among more recent works, special mention should be made of the essays in Georges Livet and Raymond Oberlé (eds.), *Histoire de Mulhouse des origines à nos jours* (Strasbourg, 1977). Biographical information is readily available in François Édouard Sitzmann, *Dictionnaire de biographie des hommes célèbres de l'Alsace* (2 vols., Rixheim, 1909–10). I have not given references to basic information contained in these volumes.

[5] On the population of Roubaix, which increased from 8,000 in 1801 to 65,000 in 1866, and of other French towns, see the tables in Paul Meuriot, *Des agglomérations urbaines dans l'Europe contemporaine. Essai sur les causes, les conditions, les conséquences de leur développement* (Paris, 1898), pp. 93–5.
The explosion of the populations of Mulhouse and Roubaix should be contrasted with the more sedate growth of most other textile towns. The population of Rouen, for example, grew by only about 15 per cent between 1801 and 1866 (from 87,000 to 100,000). Even the three-fold increases that occurred in the same period in the populations of Lyon and Lille (from 109,000 to 323,000 and from 54,000 to 154,000 respectively) seem modest by comparison with what occurred in Mulhouse.

[6] Quoted in Paul Leuillot, 'Le centenaire de Lambert (1828) dans le Mulhouse en expansion au début du XIXe siècle', in *Université de Haute-Alsace. Colloque international et interdisciplinaire Jean-Henri Lambert. Mulhouse, 26–30 septembre 1977* (Paris, 1979), pp. 75–93 (78).

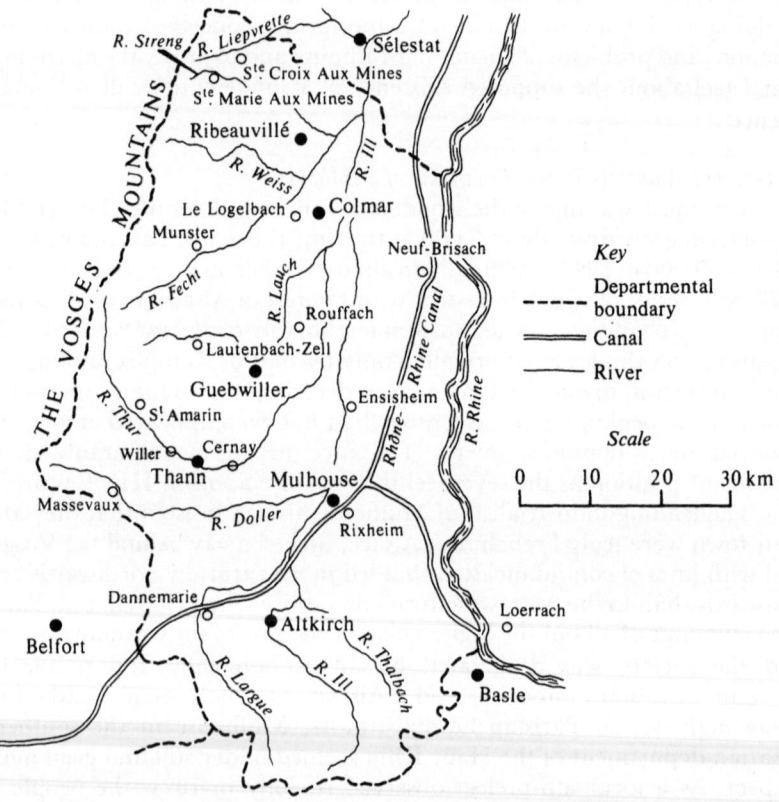

FIGURE 1. Department of the Haut-Rhin

race apart they really were, with a distinctive mentality born of a distinctive history that still weighed heavily on them in the nineteenth century.

From the fifteenth century until 1798, Mulhouse had been a conservative, independent republic, ruled by its closed community of burghers and, since 1524, by the most rigorous Calvinism. From the mid-seventeeth century, the leaders of the republic—the *bourgeois privilégiés*—had assiduously cultivated the French language (at least in public) as a way of demonstrating the superior taste which distanced them from a rustic, not to say uncouth, environment. Yet at the same time as they flaunted their immersion in French culture, they missed no opportunity of fostering other traditions that were quite alien to the France of the Ancien Régime. They preferred to educate their sons not in France but in Calvinist Switzerland—for this purpose, Neuchâtel and Lausanne were particularly favoured; and it was only at the end of their education that boys might be sent to Montbéliard or some other convenient French town in the vain hope that the last traces of their Alsatian dialect might be eliminated.[7] Marriage patterns, too, were carefully contrived to reinforce the unswerving Calvinism of the young and their attachment to their region. In this as in all the strategies of the great families of Mulhouse, both during and after the years of independence, the overriding aim of having the best of both worlds was clear. To be French without being wholly French was one of the most potent ways of securing what was, and remained until 1870, their highest priority, to maintain their local power.

During the first half of the nineteenth century, the determination to resist assimilation into the main stream of French life remained as strong as it had ever been. For one young schoolmaster, the future novelist Émile Souvestre, who was sent to Mulhouse fresh from Catholic Britanny in 1836, the shock was profound. The joyless severity of the town, perhaps even the gruff boorishness of the hotel-keepers and Alsatian-speaking tradesmen were to be expected. But the paucity of social and literary refinement was something that Souvestre had not bargained for, as he told a large and eager public in the *Revue de Paris*. His main target, inevitably, was the industrial community, obsessed with work, to the exclusion of all but the most basic human needs:

> After a full and busy day in his factories [he wrote] the *industriel* goes home simply to eat and sleep. As a result, his social contacts are limited to his closest relatives, and even at these family gatherings he says little. Tired after the day's work and anxious about the day ahead, he is usually content to do no more than digest in society.[8]

Even the rising generations offered no hope of improvement. At the age of five, according to Souvestre, a child would know the price of coal; at eight, he would understand the principles of the steam engine; by fifteen he was a foreman.[9]

For his unflattering portrait of Mulhouse, Souvestre suffered a predictable fate. Public ridicule was intolerable, and his victims duly intervened with the Minister of Public Instruction to secure his dismissal.[10] Souvestre's crime, of course, was to have perceived and published the truth. His description of the tasteless profusion of possessions in the homes of the industrial élite, of the preoccupation with comfort at the expense of elegance, and of the obsessive industriousness that left the owner of a factory with less leisure than his humblest employee, was only too accurate.

[7] Paul Mieg, 'La langue et la culture française à Mulhouse jusqu'à la fin du XVIIIe siècle', in *Les lettres en Alsace* [Publications de la Société Savante d'Alsace et des Régions de l'Est, no. 8] (Strasbourg, 1962), pp. 179–92.

[8] Émile Souvestre, 'Mulhouse', *Revue de Paris*, new ser. 31 (1836), 145–53 (147).

[9] Ibid., p. 148.

[10] It seems virtually certain that some kind of intervention occurred, though the published information on Souvestre's rapid departure from his post at the *collège communal* is limited to a note in the *Histoire documentaire de l'industrie de Mulhouse*, op. cit. (note 4), vol. 1, p. 84.

132

It is inconceivable that Souvestre overlooked the doctrinal foundations for the attitudes he described, though he kept his comments on the Calvinism of the industrial families quite separate from his general disparagement of their customs and values. In reality, such a separation never existed. In all aspects of their lives, the *industriels* of Mulhouse made a quite deliberate display of their religiously inspired austerity. Calvinism and earnestness were both very much part of the public man, and they were duly invoked, in generation after generation, as the hallmarks of superiority. Obituaries throughout the nineteenth century made it plain that the success of the industrial clans owed everything to their simplicity, their dedication to work, and, above all, the manifest piety which they displayed not only in formal religious observances but also in compassionate acts of charity. These were the qualities which justified wealth and authority and which made the *industriels* fit objects for emulation. A comparison of the eulogies of Henri Schlumberger, Nicolas Koechlin, and Jean Zuber *père*, all of whom died in 1852, makes the point very plainly.[11] In all three cases, a story of early struggle, resilience in the face of personal tragedy, generosity to family and employees alike, and lightly borne distinction was united with references to providence which implied unmistakably that here, in a remote corner of Alsace, something resembling a divine plan was being unfolded.

In suggesting that the maintenance of local power was the highest priority of the Mulhousien élite, I am making a deliberately undifferentiated statement. For while the character and objectives of the élite remained the same in the two hundred years or so up to 1871 the means of power varied greatly. Until the mid-eighteenth century, the means had been straightforwardly political: a few families—those of Dollfus, Koechlin, and Hofer, in particular—had hogged the main offices of the republic, including the all-important position of burgomaster, and steered the republic's Grand Council in whatever direction best served their interests. But from 1746, when the first calico-printing works were established in Mulhouse, the context for the exercise of power began to change, albeit with no significant shift in the *location* of power. All that happened in the later eighteenth century was that the political ascendancy of the ruling families came to be buttressed by economic success in the expanding world of manufacturing. The involvement of Samuel Koechlin (the son-in-law of one of the most distinguished eighteenth-century burgomasters, Jean Hofer) and of Jean-Henri Dollfus (who later became burgomaster himself) in the first calico-printing venture is entirely typical of the way in which the arrival of industry reinforced, rather than weakened, the established oligarchy.[12]

[11] The eulogies, all delivered to the Société Industrielle de Mulhouse, are published in *Bulletin de la Société Industrielle de Mulhouse*, 24 (1852) 115–29 (Auguste Scheurer-Rott on Schlumberger), 193–217 (Achille Penot on Koechlin), and 269–81 (Jean Weber on Zuber).

[12] The establishment of Koechlin, Schmaltzer et Cie in 1746 marks the beginning of the history of

From the start, the Mulhouse cotton industry became predominantly the affair of the three families of Dollfus, Koechlin, and Hofer. And it long remained so, though the Hofers eventually became less prominent, and at the same time a small circle of satellite families—those of Schlumberger, Heilmann, Thierry, Mieg, Zuber, Schwartz, and Engel, in particular—was cautiously absorbed. Capital flowed between these families as freely as public offices had done for generations past; inter-marriage was practised as frequently as decency would allow; and industrial expansion was made possible (and in some degrees made necessary) by a succession of huge progenies (see Figures 2 and 3). Samuel Koechlin began the tradition by producing seventeen children. Two of them, in turn, had fourteen children each, while another son (Jean Koechlin, the greatest of the early calico-printers) married a Dollfus and had twenty. Of these twenty children, all eleven of the boys who survived to adulthood entered the textile industry, and all the girls married textile industrialists and bred more of them the same.[13] The Koechlins, in particular, were so relentlessly prolific that in 1881, a hundred years after the death of Samuel Koechlin, it was estimated that he had more than 2,250 living descendants.[14]

In the last four decades of the Ancien Régime, the carefully managed strategy of family control and prudent investment allowed the new textile industrialists of Mulhouse to refashion the economy not only of the republic itself (the area of which was no more than eight square miles) but also of a wider region, in French territory, extending eastwards to the Rhine, westwards to the precipitous valleys of the Vosges, and northwards in the direction of Colmar, twenty-five miles away.[15] In Mulhouse alone, the number of establishments engaged in calico-printing had grown to fifteen by 1768, and hand-weaving (at this time almost entirely of imported

Mulhouse as a significant industrial town. Two of the partners—Jean-Jacques Schmaltzer and Koechlin—had had industrial or commercial experience; the role of Dollfus was chiefly as a designer.

[13] See the list of children and their occupations and marriages in André Brandt, 'Une famille de fabricants mulhousiens au début du XIXe siècle. Jean Koechlin et ses fils', *Annales ESC*, 6 (1951), 319–30 (321n).

[14] Auguste Dollfus, 'La famille Koechlin', *Bulletin du Musée Historique de Mulhouse*, 6 (1881), 108–10 (108).

[15] The industrial communities beyond the boundaries of the republic were established to help in the securing of markets in France and to avoid customs duties and some restrictive legislation within the republic. Achille Penot explains this legislation as an attempt to protect the older, small-scale manufacturers of woollen cloth and the associated traders; see *Histoire documentaire de l'industrie de Mulhouse*, op. cit. (note 4), vol. 1, p. 298, and cf. the similar analysis given in Xavier Mossman, *Les grands industriels de Mulhouse* (Paris, 1879), pp. 8–9 and 17–18. However, it seems necessary to draw a distinction between the obstructiveness often displayed by the six 'tribes', or trade corporations, into which the population of Mulhouse was divided, and the attitudes of the civic leaders of the republic, most of whom were cautiously favourable to the new industry of calico-printing; see Frédéric Engel-Dollfus, 'Rapport sur un mémoire traitant de l'industrie du coton du Haut-Rhin', *Bulletin de la Société Industrielle de Mulhouse*, 32 (1862), 527–33 (530), and Laufenburger and Pflimlin, *Cours d'économie alsacienne*, op. cit. (note 4), vol. 2, pp. 185–216.

It should be noted that Mulhousien influence never embraced Colmar. The difference between the economic and social development of the two towns is marked. The greater openness of Colmar to outside influences is suggested by the fact that the only Catholic textile manufacturers in southern Alsace, Antoine Herzog (father and son), were established just outside the town, at le Logelbach, from 1818 to 1870.

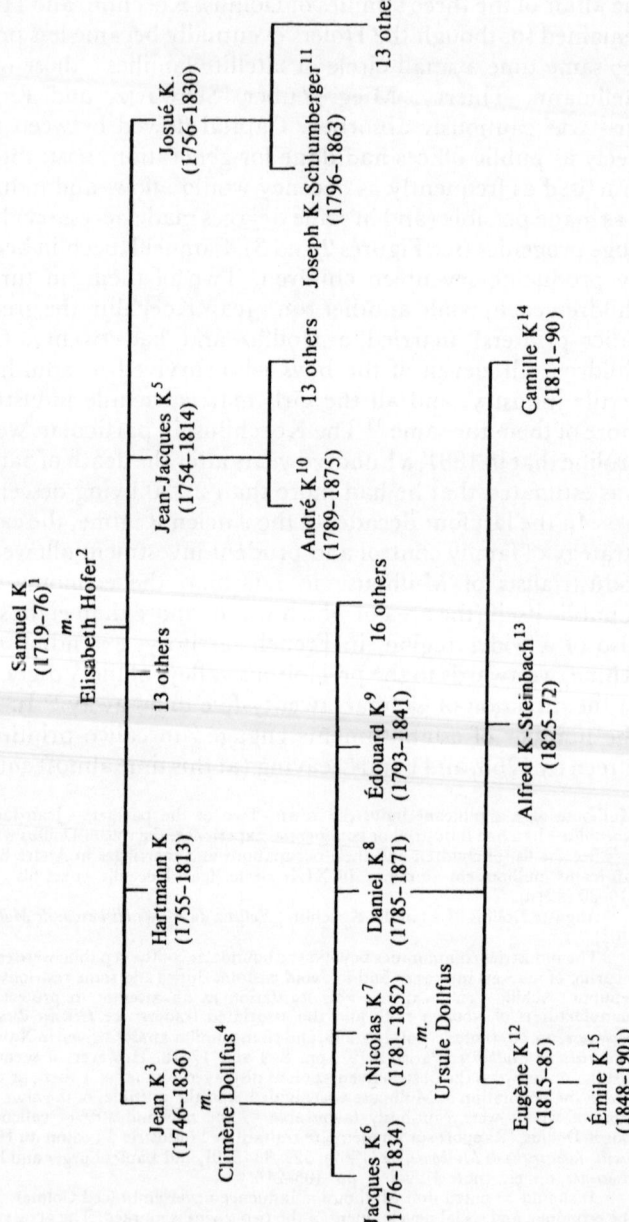

FIGURE 2. The Koechlin family

Notes to Figure 2

[1] Descendant of Hartmann Koechlin of Zurich, who came to Mulhouse in 1596.

[2] Daughter of Jean Hofer, burgomaster of Mulhouse (1748–81).

[3] Established, with his brothers Josué and Hartmann, Koechlin frères, calico-printers, in 1777. Later associated with Nicolas Koechlin et frères.

[4] Sister of Daniel Dollfus-Mieg (1769–1818) and daughter of Jean Dollfus (1729–1800); see Figure 3.

[5] Doctor.

[6] Mayor of Mulhouse (1819–21) and deputy for the Haut-Rhin. Financed early railway building in Alsace.

[8] Studied chemistry in Paris with Fourcroy and Vauquelin. Head of Frères Koechlin.

[9] Associated with Nicolas Koechlin (6) in development of railways in Alsace.

[10] Mayor of Mulhouse (1832–43). Secured by marriage a major interest in Dollfus-Mieg et Cie. Locomotive builder. Deputy for the Haut-Rhin (1832–48).

[11] Spinner and calico-printer. Mayor of Mulhouse (1852–63). Geologist in later life.

[12] Succeeded his father as head of Frères Koechlin.

[13] Leading figure in republican opposition to the Second Empire.

[14] Eminent colourist.

[15] Succeeded his father at the head of Frères Koechlin.

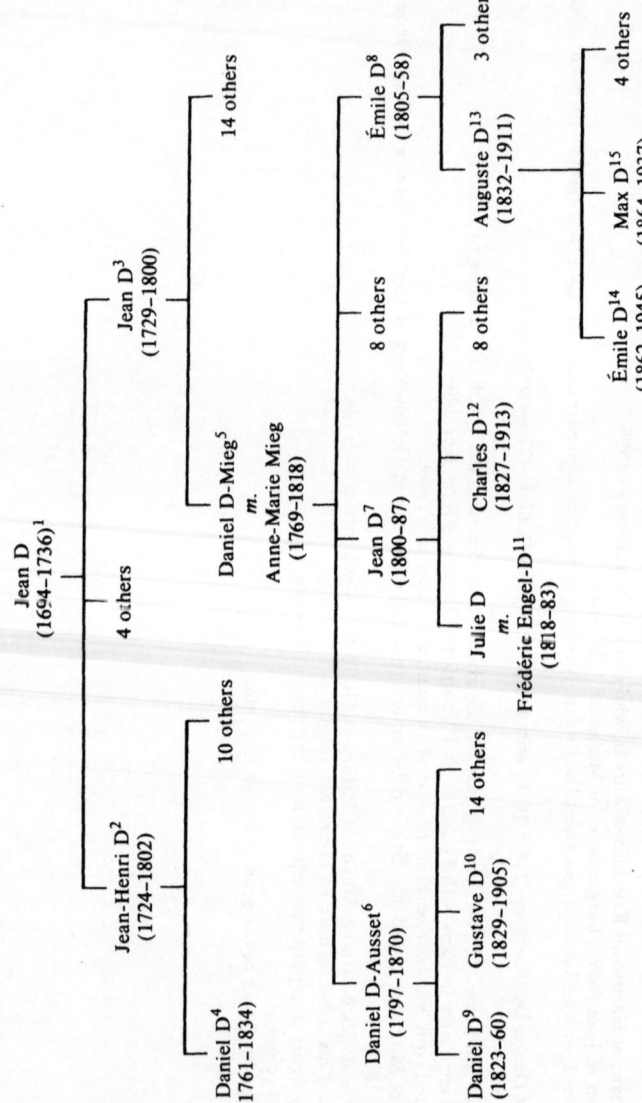

FIGURE 3. The Dollfus family

Notes to Figure 3

[1] Merchant.

[2] Designer and partner in Koechlin, Schmaltzer et Cie (1746).

[3] Calico-printer. Grandson of the mathematician Jean Bernoulli. One of the three last burgomasters of Mulhouse.

[4] Calico-printer. Son-in-law of Jean-Michel Haussmann.

[5] Founder of Dollfus-Mieg et Cie (1800).

[6] Chemist, geologist, alpinist. Technical director of Dollfus-Mieg et Cie.

[7] Calico-printer. Senior partner in Dollfus-Mieg et Cie. Mayor of Mulhouse (1863–9). Deputy to the Reichstag (1877–87). Promoter of the Cités ouvrières from 1853.

[8] Partner in Dollfus-Mieg et Cie. President (1834–58) and benefactor of Société Industrielle. Mayor of Mulhouse (1843–9). Deputy and active in national politics (1846–52).

[9] Partner in Dollfus-Mieg et Cie. President of the Société Industrielle (1858–60). Collector and connoisseur.

[10] Graduate of École Centrale des Arts et Manufactures. Active in Dollfus-Mieg et Cie. Agriculturalist.

[11] Partner in Dollfus-Mieg et Cie. Supporter of Jean Dollfus (7). in campaign for free trade in 1860s.

[12] Writer and journalist.

[13] Textile manufacturer. Member of Town Council of Mulhouse (1865–1902). President of Société Industrielle (1864–1911).

[14] Chairman of the Board of the Société Alsacienne de Constructions Mécaniques. President of the Société Industrielle.

[15] Historian of the Dollfus family.

138

cotton thread) had begun to be organized on a factory basis. An unusually favourable system of tariffs and the accessibility of markets in Germany, Italy, Holland, and, above all, France meant that profits came easily. It is true that by the 1770s and 1780s there were some signs of competition: the most notable challenge came from Jouy, near Versailles, where Christoff-Philipp Oberkampf was beginning to exploit the removal of the ban which had prevented the production of printed cottons (*toiles peintes*) in France from 1686 to 1759.[16] But on the eve of the Revolution, even Jouy presented no serious threat to Mulhouse, at least in fashion fabrics, the *indiennes fines*,[17] for which original designs and colours (already emerging as the main Mulhousien specialities) were all-important.

Although the events of 1789 had few political repercussions in Mulhouse, the increasingly hostile tariff policies of successive French governments and the revolutionary wars that began in 1792 eventually took their toll on trade. In particular, unfavourable duties and disruption virtually deprived the republic of its most lucrative market, in the Midi. And so what had been highly propitious geographical and political circumstances now became so great a handicap that in 1798 there was no realistic alternative but to accept integration with France. On 15 March, the keys, flag, and other trappings of Mulhousien independence were buried with great ceremony and a tree of liberty symbolizing the nascent union of the two republics was planted. There were those—some members of the Hofer family and other so-called *aristocrates*—who opposed the union and derived a perverse satisfaction from seeing the tree quickly die.[18] But in the debate preceeding the union, the rival party of *patriotes*, led by the formidable clan of Koechlins, held and played very effectively the powerful trump card of commercial necessity. The vote of 97 to 5 taken by part of the Grand Council and the republic's forty elders is an indication of the dominant position which the industrial interest had come to occupy by 1798.

[16] On Oberkampf, whose factory at Jouy began operating in 1760, see Serge Chassagne, *Oberkampf. Un entrepreneur capitaliste au siècle des lumières* (Paris, 1980), and, for a briefer treatment in English, S. D. Chapman and Serge Chassagne, *European textile printers in the eighteenth century. A study of Peel and Oberkampf* (London, 1981).

[17] *Indiennes*, or *indienneries*, was originally the name given to the printed cotton fabrics which had been produced in India since ancient times. But it was quickly applied to the similar, though invariably coarser, products which began to be manufactured in Europe from the seventeenth century.

[18] In most of the standard histories, the conflict between the *aristocrates* (led by Josué Hofer) and the *patriotes* (led by the families of Koechlin and Thierry) is played down. But see Max Dollfus, *Histoire et généalogie de la famille Dollfus de Mulhouse 1450–1908* (Mulhouse, 1909), pp. 9–10, on what was in reality a bitter confrontation. The Hofer and Dollfus families were both split on the issue, and while the members of those families which had strong industrial interests never doubted that the union was necessary, laments about the passing of the old order continued to be voiced until well into the nineteenth century. The most vociferous of those who deplored the disruptive effect of industry was Mathieu Mieg ('the chronicler'). His coolness towards the industrialists is very plain in his main historical works: *Der Stadt Mülhausen Geschichte bis zum Jahr 1816 [1817]* (2 vols., Mulhouse, 1816–17) and *Relation historique des progrès de l'industrie commerciale à Mulhausen et ses environs* (Mulhouse, 1823).

The Strategies of Industrial Success, 1798–1830

Despite the triumph of the *patriotes*, it was only eight years later, with the establishment of the Continental System, that the advantages of the union began to be fully realized. Then, in the absence of English competition, the cotton fabrics of Mulhouse penetrated the eastern parts of the French Empire with ease. Prices for *indiennes* were high, and substantial profits earned in virtually captive markets provided a steady flow of capital for investment on a scale that gives the lie to any notion of French industrial stagnation during the Napoleonic wars. Some of the capital was used simply to expand existing activities in dyeing, calico-printing, and (a relative newcomer) wallpaper manufacture. But most of it seems to have been directed to financing a totally new departure into spinning, with the aim of achieving self-sufficiency at a time when spun cotton from traditional sources abroad was either unavailable or, because of heavy duties and transport costs, prohibitively expensive.

The mixture of growth and restructuring had impressive consequences. By 1812, only ten years after the opening of the first spinning mill at Wesserling, there were eleven such mills in the Mulhouse area. With the mills, which marked the beginning of advanced manufacturing technology in the region, there emerged a new breed of owner-managers provided by the rising generations of the great families. In view of the undiminished resolve of these families to protect their economic interests and to resist any dilution of their power, it was no coincidence, but rather the start of a regular pattern of development, that Nicolas Koechlin (at Massevaux) and Nicolas Schlumberger (at Guebwiller) established themselves in these years as two of the region's leading cotton-spinners, even though both men were still in their twenties.

Between 1806 and 1814, the cotton industry of Mulhouse assumed a character for which the desire to maintain an existing social structure, the protectionism of the Continental System, and a location in on the rather thinly populated periphery of France all had their share of responsibility. So long as Mulhouse maintained its reputation for quality and sensitivity to fashion (responding, for example, to the vogue for light, coloured 'shawls' after the Egyptian campaign), the pickings were easy. In certain respects, they may even have been too easy, for profits were high, even with a relatively low level of production, and, as a result, mechanization and the reduction of manufacturing costs became secondary objectives. The priorities in technology lay unmistakably in the field of high-quality colour printing. Hence at a time when machinery remained (by English standards) primitive, new techniques for the preparation and application of dyestuffs were developed in ways that even the specialist dyers of Accrington would have found impressive. Using methods which had distant roots in England and Augsburg, where he had worked as a chemist

140

in J. H. de Schulé's calico-printing works in the 1760s and 1770s, Jean-Michel Haussmann fostered a particularly distinguished tradition of research and development at his own works at le Logelbach near Colmar; his successes, ranging from improved methods for the fixing of well-known dyes such as Prussian blue and the brilliant *rouge d'Andrinople* to the introduction of organic mordants, were outstanding, but they provoked emulation in most of the calico-printing enterprises of the area.

The history of research and development in the technology of dyestuffs and their application in this period can only strengthen the view that the gap between British and French technology under the Empire is due for reassessment. Perhaps, in Britain at least, we have been too swayed by the disparaging tone of most British accounts of French industry after 1814 and by some calculated scaremongering on the part of Charles Dupin and other French visitors who came to England in the early years of the Bourbon Restoration.[19] My own preference, in fact, is for an analysis of British and French technical achievements that would dwell on clear distinctions between different technologies and between different regions of France, rather than on the well-worn track of a supposed British superiority across the board. In this respect, I am inclined to follow the lead taken in the Comte de Chaptal's justly unapologetic account in his *De l'industrie française* (1819).[20] The evidence of Mulhouse would certainly endorse Chaptal's assessment of France's high standing in the 'chemical arts',[21] as it also supports the very important recurring theme of Graham Smith's study of the heavy chemical industry in the regions of Paris, Marseille, and Rouen.[22] Clearly, as Smith shows, in the revolutionary and Napoleonic periods, France had areas of real technical superiority.

Still, the traditional point about the backwardness of Mulhouse in mechanization and power technology remains. The correspondence of Oberkampf's nephew, Samuel Widmer, who visited several factories in southern Alsace in 1809, suggests that machinery in the region was more primitive and less well used than in Jouy.[23] Roman's spinning mill at Wesserling was likened to a dirty stable; printing with copper rollers was rarely practised, and only then very inefficiently; and even at Dollfus-Mieg et Cie, always regarded as a technological pacemaker in Mulhouse,

[19] See, in particular, F. P. Charles Dupin, *Voyages dans la Grande-Bretagne, entrepris . . . en 1816, 1817, 1818, 1819, et 1820* (6 vols., Paris, 1820–4).

[20] J. A. Chaptal, *De l'industrie française* (2 vols., Paris, 1819).

[21] In his 'Discours préliminaire', Chaptal wrote, with only slight exaggeration, that by 1819 France had established herself 'in the first rank of manufacturing nations' and that she was 'unrivalled in the chemical arts'. See Chaptal, *De l'industrie française*, op. cit. (note 20), vol. 1, p. xlv.

[22] John Graham Smith, *The origins and early development of the heavy chemical industry in France* (Oxford, 1979), especially p. 312. Smith's book is a notable exception to the point I make at the end of note 3, above. The local context of the industrial science he describes is treated in great detail.

[23] The relevant correspondence is quoted in Chassagne, *Oberkampf*, op. cit. (note 16), pp. 218–19, and less extensively in Chapman and Chassagne, *European textile printers*, op cit. (note 16), pp. 142–3. For a full transcription, see also 'Lettres écrites d'Alsace par S. Widmer (1788–1809)', *Bulletin du Musée Historique de Mulhouse*, 34 (1910), 105–17 (107–17).

Widmer 'did not see anything special'. The contrast would have been even more striking if the comparison had been made with Lancashire rather than Jouy, as witnesses at the time of the renewal of contract with Britain in 1814 make abundantly clear. Britain's undoubted lead in mechanization provides the starting-point for Charles Ballot's classic study, *L'introduction du machinisme dans l'industrie française* (1923), and it is referred to time and again in the evidence given in 1824 in London before the House of Commons Select Committee on Artisans and Machinery. According to the much-quoted testimony of Adam Young, a Manchester carder who worked for Nicolas Schlumberger at Guebwiller from 1818 to 1820, the Alsatians were at least twenty years behind the British in their technology for the spinning of fine thread.[24] His point is supported by the fact that, at the time when Young was there, Schlumberger employed six skilled operatives from England. Clearly, English carders, spindle-makers, spinners, and stretchers were prized, despite their reputation for intemperance and their unreasonableness as employees, and they were paid accordingly. Young's wage was 12 francs (about ten shillings) a day, roughly six times the amount paid to the Alsatians who worked under him.

Yet even this telling and seemingly reliable evidence has to be read with caution. Three qualifications are necessary. First, the total number of British workmen in Alsace was not large.[25] The six who were employed at Guebwiller were part of a workforce of over 600, and it has been estimated that between 1814 and 1830 no more than a hundred English immigrants ever worked in Alsace.[26] Moreover, the great majority of them stayed for only a year or two and then returned to Britain, probably for reasons similar to those which brought Young back to Manchester: 'I did not like the diet, nor the people, nor anything they had' was his comment.[27] Secondly, the very selective nature of the integration of immigrants in the textile industry is plain. The demand in Alsace was for a small range of

---

[24] Young's evidence appears in the *Fifth report from Select Committee on Artizans and Machinery*, Parliamentary Papers (hereafter P.P.) 1824, vol. 5, pp. 579–82.

[25] It was certainly far smaller than the number engaged in the region of Paris, chiefly by manufacturers of steam-engines and other industrial machinery. At Humphrey Edwards's Chaillot works, for example, 500 English workmen were said to be employed in 1824; see John Martineau's evidence in *First report from Select Committee on Artizans and Machinery*, P.P. 1824, vol. 5, p. 9. Cf. also the figures of 200 and 300 English workmen said to be employed at the iron works of Manby and Wilson at Charenton; it seems that all positions at Charenton, except those of unskilled labourers, were filled by Englishmen. The figure of 200 is given by William Turner, a steam-engine fitter, in *Second Report from Select Committee on Artizans and Machinery*, P.P. 1824, vol. 5, p. 110; the figure of 300 is Alexander Galloway's, given in *Report from the Select Committee on the Laws relating to the Export of Tools and Machinery*, P.P. 1825, vol. 5, p. 43. According to Galloway, between 15,000 and 20,000 British artisans were employed in what he described as 'the French Empire'. Roughly a tenth of this number were employed in the manufacture of iron; about 1,000 of the workmen were in Paris.

[26] André Brandt. 'Travailleurs anglais dans le Haut-Rhin dans la première moitié du XIXe siècle', in *Actes du 92ᵉ Congrès National des Sociétés Savantes. Strasbourg et Colmar 1967. Section d'Histoire Moderne et Contemporaine* (2 vols., Paris, 1970), vol. 2, pp. 297–312 (300).

[27] Cf. the equally disenchanted comment of James Lever, formerly a textile worker in Saint-Quentin, who complained, on his return to England, that he could not live as comfortably in France as he could in Manchester and that he could obtain 'no good ale' and only inferior beef and mutton. Laver's evidence is in *Fifth report*, op. cit. (note 24), pp. 336–7.

highly specialized skills, like Young's, that were relevant to the spinning of fine yarns (an activity which Nicolas Schlumberger brought to Alsace in 1819) and to the construction and maintenance of the appropriate machinery. Thirdly, it is clear that as local workmen learned their skills, the senior British operative became a rarity. Paradoxically, the repeal of the British restrictions on the emigration of labour in 1825 only hastened the process, since it allowed workmen to return to Britain without fear of a penalty, and many of them who were in France at the time seized the opportunity.

The determination that textile production should draw selectively on British expertise and then move resolutely towards autonomy guided industrial policy in southern Alsace throughout the Restoration. Yet autonomy was never seen as synonymous with isolation. Even at the height of the war with Britain, Nicolas Koechlin and Nicolas Schlumberger had visited Manchester,[28] and after 1814 such exchanges (now in both directions) became common, with obvious advantages for the transfer of technical information. For the Mulhousien visitors to north-west England, it was, predictably, the mechanization of the British textile industry which was most striking, especially in the early years of the century. In 1814, the gap between a region that introduced its first steam-engine (a ten-horse power engine of Parisian origin but unknown design used to drive spinning machinery at Dollfus-Mieg et Cie[29]) only two years earlier, and Manchester, with its scores of engines, was glaring. But over the next fifteen years, even that gap narrowed, at least in terms of the quality and modernity, if not the number, of the machines in use in Alsace. The narrowing owed something to the (largely clandestine) importation of machinery from England: four steam-engines from Manchester were said to be in use in the Haut-Rhin in 1826, for example. But it owed more, especially in the later years of the Restoration, to the growth of an indigenous machine construction industry in and around Mulhouse. The success of that industry is reflected in improvements that affected both productivity and quality. Whereas in 1816–17 a mule with 240 spindles would spin 3 kg of

[28] Nicolas Schlumberger's contacts with Manchester were particularly close. He worked in England for over three years between 1802 and 1805 , corresponded with Benjamin Kennedy and William Fairbairn, and seems, after the Empire, to have been very effective in persuading British workmen to to emigrate to Alsace (see below, note 31, for example). On the ease of contacts between the Haut-Rhin and south Lancashire in the early decades of the nineteenth century, see André Brandt, 'Apports anglais à l'industrialisation de l'Alsace au début du XIXe siècle', *Bulletin de la Société Industrielle de Mulhouse*, no. 1 (1967), 27–41, and 'Travailleurs anglais dans le Haut-Rhin', op. cit. (note 26).

[29] The engine, which was probably of the kind designed by Watt in the 1780s, was constructed by Salneuve in Paris and used to drive spinning machinery. See 'Résumé des notes laissées par M. Hartmann-Liebach sur l'histoire industrielle du Haut-Rhin, depuis les premières années du XIXe siècle', *Bulletin de la Société Industrielle de Mulhouse*, 47 (1877), 218–35.

The extent of Mulhousien backwardness in power technology is also conveyed by Émile Dollfus's observation that water power only began to replace horses and manual labour for the driving of machinery in 1809–10. See Dollfus, 'Notes pour servir à l'histoire de l'industrie cotonnière dans les départements de l'Est', *Bulletin de la Société Industrielle de Mulhouse*, 27 (1855–7), 435–61.

cotton in one day, between 8 and 9 kg would be spun (to a higher standard) on a machine of comparable size in 1831.[30]

The changing nature of the relations between Alsace and Britain in the field of machine construction is illustrated clearly by a case-history that begins with an obscure foundryman and engineer by the name of Job Dixon. It was Dixon, coming penniless from Manchester to Cernay in 1820, who provided the technical skills for the firm of Risler frères et Dixon, where mechanical engineering in southern Alsace effectively began.[31] Until its bankruptcy amid the economic crisis of 1827, Risler frères et Dixon supplied the region with machinery of the latest design for spinning and weaving, and served as a training-ground for a new generation of Alsatian engineers, including Émile Koechlin, a nephew of one of the senior partners, Jérémie Risler. Another, even more important route by which English influence stimulated engineering in Alsace was through the association of André Koechlin et Cie with the Manchester engineers, Sharp, Roberts, and Co. Between 1826, when the association began, and 1828, Richard Roberts made three visits to Mulhouse and, in return, received Alsatian engineers at his works in Manchester.[32] As in the case of Dixon's activity at Cernay, the Alsatian partners in the relationship learned quickly. An initial arrangement whereby André Koechlin et Cie constructed textile machinery under licence, using designs supplied by Sharp, Roberts, and Co., soon evolved into one of far greater, though never total independence. It is a mark of the incompleteness of Alsatian emulation and of the smaller size of the industry in Alsace that when new technologies and construction on a much larger scale were called for—at the time of the building of the railway lines from Mulhouse to Thann (1839) and between Strasbourg and Bâle (1841)—Sharp, Roberts, and Co. were called upon once again, with the Koechlin company assuming, at first, a secondary role.[33]

However, in the Restoration as in the Empire, dyeing remained one technology in which Mulhouse was unquestionably the pacemaker and in which self-sufficiency was a reality. As early as 1815, the Lancashire calico-printer James Thomson had reversed the more familiar direction of technological indebtedness by securing the exclusive right to import and sell in Britain certain of the finer printed cottons (in particular the

[30] Achille Penot, *Statistique générale du département du Haut-Rhin* (Mulhouse, 1831), pp. 322 3.
[31] Brandt, 'Apports anglais', op. cit. (note 28), pp. 30–1. According to Leuillot, *L'Alsace au début du XIXe siècle*, op. cit. (note 4), vol. 2, p. 347, Dixon was recruited in the first place by Nicolas Schlumberger. Risler frères had been established as recently as 1818. They were the first machine-builders of any consequence in the region.
[32] Brandt, 'Apports anglais', op. cit. (note 28), pp. 32–3. For a list of fourteen British engineers engaged by André Koechlin et Cie in 1827, see Brandt, 'Travailleurs anglais dans le Haut-Rhin', op. cit. (note 26), pp. 308–9.
[33] Brandt, 'Apports anglais', op. cit. (note 28), p. 33. Nicolas Koechlin, who was chiefly responsible for the construction of the railway system in southern Alsace, ordered the first three locomotives for the new lines from Sharp, Roberts, and Co. Thereafter, most locomotives were constructed by engineers in Mulhouse, who used the imported locomotives as their prototypes.

celebrated *rouges turcs*) of Nicolas Koechlin et frères.[34] Twenty years later, despite being more costly than comparable British products,[35] Alsatian fashion fabrics were still sought after. They were of outstanding design, and, above all, benefited from a technical expertise developed by a new and very distinctive industrial profession, that of the textile chemist or colourist. Whereas Haussmann, Nicolas Koechlin, and the other calico-printers of the early years of the century had regarded dyeing as just one of a range of skills which they had to master, by the 1820s calico-printers were more likely either to specialize in dyeing themselves or to employ one of their younger relatives for the purpose. The careers of two of the most distinguished colourists of the Restoration and July Monarchy—Daniel Koechlin-Schouch and Henri Schlumberger—show how international reputations could be won in either way. Koechlin-Schouch was a calico-printer who devoted himself increasingly, in his later career, to dyestuffs and their application; his pupil, Henri Schlumberger, was typical of a slightly later generation in that he began his career as an apprentice colourist, entering the firm of Nicolas Koechlin et frères in 1818 at the age of fifteen and rising eventually to the top of his profession as chief chemist at Dollfus-Mieg et Cie.[36]

It is not hard to unravel the economic and social priorities that gave industrial development in southern Alsace its distinctive character. Poor communications and the high price of coal and raw materials (on both of which counts Normandy had a marked advantage) made it natural to concentrate on quality and design rather than on mass production. An important influence was also exerted by the undiminished sense of separateness in religion and culture and by the perennial problem of finding suitable local employment for the younger members of the great families. The practice of recruiting the main colourists from these families undoubtedly contributed to the high status and salaries associated with the profession as it existed in Mulhouse.[37] Likewise, family bonds guided the pattern of diversification from calico-printing into spinning, weaving, wallpaper manufacture, and machine construction, and the trend by which these activities became the responsibility of separate firms with

[34] Daniel Koechlin-Schouch, 'Notice nécrologique sur M. James Thomson', *Bulletin de la Société Industrielle de Mulhouse*, 23 (1850–1), 182–5.

[35] At the Frankfurt fair of 1818, for example, Mulhousien printed cottons were 40 per cent dearer than their British rivals, yet they were preferred by buyers because of their superior design; see Leuillot, *L'Alsace au début du XIXe siècle*, op. cit. (note 4), p. 389. Over the next half century, the preference was not sustained, as I point out in note 102, below.

[36] See the obituary cited in note 11, above.

[37] At Dollfus-Mieg et Cie, for example, *coloristes* would commonly receive an annual salary of 12,000 francs (about £500). This should be compared with the salaries of professors in the provincial faculties of science, which seldom exceeded 5,000 francs. Even at the end of a long and distinguished academic career (spent almost entirely in Mulhouse as a close associate of the great industrial families), the chemist Achille Penot earned less than 6,000 francs p.a. in the early 1860s; see his personal file in Archives Nationales, F[17] 21456. It is also instructive to compare the salary of £400, rising to £600, that was offered to Lyon Playfair when he accepted his appointment as 'chemical manager' with James Thomson at Clitheroe in 1841; see T. Wemyss Reid, *Memoirs and correspondence of Lyon Playfair* (London, 1899), p. 44. Such a salary was quite exceptional in England at the time.

personal ties to a larger parent company.[38] As a result, the industrial structure of southern Alsace changed but it did so in a manner which left economic power in precisely the hands which had always wielded it.

Of the numerous institutions and activities through which this strategy of meticulously controlled expansion was pursued, the most significant was the Société Industrielle de Mulhouse.[39] From its foundation in 1826 until the annexation in 1871, the society served as a mouthpiece for the expanding industrial élite and as a way of profitably engaging its intellectual energies, especially those of its younger members. The society's declared objectives were predictable enough: to provide a scientific basis for industrial practice, to encourage the spirit of enterprise, and to advance public welfare. In all these respects, it was successful. Specialized committees on chemistry and machinery, substantial prizes for technical improvements, and a heavily subsidized *Bulletin* fostered discussion and the dissemination of industrial research that transcended such frail boundaries as existed between individual firms; a very effective system of *plis cachetés* encouraged the protected inventions and innovation; and, especially from the mid-century, the committee on 'économie sociale' developed an important role in the promotion of philanthropy and education.

Although the enthusiasm for the economic activities and good works was no empty charade, it cannot be understood in isolation from the unspoken motives and hidden bonds that also fired Mulhousien paternalism. Thirteen of the twenty-two founder-members of the society bore the names of Koechlin, Dollfus, Schlumberger, Thierry, or Heilmann, or combinations resulting from inter-marriage. And twelve of the twenty-two belonged to the masonic lodge, La Parfaite Harmonie, whose activities had assumed a new vigour since 1824, when its head, Jacques Koechlin, had returned as a liberal hero from a much-publicized sentence in the Sainte-Pélagie prison.[40] The particular brand of Freemasonry that

[38] These two tendencies are evident in the history of Nicolas Koechlin et frères between 1802 and 1836. The firm began in 1802 when, at the age of twenty, Nicolas Koechlin set up as a spinner at Massevaux. A quarter of a century later, at the peak of their prosperity, Nicolas Koechlin et frères were engaged in calico-printing and spinning in Mulhouse; spinning, weaving and bleaching at Massevaux; and calico-printing and weaving at Loerrach. By then, there were, in all, 5,000 employees. When the firm was wound up in 1836, the various activities continued to be pursued, but under a number of separate firms. By the 1830s, the huge firm of Dollfus-Mieg was unusual in maintaining strong interests in all three main branches of the textile industry: spinning, weaving and calico-printing. On the trend, which tended inevitably to undermine the community of interests among the *industriels*, see Laufenburger and Pflimlin, *Cours d'économie alsacienne*, op. cit. (note 4), vol. 2, pp. 270–1.

[39] The standard history of the Société Industrielle in its first fifty years is Achille Penot, 'La Société Industrielle de Mulhouse', on pp. 1–136 of *Travaux et mémoires présentés à la Société Industrielle lors de la célébration du cinquantième anniversaire de sa fondation*, a supplement to volume 46 (1876) of the *Bulletin* of the society. See also *Centenaire de la Société Industrielle* (2 vols., Mulhouse, 1926), vol. 1, pp. 11–187.

[40] Jacques Koechlin was imprisoned specifically for the pamphlet, *Relation historique des événemens qui ont eu lieu à Colmar, et dans les villes et communes environnantes, les 2 et 3 juillet 1822* (Paris, 1822), in which he criticized the provocative behaviour of the civil and military authorities in the arrest and execution of the Bonapartist conspirator, lieutenant-colonel Joseph-Augustin Caron.

In fact, the masonic associations of the Société Industrielle were even stronger than I indicate in the text, for another three of the founder-members subsequently joined the Parfaite Harmonie lodge. See Max Koehnlein, 'Un inspirateur de la Société Industrielle treize ans avant sa fondation', *Bulletin de la Société Industrielle de Mulhouse*, 99 (1933), 453–62 (458).

pervaded the Parfaite Harmonie lodge helped to invest the Société Industrielle as a whole with an aura of liberalism and Bonapartism which condemned it in the eyes of the Bourbon authorities, while allowing it to appear in Alsace as a champion of local, as opposed to national, interests. But the broader ideology always remained implicit, buried beneath a commitment to acts of public utility which gained in strength throughout the remaining decades of French rule. As I show in the next two sections, this sustained seriousness can be interpreted (in a manner with which historians are now very familiar) as a way of reinforcing a threatened social order; but it also produced lasting material benefits for industry and the community.

The Challenge to Authority

The reputation of the printed cottons of Mulhouse was such that by the early 1830s they had largely eliminated their British rivals from the French market and made some modest inroads on foreign markets as well. The quality of the *toiles peintes* of Dollfus-Mieg et Cie was such that in 1834 half of the company's production was sold abroad.[41] In order fully to appreciate the magnitude of this achievement, the natural disadvantages of the Mulhousien situation after the collapse of the Empire have to be borne in mind. Imported cotton which reached Rouen within hours of its arrival at Le Havre took three weeks on a difficult overland route to reach the Haut-Rhin. The improvement in communications which came with the opening of the Rhône-Rhine canal through Mulhouse in 1832 did little, if anything, to alleviate the problem, and it was only in the 1850s, when the railway lines from the west via Strasbourg and later via Belfort were opened, that significant reductions in time and expense were achieved.[42]

So the expansion that occurred in the twenty years or so following the return of the Bourbons must be regarded, by any standards, as impressive. It is not surprising that the *industriels* of the regions seized every opportunity of flaunting their success before their workers and the governments whose policies could do so much to reinforce or impede their efforts. When Charles X visited Alsace in 1828, the Société Industrielle resolved on a particularly extravagant display of regional pride, through an exhibition demonstrating the industrial strength of the Haut-Rhin. For the occasion, the most entrenched political principles were laid aside. Even the

The establishment of the Parfaite Harmonie lodge was part of a revival of masonic activity which occurred widely in Alsace. The lodge was an important focus for liberalism and bonapartism in the Restoration, though in later years, especially after 1848, it came to be more closely associated with republicanism. See Paul Leuillot, 'Bourgeoisie d'Alsace et Franc-Maçonnerie aux XVIIIe et XIXe siècles', in *La bourgeoisie alsacienne. Études d'histoire sociale* [Publications de la Société Savante d'Alsace et des Régions de l'Est, no. 5] (Strasbourg, 1967), pp. 343–76 (362–5).

[41] Lévy, *Histoire économique de l'industrie cotonnière en Alsace*, op. cit. (note 4), p. 234.

[42] Claude Fohlen, *L'industrie textile au temps du Second Empire* (Paris, 1956), p. 139. Fohlen does note, however, that the opening of the canal did help to alleviate (though it never solved) the very serious problem of obtaining cheap coal in the Mulhouse area; see note 88, below.

declared Bonapartist Nicolas Koechlin brought himself to make a respectful public address to the king and subsequently to accept the rank of Chevalier of the Legion of Honour (though in 1830 he duly reverted to type by voting in the Chamber of Deputies for the overthrow of the Bourbon line).[43]

Despite their understandable pride, the readiness of the Mulhousien industrialists to be involved in the civic junketings that accompanied the royal visit is striking. It seems that a concern about recent developments in the economy and society prevailed over a public stance, with its origins way back in the days of the republic, which in principle should have brooked no show of sympathy to the Bourbons. The fall in the price of spun cotton and printed cottons in the crisis of 1827 and 1828 had already caused unemployment in Alsace as it did in other parts of France, as well as some bankruptcies among the smaller, less versatile firms. At a more general level, there were also signs that the élite's command of local affairs might be diminishing. The long-term cause of this was demographic change on a grand scale. The growing need for labour had led, since the Empire, to an influx of Catholics from the surrounding area, including parts of Bavaria and Wurtemburg across the Rhine. In 1803, Mulhouse had had only 600 Catholics in a total population of nearly 7000: by 1834, after three decades of steady immigration, roughly half the population of 13,300 was Catholic.[44] The resulting gulf between employer and employee was further widened by the dismal conditions in which the Catholic immigrants lived. About 1830, half of the workforce lived outside the town, and journeys on foot of six miles each way between home and workplace were common. Moreover, those who resided in the town lived in a squalor that appalled even the case-hardened Louis Villermé when he went there in 1835 and 1836. These were the days when the newspapers of Mulhouse would carry announcements advertising space to let not in a house but in a bed,[45] and when, as Villermé observed, two families would commonly share one squalid room in a lodging house to avoid the debilitating trudge

---

[43] The exhibition of 1828 is described in 'Rapport sur l'exposition des produits de l'industrie, à l'occasion de l'arrivée du roi, le 11 septembre 1828', *Bulletin de la Société Industrielle de Mulhouse*, 2 (1828), 73–166. On the visit of the king to Mulhouse see P. J. Fargès-Méricourt, *Relation du voyage de Sa Majesté Charles X en Alsace* (Strasbourg, 1829), pp. 145–60. Despite the fuss, the visit to Mulhouse lasted a mere 4½ hours.

It cannot have been easy for Koechlin to show public enthusiasm for the royal visit, but for him, the fall of Villèle's reactionary ministry in January 1828 would certainly have resurrected hopes of some liberalization of the Bourbon régime which he professed to despise. Koechlin also became involved in the visit as the most generous of the *industriels* who financed the elegant residential and commercial development known as the New Quarter. Both the king's visit and the exhibition of industry were organized in ways that drew attention to the magnificence of the new buildings and so, indirectly, celebrated its sponsors. It is typical of Koechlin's opportunism that after making his address, he presented the king with a written statement of the needs of Mulhousien industry; see Fargès-Méricourt, *Relation du voyage de Charles X*, p. 153.

[44] See Table 3. As the Table shows, the trend continued. Thirty years later, Catholics outnumbered Protestants in the ratio of three to one.

[45] See Penot's obituary of Nicolas Koechlin, cited in note 11, p. 202.

between Mulhouse or the industrial suburb of Dornach and an outlying village.[46]

The publication of Villermé's *Tableau de l'état physique et moral des ouvriers* in 1840 gravely embarrassed employers whose public face of Calvinist piety and charity could not be made to square with the revelations. Only Lille, it was said, could match the Haut-Rhin in the degradation suffered by its cotton-workers.[47] In both towns, the hours were long. In summer, when work was plentiful, the working day in spinning and weaving would begin at 5 a.m. and end at 8 or 9 p.m., fifteen hours or more, with breaks amounting to no more than $1\frac{1}{2}$ hours. The consequences were entirely predictable. Inadequate food and housing went hand in hand with rampant illegitimacy and a pitiful standard of health that explained the expression 'Nègres-Blancs' used to describe the pallid inhabitants of Thann and Mulhouse. Villermé's indictment confirmed the suspicions which Frenchmen outside Alsace had harboured and which loyal Mulhousiens had vehemently denied for many years past.[48] As early as 1809 Samuel Widmer had been shocked by the appalling conditions in which calico-printers at Wesserling were expected to work, with temperatures in the printing shed rising to 40°C.[49] And in 1824 the prefect of the Haut-Rhin had stated (with confidence and obvious satisfaction) that the Catholic, German-speaking population of his department was totally antipathetic to the Koechlin clan.[50] All this may well have been true, but the sense of misery and discontent did not begin to harden into systematic disaffection before the mid-1830s, when a period of particularly rapid

[46] Louis Villermé, *Tableau de l'état physique et moral des ouvriers employés dans les manufactures de coton, de laine et de soie* (2 vols., Paris, 1840), vol. 1, p. 27.

[47] Ibid., vol. 1, p. 439. The information that I give on Mulhouse is taken from vol. 1, pp. 14–61 and 437–46.

[48] For a typically indignant riposte to the charges, see Penot, *Statistique générale du Haut-Rhin*, op. cit. (note 30), pp. 316–17:

The criticisms of those industries which employ large numbers of workers in one building . . . have been directed chiefly at spinning. The criticisms have been exaggerated. No, our workers are not the pinched, stunted creatures that they are said to be . . .

Needless to say, Penot did not mention that, in 1827, a committee of the Société Industrielle de Mulhouse had declined to take action on a proposal by one of the more compassionate employers, the spinner Jean-Jacques Bourcart of Guebwiller, for the imposition of a limit of twelve hours on the working day in spinning mills and a ban on the employment of children under the age of nine. For Bourcart's proposal, which was modelled on the British legislation of 1825 on working hours and the employment of children, see *Bulletin de la Société Industrielle de Mulhouse*, 1 (1826), 373–86. The proposal was rejected on the grounds that it would infringe the personal freedom of employer and employee alike.

[49] Chassagne, *Oberkampf*, op. cit. (note 16), p. 218, and 'Lettres écrites d'Alsace', op. cit. (note 23), p. 111.

[50] Leuillot, *L'Alsace au début du XIXe siècle*, op. cit. (note 4), vol. 1, p. 444. The statement by Puymaigre, a prefect of unimpeachable loyalty to the Bourbons, was made at the time of the elections of 1824. His belligerence towards the Koechlins is reflected very clearly in his comment: 'nous allons voir si c'est la famille des Bourbons ou la famille Koechlin qui gouverne le Haut-Rhin'. In the event, Jacques Koechlin was re-elected, but the larger vote for the other deputy, a legitimist, justified Puymaigre's view that the liberal cause had suffered since the previous elections in 1821.

immigration was followed by a sudden crisis in the spring of 1837, which caused sackings, short-time working, and even some closures.[51]

Despite the long history of the underlying causes, the publication of Villermé's book coincided with, and almost certainly contributed to, an unprecedented level of social unrest. Whereas in the 1830s there had been no disorder remotely comparable with that in Lyon in 1831 and 1834, new economic problems in 1846–8 provoked major disturbances (including a notorious bread riot in 1847) and, in response, shows of military strength which on one memorable occasion brought a leading employer, Jean Koechlin-Dollfus, face to face with the mob in his capacity as commander of the town's National Guard.[52]

A quite different challenge to the authority of the *industriels* arose from another protracted process which threatened their carefully contrived command of all the main seats of power in the region. Their dominant position in the Town Council, the Parfaite Harmonie lodge, the Société Industrielle, and the Chamber of Commerce remained secure enough until well into the Second Empire, but in the middle decades of the nineteenth century rival élites did begin to emerge, chiefly as a result of the expansion of governmental bureaucracy. From as early as the 1820s, pretensions to autonomy, in accordance with the old tradition of independent rule by a closed community of burghers, were increasingly seen, from Paris, as impediments to the ideals of tidy centralization and political conformity. Conflicts, amounting on occasions to a running battle, were inevitable.

One early skirmish, which illustrates very clearly the icy welcome awaiting the government officials who were unlucky enough to find their way to Mulhouse, concerned the building of the Rhône-Rhine canal. Plans for this work had been gestating since the eighteenth century, and construction had begun in 1785. But it was only after the Empire, when the slowness and the high cost of transport were recognized as major constraints on local manufacturing, that the *industriels* of the Haut-Rhin began to press for completion of the canal, on the understanding that it would pass through Mulhouse. Military considerations helped to engage the government's interest, but they also dictated a route through quite unsuitable terrain which their engineers, if given a free hand, would probably never have chosen. Seepage proved to be a constant problem and a source of delays well calculated to elicit the derision of men who were convinced that the competence of outsiders could never match that of the home-grown Mulhousien product. At last, the point was made explicitly.

[51] Villermé, *Tableau de l'état des ouvriers*, op. cit. (note 46), vol. 1, p. 24n. Villermé (ibid., vol. 1, pp. 14–18) noted the marked demographic change that occurred in a period of less than two years in the mid-1830s. In April 1834, 4,960 of the 9,860 workers in the cotton mills of Mulhouse lived in the town, 4,900 of them in surrounding villages; by the end of 1835, of a total of 11,637 workers, 6,573 lived in the town, with only 5,064 coming from outside.

[52] Achille Penot, 'Notice sur M. Jean Koechlin-Dollfus', *Bulletin de la Société Industrielle de Mulhouse*, 41 (1871), 52–61 (57).

150

In 1828, in what was ostensibly a purely academic paper on local geology, Édouard Koechlin delivered a devastating criticism of the Ponts et Chaussées engineers who had undertaken (and bungled) the preliminary surveys for the canal.[53] Coming from someone who had no formal qualifications in either geology or civil engineering, it was an audacious attack, but it almost certainly represented the collective opinion of the *industriels*. It is hard to imagine that a pained defence of the Corps des Ponts et Chaussées, which was published soon afterwards,[54] was given much of a hearing.

An even more intrusive form of bureaucratization was apparent in education. Here, to a remarkable degree in the early nineteenth century, Mulhouse had stood apart from the rest of France. In the characteristic manner, local educational policy had been directed at independence and self-sufficiency, and that aim had been achieved through private patronage bestowed preferentially on ventures adapted to industrial employment. It was the generosity of André Koechlin, Jean Dollfus, and the firm of Nicolas Koechlin et frères, for example, which allowed the *collège communal* of Mulhouse to offer teaching in industrial chemistry, including some laboratory instruction, from 1822.[55] The innovation was an important one, and enrolments and academic standards rose steadily. From 1854, both the lectures and the practical classes were integrated in the curriculum of the new municipal École Professionnelle, with the young Paul Schützenberger taking charge of the laboratory; and in 1866 the teaching of chemistry was removed to a new École Superieure de Chimie which, especially after the annexation, attracted a dazzlingly international body of students, chiefly from Alsace but also from Russia, Germany, France, Austria, Switzerland, and Italy.[56]

Although Mulhouse earned its reputation in industrial education chiefly for the teaching of chemistry, there were other important developments for which the Société Industrielle and the town council, in varying degrees though with a clear community of interests, were responsible. These included schools of design (1829), weaving (1861), spinning (1864), and commerce (1866). Invariably, the emphasis in these local ventures was on strictly vocational training for what the promoters defined as the needs of Mulhousien industry, with national examinations and non-vocational subjects having virtually no place. It is all too easy, and

[53] Édouard Koechlin, 'Aperçu géologique sur les environs de Mulhouse', *Bulletin de la Société Industrielle de Mulhouse*, 2 (1828), 258–76 (276).

[54] *Bulletin de la Société Industrielle de Mulhouse*, 3 (1829), 1–21.

[55] On the history of the teaching of industrial chemistry in Mulhouse, see (in addition to the standard works cited in note 4) *Histoire de l'École de Chimie de Mulhouse publiée à l'occasion du 25e anniversaire de l'enseignement de M. le Dr Emilio Noelting 1880–1905* (Strasbourg, 1905), especially pp. 1–45, and Raymond Oberlé, *L'enseignement à Mulhouse de 1798 à 1870* (Paris, 1961), pp. 215–17. Oberlé's book is an invaluable source for all aspects of the history of education in Mulhouse.

[56] See the Tables on pp. 31–5 of *Histoire de l'École de Chimie*, op. cit. (note 55), pp. 31–5. The largest categories of students in the period 1879–1905 were: Alsatian (37·93%), Russian (15·60%), German (9·16%), French (8·32%), Austrian (8·06%), Swiss (7·47%), and Italian (6·24%).

profoundly misleading, for these schools to be disregarded as minor appendages to the national system of education. In fact, the schools of Mulhouse were typical of a world of locally supported vocational training over which the Ministries of Public Instruction and Commerce had no jurisdiction. There is an obvious parallel, for example, in the school for the instruction of the operatives of steam-engines, established in 1858 by the Société des Sciences, de l'Agriculture et des Arts in Lille.[57] Just as the École des Chauffeurs in Lille responded to the startling growth in the number of steam-engines in the region (nearly 2,000 were in use in the department of the Nord in 1858), so the schools of weaving and spinning in Mulhouse followed on the heels of the introduction of the technically advanced power looms and self-acting mules which I discuss later in the paper.

The enthusiasm of the *industriels* for their own schemes of vocational education contrasts unmistakably with their relative indifference to initiatives emanating from Paris. For example, they seem never to have campaigned for the establishment of a *lycée* in the town; the *collège communal*, despite its formally lower status, served their purposes adequately.[58] And they did little more than acquiesce in the creation of the École Préparatoire à l'Enseignement Supérieur des Sciences et des Lettres in 1855.[59] The explanation for this coolness is simple. For although the school was a municipal one, it was established in accordance with a plan for the extension of higher education in the industrial areas which emanated from the Ministry of Public Instruction, specifically as the brainchild of the Minister, Hippolyte Fortoul. The industrial employers of Mulhouse could not have overlooked the general drift towards academic centralization which the new Écoles Préparatoires were intended to promote,[60] and they can have had little sympathy for a curriculum which led, after two years of study, to an examination embracing geography, French history, and literature, as well as scientific and technical subjects, and to the award of a ministerial *certificat de capacité* which in the event proved worthless. It is hardly surprising that, in the absence of active encouragement on the part

[57] On this school, see 'Séance d'installation de l'École gratuite des Chauffeurs', in *Mémoires de la Société Impériale des Sciences, de l'Agriculture et des Arts de Lille*, 2nd ser. 5 (1858), v–viii.
[58] As mayor from 1836 to 1843, André Koechlin was a powerful opponent of any attempts to reduce the emphasis of the *collège* on scientific and industrial studies and to extend its very limited teaching in Greek and Latin. During the mayoralty of Émile Dollfus (1843–8), a different philosophy prevailed, and the *collège* assumed increasingly the character of a *collège royal* without ever being formally designated as such. See Oberlé, *L'enseignement à Mulhouse*, op. cit. (note 55), pp. 132–47.
[59] On the establishment of the École Supérieure, see Oberlé, *L'enseignement à Mulhouse*, op. cit. (note 55), pp. 197–205. The pamphlet *Inauguration de l'École Préparatoire à l'Enseignement Supérieur des Sciences et des Lettres de Mulhouse*, published to mark the opening of the school on 17 November 1855, is also helpful. Comparable schools were established at the same time in Rouen, Angers, and Nantes, in an attempt to meet the new demands being made on the educational system as a result of economic and demographic change. The instruction, which lasted two years, was practical in orientation, adapted to the needs of young people entering industrial and commercial careers for whom the more advanced and 'purer' curriculum of the faculties was inappropriate.
[60] For a comment on Fortoul's desire for centralization and its consequences see Robert Fox, 'Science, the university, and the state in nineteenth-century France', in Gerald L. Geison (ed.), *Professions and the French state, 1700–1900* (Philadelphia, 1984), pp. 66–145 (86–92).

of the *industriels*, enrolments in Mulhouse fell to an extremely low level: in the late 1860s the total number of students was even in single figures.[61]

For my present purpose, the main significance of the École Préparatoire lies in its contribution to the accelerating erosion of the old social order of Mulhouse. Like the *collège communal* and the École Professionnelle, the École Préparatoire brought to Mulhouse educated Frenchmen who often had no allegiance to Alsace, still less to her Calvinist textile manufacturers. Predictably, there were those among the newcomers who found the Alsatian speech and the drab utilitarianism of the town as hard to take as Émile Souvestre had done in 1836. One new arrival who recorded some particularly vivid impressions was Émile Boissière, who was appointed in 1855 to teach literature in the École Préparatoire and the *collège*. After two hours in Mulhouse, Boissière was on the point of returning hot foot to Paris,[62] though eventually he stayed for twenty years, in the course of which he saw a modest degree of gaiety and sophistication injected into Mulhousian polite society, chiefly through the influence of the professional men and administrators from outside Alsace of whom he was typical. The removal of the Sous-Préfecture and some associated legal officials from Altkirch to Mulhouse in 1857 was, in this respect, an important new departure.[63] It was a departure which the tight-knit industrial and municipal élite cautiously welcomed as signalling the growth of their town, and as an inevitable rationalization. Yet its contribution to the growing pluralism among the *notables* of Mulhouse can hardly have been missed. The price to be paid was certainly not negligible.

Confronted, on the one hand, with an increasingly restless workforce and, on the other, with new contenders for authority, the *industriels* resorted once again to paternalism. Earlier in the century, the encouragement of savings banks and mutual aid societies, subsidized pharmacies, lending libraries in factories, and piecemeal masonic philanthropy had been adequate instruments of control. But by the 1850s, with a population almost three times what it had been thirty years before and with Saint-Simonian, Fourierist, and socialist ideas beginning to exert a tardy influence on the nature of workers' aspirations and on their forms of collective action, the response was necessarily on a rather grander scale. Now, in a scheme that was initiated by the Société Industrielle, the workers whose pitiful conditions Villermé had deplored only a few years before

[61] Oberlé, *L'enseignement à Mulhouse*, op. cit. (note 55), pp. 207–10. As Oberlé notes, the shortage of students was equally marked in the other Écoles Préparatoires.

[62] Jean-Louis-Émile Boissière, *Vingt ans à Mulhouse 1855–1875* (Mâcon, 1876), p. 7.

[63] Boissière noted the importance of this administrative change; see ibid., p. 121. It seems that the *salon* organized by the new sub-prefect was an object of particular interest, though *salons* were by no means unknown in Mulhouse by the mid-century, as Boissière's comments on gatherings presided over by Madame Nicolas Koechlin make clear.

According to the *Histoire documentaire de l'industrie de Mulhouse*, op. cit. (note 4), vol. 1, p. 130, the town council of Mulhouse had petitioned on several occasions since 1814 for the transfer of the Sous-Préfecture from Altkirch to Mulhouse. The transfer necessarily entailed that of the *tribunal de première instance* (the main regional court) as well.

were provided with a model town, close to some of the main factories, that is still impressive even today.[64] It was a scheme for which the leading promoters, Jean Zuber *fils*, Jean Dollfus, and Dolfus's son-in-law Frédéric Engel-Dollfus, expected and justly received much credit, though the conception was certainly derived from Henry Roberts's *The dwellings of the labouring classes* (1850).[65] By 1864, after twelve years of development and an investment of $1\frac{3}{4}$ million francs, the Société des Cités ouvrières had erected 616 houses, each with a garden, and many of the occupants were well on the way to full ownership of their homes.[66]

It is hard to overstate the importance of the years about 1850 as a turning-point in the history of the social structure of Mulhouse. The intrusion of central government in the affairs of the town, and the increase in population (to which the Cités ouvrières contributed by encouraging immigration from the countryside) were trends that the *industriels* were powerless to resist, even if they had wished to do so. It also seems that, among the younger leaders of Mulhousien industry who were now coming to prominence, the old traditions were not quite so sacred as they had been in the eyes of their parents and grandparents. This was in part an inevitable result of the passage of time, which had dimmed memories of the republic, even in the great families. Perhaps also the precepts of Pestalozzi and Fellenberg, which several members of the new generation had imbibed, must take some credit for a disinterested sense of social responsibility that their elders had conspicuously lacked.[67] At all events, Koechlins and Dollfuses who had once regarded it as essential to live in old family houses close to their factories were now tempted by a more gracious style of life, and there was, as a result, a growing tendency for them either to move out of the town altogether or to spend more time in their country houses, away from the smoke and simmering threat of disorder. The effect of this, allied to the growth of the Cités ouvrières, was unmistakable. Social segregation in the town was accelerated, with damaging consequences for the deteriorating relations between employer and employee to which I refer in the next section.

Science in the Industrial Context

By the mid-nineteenth century, Mulhouse vied with Rouen for the title of 'the Manchester of France'. It serves little purpose to rehearse the Mulhousien claims to this ambiguous distinction, but recent studies of

[64] On the history of the project and its realization, see Eugène Véron, *Les institutions ouvrières de Mulhouse et des environs* (Paris, 1866).

[65] Roberts's book was translated into French, at the request of the President, Louis-Napoleon, as *Des habitations des classes ouvrières* (Paris, 1850). Its effect in Mulhouse was further heightened by the publicity given to it at the time of the Great Exhibition of 1851, to which the Société Industrielle sent a deputation, and by a visit which Roberts made to Mulhouse.

[66] The scheme allowed for occupants to become the owners of their houses after paying rent (initially 22 francs a month) for twenty years.

[67] On this point, see René Martin, *La vie et l'oeuvre de Charles Dollfus (Mulhouse 1827–Paris 1913)* (Gap, 1913) pp. 17–21.

154

science in the British industrial setting inevitably provoke thoughts of a comparison. Was the Société Industrielle analogous to a Literary and Philosophical Society? Did science in Mulhouse have the ornamental character and social role which Arnold Thackray has ascribed to it in the case of Manchester?[68] Can the Koechlins be seen as the Alsatian counterparts of the Henrys or the Gregs?

Generally, I find the differences between the cultural history of Mulhouse and that of Manchester more striking than the similarities, especially when surface appearances are scraped away. I take as an illustration the drift from useful to polite culture which seems to have occurred among the élites of both towns. In the mid-1830s, the principal of the *collège municipal*, Verny, asked the Société Industrielle to take steps to remedy the cultural backwardness of Mulhouse. The tone of his address to the society was uncompromising.

> . . . there are [he said] few towns of the size and industrial and political importance of Mulhouse which display such a great need in matters concerning the general cultivation of the mind . . .[69]

The paucity of cultural provision at this time, as Verny's colleague Souvestre would have agreed, was very real. Yet Verny's scheme for public lectures to be presented 'in an easy and attractive style',[70] though briefly implemented, soon failed for lack of support.[71] Two decades later, however, Achille Penot spoke of audiences for public lectures on science and literature which for some years past it had been impossible to accommodate in the lecture-room of the Société Industrielle.[72]

At about the same time as this new polite audience emerged, there were also several *industriels* who turned to cultural pursuits of a non-utilitarian kind. One of the first to do so was Daniel Dollfus-Ausset, who gradually withdrew from his work as a calico-printer and textile chemist to devote himself to glaciology: from 1844 to 1865, his private field station on the Aar glacier was a mecca for geologists throughout Europe, and his thirteen-volume *Matériaux pour l'étude des glaciers* (1864) legitimated his claim to rank with Agassiz and others in the new speciality.[73] Later examples include Joseph Koechlin-Schlumberger, who practised as a field geologist of national standing in the 1850s and early 1860s (in addition to pursuing an

---

[68] Arnold Thackray, 'Natural knowledge in its cultural context: the Manchester model', *American historical review*, 79 (1974), 672–709.

[69] Louis-Édouard Verny, 'Proposition ayant pour objet d'encourager, sous les auspices de la Société Industrielle, le goût de la littérature et l'étude des sciences et arts', *Bulletin de la Société Industrielle de Mulhouse*, 7 (1834), 471–80 (473–4).

[70] Ibid., p. 479.

[71] Oberlé, *L'enseignement à Mulhouse*, op. cit. (note 55), pp. 239–40.

[72] See p. 7 of the address which Penot delivered at the opening of the École Préparatoire in 1855, in his capacity as director of the school, reproduced in *Inauguration de l'École Préparatoire*, op. cit. (note 59). Quite separate but equally successful lectures were organized for working men and their families in the 1850s and 1860s; see Oberlé, *L'enseignement à Mulhouse*, op. cit. (note 55), pp. 241–6.

[73] Jean Weber, 'Notice biographique sur M. Dr Dollfus-Ausset', *Bulletin de la Société Industrielle de Mulhouse*, 41 (1871), 34–44 (38–43).

active career in public life),[74] and Jean Schlumberger, who turned in his later years from spinning to botany, entomology, and ancient and medieval history.[75]

At first sight, it would appear that these cases exemplify the familiar three-generation pattern of movement away from industrial activity which Thackray describes with reference to the Henry family of Manchester.[76] But the movement in Mulhouse occurred more slowly than it seems to have done in Manchester. While the 'defections' are important, we should not overlook the vast size of the Mulhousien industrial clans and the continued involvement of the majority of their members in manufacturing, often in its most technical aspects. It would be absurd, of course, to pretend that the old families alone were responsible for the technological advances which characterized the cotton industry of Mulhouse in the 1840s, 1850s, and 1860s. They were not. I merely claim that their role remained a preponderant one.

It is true, as I argued earlier, that mechanization came late in southern Alsace. The earliest systematic use of power looms there dates only from 1826, when Isaac Koechlin introduced them at Willer,[77] and even at Dollfus-Mieg et Cie hand-looms were not finally abandoned until the 1850s.[78] In spinning, the self-acting mule was another late arrival, being first used (at Dollfus-Mieg) in 1852, a decade or more after it came into common use in England.[79] But, thereafter, the technical gaps in machinery which had long set Alsace significantly, if not very far, behind Lancashire was reduced. Major refitting in a number of the larger factories, most notably at Dollfus-Mieg, in the 1850s and early 1860s played an important part in this, as did a continuing tradition of indigenous mechanical invention. Josué Heilmann's comber was invented in Mulhouse in 1845, applied industrially in 1851, and quickly used elsewhere in Europe; and Émile Hübner's much faster circular comber followed a few years later.[80] It is noticeable that the inventions which originated in Alsace tended to be of particular benefit to the production of high-quality fabrics: the new combers, for example, were used in the

---

[74] Jean Weber, 'Notice biographique sur M. Joseph Koechlin-Schlumberger', *Bulletin de la Société Industrielle de Mulhouse*, 33 (1863), 535–53 (541–50); and Charles Grad, 'Études historiques sur les naturalistes de l'Alsace. Joseph Koechlin-Schlumberger 1796–1863', *Bulletin de la Société d'Histoire Naturelle de Colmar*, 14e et 15e années (1873–4), 283–314 (292–313).

[75] Sitzmann, *Dictionnaire de biographie des hommes célèbres de l'Alsace*, op. cit. (note 4), vol. 2, p. 691.

[76] Thackray, 'Natural knowledge in its cultural context', op. cit. (note 68), pp. 699–701.

[77] 'Resumé des notes laissées par M. Hartmann-Liebach', op. cit. (note 29), pp. 232–3.

[78] Ernest Zuber, 'Notice nécrologique sur M. Engel-Dollfus', *Bulletin de la Société Industrielle de Mulhouse*, 54 (1884), 267–95 (271–2).

[79] Dollfus, 'Notes pour servir à l'histore de l'industrie cotonnière', op. cit. (note 29), p. 444. Although the self-acting mule was an English invention, the new machinery was supplied by André Koechlin et Cie. By 1853, 30,000 spindles of the new design were in use at Dollfus-Mieg et Cie; see Dollfus, *Histoire et généalogie de la famille Dollfus*, op. cit. (note 18), p. 506.

[80] Dollfus, 'Notes pour servir à l'histoire de l'industrie cotonnière', op. cit. (note 29), pp. 445–6, and Marie-Roch-Louis Reybaud, *Le coton. Son régime, ses problèmes, son influence en Europe* (Paris, 1863), pp. 44–7.

preparation of the superior sea-islands cotton (*coton longue soie*). But the readiness to accept techniques more directly adapted to mass production and the reduction of costs is also, by the mid-century, beyond question.

It seems clear that by about 1860 the Mulhousien textile industry had access to all the technology it needed and that, in one area at least, it led the world. I refer, once again, to textile chemistry. In his envious account of French activity in this field, published in 1860, the English colourist Charles O'Neill referred to France as a whole rather than to Mulhouse specifically. 'It may safely be said', he wrote, 'that for one person of an adequate chemical education connected with dyeing or printing in England, there are ten such in France; hence their high position in these arts with regard to the finer styles and qualities'.[81] The generalized character of the comment should not deceive us, however. For the abundant references that litter his *Chemistry of calico-printing, dyeing and bleaching* make it very plain that Mulhouse was his model, and that the key to Mulhousien success lay above all in the Société Industrielle.

Ironically, the lead to which O'Neill referred may well have had damaging consequences for Mulhousien industry in the new age that dawned with the explosion of the artificial dyestuffs industry after W. H. Perkin's discovery of mauve in 1856. Until that date, the specialities of Mulhouse had been, perforce, natural dyes and their associated mordants. There was no one in the late 1820s, for example, who could match Henri Schlumberger's mastery of madder-based dyes and mordanting with oxides of iron. And, as Ernst Homburg has observed, the development of natural dyes was being pursued more vigorously and creatively than ever in the late 1840s and 1850s.[82] In this later period, Dollfus-Mieg et Cie introduced the technique of animalisation, using albumine to fix dyes normally used for wool and silk on cotton and to create the new 'pigment colours' style. It was also in this period, in 1855, that Albert Schlumberger, then a colourist at the Wesserling firm of Gros, Roman, Odier, et Cie, showed how murexide could be used commercially as a cotton dye. Within a year or two, several firms in Mulhouse, including those of Frères Koechlin, Dollfus-Mieg et Cie, and Steinbach, Koechlin et Cie, were printing with murexide; and, before the decade was out, other new natural dyestuffs, notably the highly successful French purple, were introduced. The award of a medal to Perkin by the Société Industrielle in 1859[83] and

[81] Charles O'Neill, *Chemistry of calico-printing, dyeing, and bleaching* (Manchester, 1860), p. iii. In fact, there is a slightly grudging air about O'Neill's comments on French supremacy. In his view (p. iv), the French tended to receive excessive credit for their work in calico-printing and related technologies because innovations made in Switzerland, Belgium, Northern Italy, and parts of Germany, as well as those made in France, were regularly announced in French journals, notably of course the *Bulletin de la Société Industrielle de Mulhouse.*

[82] Ernst Homburg, 'The influence of demand on the emergence of the dye industry. The roles of chemists and colourists', *Journal of the Society of Dyers and Colourists*, 99 (1983), 325–34 (329–3). Homburg's work forms part of a broader study of the development of the dye industry now nearing completion under the direction of Dr W. J. Hornix of the University of Nijmegen.

[83] *Bulletin de la Société Industrielle de Mulhouse*, 30 (1859), 225.

the research of the young Horace Koechlin, as well as Albert Schlumberger's own work, show that Mulhouse in no way turned its back on aniline dyes. Jean Gerber-Keller, now known chiefly for his central role in the Fuchsine case of 1863, was another Mulhousien who immediately espoused the new technology, developing his ideas in the familiar context of the chemical committee of the Société Industrielle.[84] Yet there can be no denying that, with the coming of the new dyes, the centre of research and production in textile chemistry shifted unmistakably from the once fertile ground of southern Alsace to Lyon and Paris.

It is tempting to suggest that the decades of Mulhousien success with natural dyestuffs inhibited an immediate recognition of the commercial superiority of the far cheaper coal-tar products, so allowing competitors elsewhere in France and in Germany and Britain to edge ahead.[85] In his report on the artificial dyestuffs displayed at the 1862 Exhibition in London, A. W. Hofmann predicted a brilliant future for aniline dyes.[86] But even if many natural dyestuffs (murexide in particular) had had their day, it was not obvious to Hofmann that all of them would necessarily be eclipsed. In a comment that helps us to understand Mulhousien conservatism, he spoke of a continuing 'struggle' between French purple and coal-tar purple in which the real advantages of the former ('fastness and resistance to the influence of light') would have to be weighed seriously.[87]

The complete congruence between the style of chemical research pursued in Mulhouse and the industrial context that gave rise to it is, I believe, beyond question. It is scarcely less so in the case of another scientific tradition, concerned with theoretical and experimental research on the steam-engine. The industrial incentive for this research (much of it centred on the economies to be obtained by the use of steam jackets) is clear. The obstinately high price of coal, still three or four times that paid by English manufacturers, assumed a new importance in the 1850s and early 1860s, as the number of steam-engines in use in the Haut-Rhin

[84] On this case, in which Jean and Armand Gerber-Keller unsuccessfully challenged the claims of Renard frères to a monopoly on the manufacture of Fuchsine (or Magenta), see L. F. Haber, *The chemical industry during the nineteenth century. A study of the economic aspect of applied chemistry in Europe and North America* (Oxford, 1958), pp. 201–2, and other standard sources.

[85] The response of the most important chemical manufacturer in southern Alsace (the firm of Charles Kestner at Thann) is instructive. Kestner's cautious response to the opportunities presented by artificial dyestuffs contrasts markedly with the firm's long-standing activity in the preparation of natural dyes. According to the *Histoire documentaire de l'industrie de Mulhouse*, op. cit. (note 4), vol. 2, pp. 578–9, Kestner did manufacture aniline violet and other products related to the new technology. But the venture was soon abandoned, and Kestner reverted to the more traditional activities of the French heavy chemical industry, specializing in particular in the production of sulphuric acid; see Laufenburger and Pflimlin, *Cours d'économie alsacienne*, op. cit. (note 4), vol. 2, p. 80n, and Charles Grad, *Études statistiques sur l'industrie de l'Alsace* (2 vols., Colmar, Strasbourg, and Paris, 1879–80), vol. 1, pp. 308–9.

[86] See Hofmann's report on Class II ('Chemical and pharmaceutical products and processes'), in *International Exhibition of 1862. Reports by the juries on the subjects in the thirty-six classes into which the Exhibition was divided* (London, 1863), p. 136.

[87] Ibid., p. 117.

increased nearly three-fold.[88] The earliest important work on steam power
in Mulhouse, by Émile Koechlin, dates from the 1830s,[89] but the most
distinguished contributions were unquestionably those which Gustave-
Adolphe Hirn, an engineer and associate of the Haussmann family, began
to make at le Logelbach in the early 1850s.

Both Koechlin and Hirn were very self-consciously Alsatian *savants*,
though with differences that arise from a regional microstructure whose
complexity the single term 'Alsatian' cannot convey. Hirn resembled
Émile Koechlin and most other Koechlins in that he was a French-speak-
ing Calvinist, born into a textile family, privately educated, and a member
of the Société Industrielle. As a native of Colmar, however, he stood
slightly apart from the Mulhousien industrial élite. While his protestantism
made it difficult for him to contemplate an academic career in the national
system of education, he deferentially courted the scientists of the capital
and secured the favourable attention of Le Verrier, among others. Hirn, in
fact, was seen in Paris as a useful auxiliary whose work was to be taken
seriously. Hence I find it entirely predictable that his experimental
demonstration that the amount of heat leaving a steam-engine in the
condenser (Q_2, in the conventional nomenclature) is less than the amount
entering it through the boiler (Q_1) and that (Q_1-Q_2) is proportional to the
work done was immediately recognized as a major contribution,[90]
contributing to Hirn's eventual election as a corresponding member of the
Académie des Sciences.

Although the science of Mulhouse never flagged or broke with its
industrial roots, there were occasional signs, even before the mid-century,
that the rising generations of the great families felt dissatisfied with the
rather hermetic, self-contained character of cultural life in southern
Alsace. When Frédéric Engel had shown an interest in the possibility of
entering the École Polytechnique and pursuing a career in one of the state
corps of engineers about 1830, his family soon convinced him of the error of
his ways; he was rescued from the corrupting influence of Paris (where he
was a pupil at the lycée Henri IV) and brought back to Mulhouse to serve
an industrial apprenticeship and to marry a Dollfus.[91] But a quarter of a

[88] According to Charles Thierry-Mieg, 'Rapport sur les forces matérielles et morales de l'industrie du
Haut-Rhin, pendant les dix dernières années (1851–1861)', *Bulletin de la Société Industrielle de Mulhouse*,
32 (1862), 431–73 (459), the number of steam-engines in the Haut-Rhin increased from 163 (a total of
3,565 H.P.) in 1851 to 473 (11,027 H.P.) ten years later. The continuing concern about the price of coal
is evidence of the limited advantages that were obtained by the opening of the Rhône-Rhine canal and
the improvement of the railway network; see note 42, above.

[89] See, for example, Koechlin's huge 'Mémoire sur les machines à vapeur, sur des expériences
comparatives à faire entre les divers systèmes de machines, et sur l'utilité que présenterait un ouvrage
complet et classique sur cette partie essentielle de l'industrie manufacturière', *Bulletin de la Société
Industrielle de Mulhouse*, 9 (1836), 79–182, and the related contributions by Joseph Koechlin and Choffel
on pp. 183–277.

[90] The first report on Hirn's experiments, which were performed on a 120 H.P. Watt engine at le
Logelbach, appeared in a letter he wrote to the President of the Société Industrielle de Mulhouse, dated
21 October 1854. See *Bulletin de la Société Industrielle de Mulhouse*, 26 (1855), 274–7.

[91] Ernest Zuber, 'Notice nécrologique sur M. Engel-Dollfus', *Bulletin de la Société Industrielle de*

century later, the lure of the capital for the young Charles Dollfus, the son of Jean Dollfus, proved too much.[92] After a conventional early education in Switzerland, an unhappy and unsuccessful year at the École Centrale des Arts et Manufactures, and one year in the law faculty at Strasbourg, Charles was totally seduced in Paris by the literary bohemianism which beguiled many another student of law. By 1851, at the age of twenty-four, he had already had the heady experience of being imprisoned for impiety for his Voltairian *Lettres philosophiques*. Thereafter, the prospect of a legal career in Colmar seemed insufferably tame, and he settled in the capital, swelling the ranks of the liberal journalists who flocked there during the Second Empire. He never returned to Alsace and did not go back on his early rejection of the industrial career which his father had envisaged for him.

The fragmentary nature of the evidence makes it difficult to identify the trends that characterized the decade or so preceding the war. But, as I have tried to show, in a variety of ways the power which a handful of families had managed to retain through nearly three-quarters of a century of French rule was at last slipping away from them. Between the 1850s and 1870, the trends which had begun to appear by the mid-century gathered pace. Not only did the old families have to contend with the challenge of rival élites in the bureaucracy and education, they also faced the first signs of insecurity within contexts which previously had been totally their preserve. Even the Société Industrielle de Mulhouse lost some of its cosy cohesiveness. This was to a large extent a straightforward result of growth: between 1850 and 1870 the membership almost doubled in size, from 272 to 512. But it also owed something to the growing prominence of new men with a technical competence to match that of the established experts. As Laufenburger and Pflimlin pointed out long ago, the Second Empire saw the lead in some of the key activities of the society being taken by professional engineers who had been trained outside Mulhouse. Émile Burnat and William Grosseteste, both of them employed by Dollfus-Mieg et Cie, were typical of this new breed.[93]

It is quite clear that, at the same time, changes were occurring inside the industrial élite as well. I have already referred to the signs of restlessness in the generation which came to maturity about 1850, the generation of Charles Dollfus. These signs coincided with other strains, notably in the solidarity which, at least in public, had always bound the employers to one another. In the early 1850s, there was the unseemly spectacle of an all too visible dispute between André Koechlin et Cie and Nicolas Schlumberger

Mulhouse, 54 (1884), 267–95 (268–9), and Xavier Mossmann, *Vie de F. Engel-Dollfus* (Paris, 1887), pp. 7–8.

[92] For a detailed biography of Dollfus, see Martin, *La vie et l'oeuvre de Charles Dollfus*, op. cit. (note 67). A contemporary of Dollfus who 'escaped' at about the same time was Charles Schlumberger. In the late 1840s, Schlumberger entered the École Polytechnique and went on to a career in state employment, as a marine engineer, and to a consuming vocation as a naturalist.

[93] Laufenburger and Pflimlin, *Cours d'économie alsacienne*, op. cit. (note 4), vol. 2, p. 286.

et Cie over the right to manufacturer Hübner's carder.[94] And worse was to follow in the 1860s, as free-traders, led by Cobden's friend Jean Dollfus, confronted protectionists, most of them the smaller spinners and weavers whose activities were gravely affected by the entry into France of cheap British products following the Chevalier–Cobden Treaty of 1860.[95]

These confrontations would not have been so damaging to the interests of the *industriels* if they had not been accompanied by a new unity among their employees (a growing proportion of whom were now engaged in the less well protected industries of spinning and weaving, rather than in calico-printing, as Table 4 shows). There can be no doubt that unity owed much to the improved living conditions which the employers had helped to promote. For, as recent work in the urban history of nineteenth-century France has shown, the emergence of organized working-class protest depended on the existence of a settled, close-knit community: the semi-migrant workers of the 1830s, for example, were far less likely to be stirred to political action than men who had lived for a decade or more in the Cités ouvrières.[96] Electoral reform, too, played its part: since 1848, an extended franchise had given the textile workers an unprecedented grip on their destinies and a new way of demonstrating the opposition to free trade which most of them felt. Between 1861 and 1865, the cotton famine accompanying the American civil war and the resulting closures and short-time working served to polarize opinion still further and to create a degree of social dislocation that would have been inconceivable in Mulhouse only twenty years before. In the later 1860s, the industrial unrest which affected many parts of France was particularly acute in southern Alsace.[97] Hence, with the benefit of hindsight, it does not seem at all surprising (though it seemed so at the time) that when the venerable Jean Dollfus stood for re-election as one of the deputies for the Haut-Rhin in 1869, the impotence of the old paternalism was ruthlessly laid bare. Dollfus was defeated by a more radical candidate, and in the following year, only days before the war with Prussia was declared, every factory in Mulhouse and most of those elsewhere in the department were on strike.[98]

In these circumstances, Bismarck had little difficulty in persuading the working population of the Haut-Rhin that their destiny lay with the new German Empire rather than as the vassals of an industrial aristocracy

---

[94] *Histoire documentaire de l'industrie de Mulhouse*, op. cit. (note 4), vol. 1, p. 230.

[95] On this issue, see, in addition to the standard sources, Mossmann, *Vie de F. Engel-Dollfus*, op. cit. (note 91), pp. 28–39. Frédéric Engel-Dollfus was deeply involved in the debate, taking the side of his father-in-law, Jean Dollfus, along with most of the large calico-printers.

[96] The correlation between a settled, concentrated community and a capacity to organize working-class protest is a recurring theme in John M. Merriman (ed.), *French cities in the nineteenth century* (London, 1982). See especially Merriman's Introduction and the contributions by Charles Tilly and Michael P. Hanagan.

[97] Fernand L'Huillier, *La lutte ouvrière à la fin du Second Empire* (Paris, 1957), pp. 59–72.

[98] On the elections, see André Brandt and Paul Leuillot, 'Les élections de Mulhouse en 1869', *Revue d'Alsace*, 99 (1960), 104–28.

that disdained their language and did not even share the Catholicism which by now the overwhelming majority of the population professed.[99] Needless to say, the response of the main *industriels* was very different. Some of them transferred at least part of their manufacturing activity over the border into France, as a way of securing their access to the French market which, at the time of the annexation, absorbed the overwhelming majority of their products.[100] Others maintained their industrial interests in Alsace but moved their homes out of the area; the younger Antoine Herzog was among those who settled in Paris. But most of them stayed on, using the Société Industrielle as a bastion of French culture in an increasingly alien environment.[101]

In reality, the maintenance of a French cultural tradition was of little more than symbolic importance. Economically, the region was quickly and completely assimilated into Germany, despite early hopes that special terms might be arranged for the entry of Alsatian goods into France.[102] Understandably, manufacturers across the Rhine were quite as dismayed by the assimilation as the Alsatians themselves. They feared the competition of an industry which in the Haut-Rhin alone consumed as much cotton as the whole of the Zollverein and which possessed a technology far superior to their own. In this respect, the modernization of many factories in an around Mulhouse, which had proceeded at a remarkable pace since the 1840s (as the indicators in Tables 5, 6, and 7 show very clearly), had only served to aggravate the problem.[103] It both widened the technological

[99] See Table 3. In 'Les élections de Mulhouse en 1869', op. cit. (note 98), p. 124, Brandt and Leuillot give the following figures for the religious affiliations of the population of Mulhouse in 1866: Catholics 45,550, Protestants 11,211, Jews 1,939, others 73 (total 58,773). In Alsace as a whole, the proportion of Catholics to Protestants was slightly lower, being of the order of three to one.

[100] For a list of nine firms that developed manufacturing activities in France in the aftermath of the annexation, see Marie-Joseph Bopp, 'L'oeuvre sociale de la haute bourgeoisie haut-rhinoise au XIXe siècle', in *La bourgeoisie alsacienne*, op. cit. (note 40), pp. 387–402 (402). André Koechlin et Cie (assimilated from 1872 as part of the very successful Société Alsacienne de Constructions Mécaniques) was a particularly notable enterprise which began manufacturing in France (at Belfort), though it did not do so until 1879.

[101] Jean Dollfus was the most notable of the public figures who stayed and remained loyal to the French traditions of Mulhouse. Despite his defeat in the elections of 1869 and his advanced age, he re-entered the world of politics as the deputy for Mulhouse in the Reichstag from 1877 to 1887.

[102] The relative importance of French and foreign markets varied very considerably between the 1830s (when the export trade in textiles from Alsace prospered) and 1871. But exporting in this period was never an easy task. On the eve of the annexation, spinning and weaving were totally dependent on the home market, and only the calico-printers exported to a significant extent, 22 per cent of their production going abroad. In these circumstances, any reduction in the ease of access to French markets was a major blow. See Fernand l'Huillier, 'Deux siècles d'exploitation textile haut-rhinoise (1750–1950)', *Société Industrielle de Mulhouse. Bulletin trimestriel*, nos. 1–2 (1950), 111–22 (119–20).

[103] The Tables show very clearly the effects of the introduction of new and greatly improved machinery in the quarter of a century before the annexation. The main changes were: a) the continued rise in the number of spindles, especially after the introduction of the self-acting mule in the early 1850s, b) the rapid adoption of the power loom in place of the hand looms on which weaving still largely depended in the 1840s, and c) the increasing productivity of calico-printers (achieved at a time when the work force and the number of factories in this industry were diminishing). See also Table 8, which makes very clear the changing structure of the cotton industry in the Haut-Rhin, with the commercial importance of spinning and weaving overtaking that of calico-printing. The faltering authority of Jean Dollfus in the 1860s probably owed something to this trend, since, despite his interests in spinning and weaving, he always spoke as the representative of the calico-printers.

gap and convinced Bismarck that the prize of Alsace was one on which there could be no compromise.

Conclusion

I began this address with prescriptions about the writing of the history of French science, and it is with prescriptions—three of them—that I shall finish. My first concerns the administrative and intellectual centralization which is commonly supposed to have inhibited the creativity of scientists in France since the early decades of the nineteenth century.[104] As I indicated in my opening remarks, Parisian ministries—in particular Public Instruction and Commerce—constantly strove to tighten their control, and I have argued elsewhere that administrative pettiness, especially in the 1850s, drove a damaging wedge between French science and the science of other countries.[105] But I hope that this study of Mulhouse shows just how much of the realm of technical education and industrially related science escaped the net. As Terry Shinn has argued with reference to the moves to make French universities more independent between 1880 and 1914, explanations of the supposed decline of French science after about 1830 that invoke centralization are likely to be frail creatures.[106]

My second prescription follows immediately from the first. It concerns the desirability of adopting a provincial as well as a Parisian perspective. As I have tried to show, most of the initiatives in industrial science and technology that emanated from the capital had little bearing on the industrial life of Mulhouse. It is significant and wholly in character that the *industriels* of the area displayed little interest in the Société d'Encouragement pour l'Industrie Nationale until the 1820s,[107] and that the three national exhibitions of French industry that were held during the Bourbon Restoration aroused only a patchy response in Mulhouse. At the first of them, in 1819, Mulhouse calico-printers won no fewer than seven gold medals, but in 1823 there were no exhibitors from the town, and in 1827 only three.[108] Clearly, in 1823 and 1827 the political objective of a public

---

[104] This explanation for the shortcomings of French science has a long history, going back to the mid-nineteenth century. By the time of the war of 1870, it had become a commonplace in the mounting demands for reform that were being voiced by Sainte-Claire Deville, Pasteur, and others; see Fox, 'Science, the university, and the state', op. cit. (note 60), pp. 105–6. For a classic statement of the ills of centralization in the modern literature on France, see Joseph Ben-David, *The scientist's role in society. A comparative sstudy* (Englewood Cliffs, New Jersey, 1971), pp. 88–107.

[105] Fox, 'Science, the university, and the state', op. cit. (note 60), pp. 84-101.

[106] Shinn, 'The French science faculty system', op. cit. (note 3), *passim*, but especially p. 326. Shinn shows how difficult it is to establish any correlation between performance in research and the degree to which French science was subjected to close central control.

[107] Leuillot, *L'Alsace au début du XIXe siècle*, op. cit. (note 4), vol. 2, p. 379.

[108] Ibid., vol. 2, p. 389n. Daniel Koechlin's indifference to the system of national exhibitions was flaunted in a characteristic way when he did not even bother to collect the gold medal (or the decoration of the Legion of Honour) that he was awarded in 1819. The more extreme royalist governments of the 1820s reciprocated the sentiment. At the 1827 exhibition, for example, the exhibits from Mulhouse were said (by Nicolas Koechlin) to have been relegated humiliatingly to a remote corner. Under the July Monarchy, which the industrialists found politically more acceptable, Mulhousien involvement in the national exhibitions was more conspicuous. In 1834, firms in the

display of opposition to the Bourbons was seen as far more important than any economic advantages which a national exhibition might have bestowed.

Thirdly, and finally, a comment on the engaging subject of the social use of science. One of my main contentions about Mulhousien science and technology has been that its pristine seriousness was maintained almost unimpaired until 1870. Ornamental learning, purveyed through public lectures, made its appearance, as I have shown. But it was an additional, not an alternative cultural form, and never a prominent one. I know that, in making this point, I run the risk of being charged with a crude functionalism, with implying that the technical needs of the town's manufacturers stimulated an appropriate, ideologically neutral industrial science and that the interaction between economic substructure and scientific superstructure constitutes the whole story. In fact, it is my contention that the relentless utilitarianism of the debates and publications of the Société Industrielle can only be understood by also taking account of the non-technical priorities to which I have repeatedly referred. The promoting of polite lectures in the Parisian style would have undermined the distinctiveness of Mulhousien society; it would also have subverted the carefully maintained public face of Calvinist austerity and probity and so given the lie to a carefully fashioned rhetoric. Hence, in the context of Mulhouse, displays of ornamental learning and elegance, far from heightening social control and winning status (in the way they seem to have done in Manchester), would have served to weaken existing authority. It is perhaps all too easy to suppose that the social uses of science are something quite separate from its economic uses; in Mulhouse, one and the same activity achieved both social and economic aims.

In all this, historians of modern British science may feel that I have been speaking to the converted: in recent years, the fine structure of provincial activity in our own country has been perceptively explored, not least by my successor as president. But the comparable field in France is almost unploughed, despite the voluminous works that French social historians have devoted to the tiniest geographical contexts. Of course, we shall never understand the place of science in French society merely by walking the byways. But I suggest that we may have spent too long on the highways, viewing French science from the centre, as nineteenth-century ministers of education obfuscatingly intended that we should.

Tables

The Tables that follow are based primarily on statistical information in: Émile Dollfus, 'Notes pour servir à l'histoire de l'industrie cotonnière

Haut-Rhin won 13 gold medals, 14 silver medals, and 9 bronze medals, and 5 *industriels* in the region were decorated with the Legion of Honour; see *Bulletin de la Société Industrielle de Mulhouse*, 7 (1834), 466–7.

dans les départements de l'Est', *Bulletin de la Société Industrielle de Mulhouse*, 27 (1855–7), 435–61.

Charles Thierry-Mieg, 'Rapport sur les forces matérielles et morales de l'industrie du Haut-Rhin, pendant les dix dernières années (1851–1861)', *Bulletin de la Société Industrielle de Mulhouse*, 32 (1862), 431–73.

Achille Penot, 'Notes pour servir à l'histoire de l'industrie cotonnière dans le département du Haut-Rhin', *Bulletin de la Société Industrielle de Mulhouse*, 44 (1874), 145–260.

Charles Grad, *Études statistiques sur l'industrie de l'Alsace* (2 vols., Colmar, Strasbourg, and Paris, 1879–80).

Marie-Roch-Louis Reybaud, *Le coton. Son régime, ses problèmes, son influence en Europe* (Paris, 1863).

The information given relates, except where stated, to the whole of the department of the Haut-Rhin.

TABLE 1. The chronology of industrial growth and innovation in the Haut-Rhin

1746	Establishment of the first calico printing works in Mulhouse by Samuel Koechlin, Jean-Jacques Schmaltzer, and Jean-Henri Dollfus
1790	Wallpaper manufacture begun by Frères Dollfus et Cie
1800	Establishment of Dollfus-Mieg et Cie
1802	Opening of first spinning mill at Wesserling by Gros, Davillier, Roman et Cie
	Nicolas Koechlin et frères established by Nicolas Koechlin, following an apprenticeship served with his uncle, Daniel Dollfus-Mieg
	Jean Zuber et Cie, wallpaper manufacturers, established at Rixheim
1803	Flying shuttle in use at Wesserling
1804	Calico-printing with copper rollers practised at Dollfus-Mieg et Cie and Nicolas Koechlin et frères
1808	Establishment of Nicolas Schlumberger et Cie, cotton-spinners, at Guebwiller, later to become the largest spinning mill in France (37,500 spindles in 1826)
1812	First steam-engine employed in spinning, by Dollfus-Mieg et Cie
c.1818	Henri Schlumberger and Daniel Koechlin-Schouch appointed as colourists by Mulhouse calico-printers
1818	Mathieu and Jérémie Risler establish machine construction works at Cernay, with Job Dixon
c.1820	Chlorine bleaching in general use, though first used by J. M. Haussmann at le Logelbach in 1791
1826	Establishment of André Koechlin et Cie, engineers
1827	Power looms first used, by Isaac Koechlin at Willer (designed by Josué Heilmann, constructed by André Koechlin et Cie)

TABLE 1 (*cont.*)

c.1830 Machinery for calico-printing in two colours introduced at Dollfus-Mieg et Cie by Daniel Dollfus-Ausset
  1839 Manufacture of worsted wool begun
  1845 Invention by Josué Heilmann of Heilmann combing-machine, subsequently improved by Jean-Jacques Heilmann and Henri Schlumberger and manufactured by Nicolas Schlumberger et Cie
  1852 Self-acting mule introduced in spinning
  1853 Manufacture of the Hübner comber by André Koechlin et Cie. Widely used from the late 1850s, notably by Dollfus-Mieg et Cie
  1856 Murexide introduced as a dyestuff for cotton
  1858 Mauve (aniline violet) in industrial use
  1859 First use of French purple and Fuchsine

TABLE 2. The population of Mulhouse, 1798–1870

Date	Total
1798	6,018
1800	6,618
1805	8,021
1810	9,353
1815	9,350
1820	9,598
1825	12,038
1830	13,231
1835	13,804
1840	17,250
1845	23,393
1850	29,268
1855	29,574
1860	45,981
1865	56,541
1870	65,000

TABLE 3. Religion in Mulhouse

Date	Protestants (%)	Catholics (%)	Jews (%)
1803	91	9	?[1]
1834	50	50	?[1]
1851	43	57	?[1]
1865	25	75	?[1]
1875	26	74	?[1]
1888	27	70	3%
1899	23	72	5%

[1] Information about the Jewish community in Mulhouse for these dates is sparse, but the population was probably of the order of 2–3%, representing a population numbering a few hundred.

TABLE 4. Number of operatives employed in the cotton industry

Date	Spinning	Weaving	Calico-printing
1827	10,240	23,352	11,248
1846			c.10,000
1851	c.14,000	c.19,000	
1856			9,765
1861	c.14,000	c.22,000	
1871	12,245	33,243	8,611

TABLE 5. Number of spindles (cotton)

Date	No.
1810	24,000
1812	47,508
1827	466,363
1834	540,000
1839	683,000
1844	763,734
1849	786,312
1855	912,000
1859	1,154,220
1864	1,234,626
1866	1,428,850[1]
1871	1,411,011

[1] The level of activity in spinning in each of the two other main centres of the cotton industry in France at this time—Lille-Roubaix-Tourcoing and Seine-Inférieure (chiefly Rouen, Elbeuf, and Le Havre)—was roughly comparable. To obtain an impression of the importance of these activities in the general European context, the figures given have to be compared with the 40 million spindles that were in use in the United Kingdom in the late 1860s and the 3 million of the German Zollverein. The corresponding figure for the whole of France was 6,800,000.

TABLE 6. Mechanization of cotton weaving

Date	Number of power looms	Number of hand looms
1831	426	21,651
1834	3,090	31,000
1839	6,000	
1844	12,000	19,000
1856	18,139	8,657
1864	24,133	3,000–4,000
1865	24,646	3,000–4,000
1866	30,421	3,000–4,000

TABLE 7. Production of printed cotton

Date	Factories	Metres of fabric produced
1798		2,500,000
1828	27	17,949,790
1836	35	
1847	20	37,800,000
1856	21	49,000,000
1862	18	50,000,000
1867	14	65,000,000
1871	18	82,537,934

TABLE 8. Turn-over in the major industries

Date	Calico-printing (M francs)	Weaving (M francs)	Spinning (M francs)
1828	33[1]	20	16
1862	50[1]	70	60

[1] On p. 447 of his 'Rapport sur les forces matérielles et morales de l'industrie du Haut-Rhin' (see list of sources for these Tables), Charles Thierry-Mieg contrasts the modest growth in this industry with that in England, where the comparable figures were 100 million francs for 1828 and 300 million francs for 1862.

VI

FROM CORFU TO CALEDONIA: THE EARLY TRAVELS OF CHARLES DUPIN, 1808–1820

It is a common conceit of the English that France was cripplingly isolated during the revolutionary and Napoleonic wars that dragged on, with only one short break, from 1792 to 1814. In this way, comforting sense can be made of French backwardness in some key branches of industrial technology (notably in the use of steam power) and of the intense curiosity about Britain which French writers displayed after the fall of Napoleon. When viewed from France, however, the picture is somewhat different. For although France was indeed isolated from Britain during the wars, she was by no means isolated from lands to the east and the south. In fact, at various times during the Consulate and the Empire, Egypt, Germany, Italy, Spain, Russia and the islands and coast of the Adriatic were all unprecedently accessible as a result of military conquest. It would be hard to overestimate the importance for French culture of this new accessibility. The Egyptian style in dress and furniture under Napoleon was a vogue directly encouraged by travels and travellers' tales, while in literature, Stendhal's largely autobiographical *Chartreuse de Parme* (1839) was a particularly well known if tardy product of a love of Italian culture first conceived under the Empire.

It could never be suggested, of course, that French interest in foreign lands and unfamiliar civilizations was a product of the Revolution. In the later Ancien Régime, Louis-Antoine de Bougainville's *Voyage autour du monde* (1771) and Claude-Étienne Savary's *Lettres sur l'Égypte* (1785) and *Lettres sur la Grèce* (1788) enjoyed a success which displays very plainly a taste for the exotic fostered by an unprecedented adventurousness that emerged in the later Enlightenment. My claim, therefore, is not that the decade of the 1790s marked a new departure, but rather that it saw the intensification of a trend. Between then and 1814 the number of Frenchmen who travelled over the Rhine and down to and across the Mediterranean was far greater than ever before.

In this paper, I develop a particular perspective on the cultural phenomenon I have described by discussing the experiences of Charles Dupin. As a young man, almost exactly a contemporary of Stendhal, Dupin was taken by his naval duties under Napoleon's Empire to the Ionian Islands and Italy, and there he was fired with an enthusiasm for travel which he pursued, after Waterloo, in a remarkable series of five visits to Britain undertaken between 1816 and 1820.

J.D. North and J.J. Roche (eds.), The Light of Nature. ISBN 90-247-3165-8.
© *1985, Martinus Nijhoff Publishers, Dordrecht. Printed in the Netherlands.*
Reprinted by permission of Kluwer Academic Publishers.

As I shall argue, both in his exposure to the east and in his subsequent infatuation with Britain, Dupin was not merely indulging a personal whim. He was experiencing a widening of cultural horizons that left its indelible mark on a whole generation of Frenchmen whose youth was coloured by the excitement of the Empire and then by the economic, social, and political changes that accompanied the return of the Bourbon line.

Dupin's talents and the long and distinguished public life in which he applied them were dazzling. [1] Born at Varzy in the Nivernais in 1784, the son of a lawyer with literary inclinations and a Rousseauite mother besotted with *Émile*, he was one of the most brilliant products of the short-lived system of *écoles centrales*. Progressing from the *école centrale* of the department of the Loiret in Orléans, he entered the more advanced classes that were available in Paris and finally was admitted to the École Polytechnique in 1801, at a particularly distinguished period in the institution's history. Two years later, he was one of three graduates admitted to the prestigious corps of naval engineers, and for the next four years he served with the Channel and Mediterranean fleets, chiefly in Antwerp, Toulon, and Genoa. By this time, he had already established a personal friendship with the family of Lazare Carnot[2] and attracted the admiring attention of Gaspard Monge through his precocious mathematical talents, displayed in some early work on the radii of curvature of surfaces of the second order.[3] Thereafter, he remained fiercely loyal to Monge and to the style of descriptive geometry which Monge taught at the Polytechnique from its inception in 1794 until 1810.

It was the Treaty of Tilsit in July 1807 which allowed Dupin to break with the routine of refitting and modest coastal exercises to which the French navy had been condemned since England's victory at Trafalgar nearly two years before. By the Treaty, the Ionian Islands, on which Dupin and thousands of other young Frenchmen were to spend several years of their lives, were ceded by Russia to France.[4] From the start, the islands were a mixed blessing, having been

1. For biographical information on Dupin, see A. Victor Lacaine and H. Charles Laurent, *Biographies et nécrologies des hommes marquants du XIXe siècle*, 7 vols., Paris, 1844–50, vol. 4, pp. 273–312, and Joseph Bertrand, *Éloges académiques*, Paris, 1890, pp. 221–46.

2. It is clear that by about 1803 Dupin was already a friend of the Carnot family. See [Lazare-Hippolyte Carnot], *Mémoires sur Carnot par son fils*, 2 vols., Paris, 1861–3, vol. 2, pp. 277 n., where it is recorded that Lazare Carnot had treated Dupin "like a son".

3. For examples of his early work in descriptive geometry, see the two short papers in the *Correspondance sur l'École Impériale Polytechnique*, 1 (issues for July 1806 and January 1807), pp. 183–4 and 218–25, and his "Essai sur la description des lignes et des surfaces du second degré", *Journal de l'École Polytechnique*, 1808, 7, 14e cahier, pp. 45–83.

4. On the history of the Ionian Islands in the early nineteenth century, the most useful sources are: Jean-Pierre-Guillaume Pauthier, *Les îles ioniennes pendant l'occupation française et le protectorat anglais d'après des documents authentiques la plupart inédits tirés du général de division Comte Donzelot*, Paris, 1863; Emmanual Rodocanachi, *Bonaparte et les îles ioniennes. Un épisode*

marked equally by the four centuries for which they had existed as a backward part of the Venetian Republic, and by the previous decade, in which they had been a pawn in the wider European conflict.

The islands had become involved in the main stream of European affairs in 1797, when they were first occupied by the French. Two years later, they were liberated by a combined force of Turks, Albanians, and Russians, and in 1800 they were declared an independent republic, the Septinsular Republic,[5] under Turkish and Russian protection. The solution proved an unstable one. After the Russian forces left the islands in 1801, the republic quickly began to disintegrate, with two of the seven islands (Cephalonia and Ithaca) seceding and another, Zakinthos, admitting an English occupying force. On the main and northern-most island, Corfu, there was a wave of rebellion and assassination which culminated in a victory for the democratic faction, the establishment of a new constitution, and the abolition of the privileges and titles of the nobility. For the next six years, the internal strife between nobles and peasantry continued unabated, with the old ruling families, the Signori, pining for Venetian rule and looking to Turkey and then to England for support, and the Russians, the traditional enemies of Venice, eventually intervening to form the dependency which lasted until the Treaty of Tilsit. The compromise of a "constitutional" nobility which the Russians imposed was a promising one, but it seems, in the event, to have done nothing to remove the threat of civil war or to ease the perpetual turmoil which had existed since 1797.

There is no evidence that the French administration which took possession of the islands in 1807 had any clear political or social mission. Napoleon's public statements and, still more obviously, his correspondence make it plain that, for him, the islands were of purely strategic importance and that, even strategically, only Corfu, the "key to the Adriatic", as he called the island, was of real consequence.[6] The first task, that of placing a garrison on Corfu, underlined

---

des conquêtes de la République et du Premier Empire (1797–1816), Paris, 1899; Guillaume de Vaudoncourt, *Memoirs on the Ionian Islands, considered in a commercial, political, and military point of view*, transl. by William Walton, London, 1816; and Captain Henry White-Jervis [correctly Jervis-White-Jervis], *The Ionian Islands during the present century*, London, 1863. For a very full account of the topography, condition, and history of the islands up to 1800, see André-Grasset Saint-Sauveur, *Voyage historique, littéraire et pittoresque dans les îles et possessions ci-devant vénétiennes du Levant*, 3 vols. plus atlas, Paris, an VIII [1800]. By comparison, [Virgile-Antoine Schneider], *Histoire et description des îles ioniennes depuis les tems fabuleux et héroïques jusqu'à ce jour*, 1 vol. plus atlas, Paris, 1823, is rather thin, but the atlas contains useful statistical information. A modern account of the two periods of French occupation is Jacques Baeyens, *Les Français à Corfou (1797–1799 et 1807–1814)*, Athens, 1973.

5. So called because of the seven islands which composed the Republic: Corfu (or Corcyra), Paxos, Cephalonia, Ithaca, Zakinthos (or Zante), Kithira (or Cerigo), and Levkas (or Santa Maura).

6. The letters in the *Correspondance de Napoléon Ier publiée par ordre de l'empereur Napoléon*

the Emperor's point. With the English commanding the sea, it was no small achievement to land the first French detachment in August 1807, just six weeks after the Treaty was signed, and then to have over 8,000 men on the islands by November. Despite the size of this force, however, it was only in the spring of 1808 that the occupation began to assume real stability. By then the vain and self-willed first governor, César Berthier, had been replaced by a man of charm, ability, and tact, Baron François-Xavier Donzelot, and a small fleet of ships under Vice-Admiral Honoré Ganteaume had triumphantly evaded the English to land men and supplies. By early April, Napoleon could write that the seas around Corfu had been cleared of British ships.[7] This soon proved to have been an optimistic statement, but the English blockade was never rigorously imposed, and, in any case, the fact that Corfu now had ample weapons and two years of provisions meant that the French presence was secure.[8]

It was Ganteaume's fleet which brought Dupin to Corfu in March 1808. He came, it seems, with freshly acquired skills in Latin, Greek, and Italian, a burning antiquarian interest in France's Italian possessions, and a youthfully romantic Hellenism that responded readily to the beauties of the place and the degradation of the people. The pressures of his first duties, which included the organization of some rapid repairs to the French ships before their return to Toulon, quickly subsided, and he then found time for scientific pursuits, ranging from his own continuing research in descriptive geometry to experiments on the strength of materials which he conducted on masts in the naval shipyards,[9] and for the advancement of Greek literature and national spirit.

The setting for Dupin's cultural mission was the Ionian Academy, which he founded in 1808 and led until he left Corfu some four years later. The model for the Academy was clearly the far better-known (and bigger) Institut

---

III, 32 vols., Paris, 1858–70, reflect Napoleon's keen interest in the islands between 1807 and 1811, but his interest clearly declined after 1811. The reference to Corfu as "la clef de l'Adriatique" appears in an address to a deputation from the islands on 18 August 1811; see *Correspondance de Napoléon Ier*, vol. 22, p. 417. It is a mark of Napoleon's indifference to the islands other than Corfu that control of Zakinthos, Cephalonia, Ithaca, and Kithira (in 1809) and of Levkas (in 1811) was relinquished without any serious struggle in the face of the British fleet.

7. Napoleon to Ganteaume, 18 April 1808, in *Correspondance de Napoléon Ier, op. cit.*, note 5, vol. 17, p. 21.

8. *Ibid.*, vol. 17, pp. 21–2, where Napoleon refers to a garrison of 10,000 men on the islands, the overwhelming majority of them on Corfu.

9. The experiments in the naval shipyards were described in Dupin, "Expériences sur la flexibilité, la force et l'élasticité du bois . . . faites dans l'Arsenal de la marine, à Corcyre, en 1811", *Journal de l'École Polytechnique*, 1815, 10, 17e cahier, pp. 137–211; also in Dupin, *Mémoires sur la marine et les ponts et chaussées de France et d'Angleterre*, Paris, 1818, pp. 384–410. His most important work in descriptive geometry eventually appeared as *Développements de géométrie . . . pour faire suite à la géométrie descriptive et à la géométrie analytique de M. Monge*, Paris, 1813.

d'Égypte, established in Cairo almost exactly ten years before. [10] In this respect, it was helpful to have in Donzelot a governor who was at once highly cultivated and a veteran of the Egyptian campaign: certainly the parallels between the Egyptian and the Ionian institutions are not hard to find. In Corfu as in Cairo, for example, there were marks of a rather heavy cultural imperialism. Dupin assured his audiences of the "paternal hand" which the French administration would extend over their interests. [11] One delayed product of this paternalism was a scheme, briefly implemented in the winter of 1813–14, for thirty young islanders to be sent to France each year to receive a technical education at one of the *écoles d'arts et métiers*. [12] Another face of paternalism was displayed in the plan for two *prix olympiadiques* which the Academy proposed to award every four years at the beginning of each new Olympiad: the winners were to receive medals (in iron, the symbol of honour and virtue and the currency of ancient Sparta) bearing the head of Napoleon and the inscription "Napoléon, bienfaiteur et protecteur". [13] Despite these examples, however, the cultural imperialism was always tempered by the palpable lack of interest in this aspect of the occupation displayed by Napoleon and his ministers and, more obviously, by the nature of Dupin's own aspirations, which were concerned far more with Greece, her people, and, above all, her language than with the extension of French influence. It is entirely in keeping with this priority that entries for the two *prix olympiadiques* were to be in modern Greek: one prize was for a translation of a foreign work, the other for an original work in Greek, distinguished for its linguistic purity. [14]

Hellenism of the kind which Dupin and many other of his generation espoused was emphatically concerned with Greece as a whole. But there can be no doubt that the Ionian Islands presented a special challenge. Observers were unanimous in deploring the condition of the islands and their people. [15] Both at

10. The original statutes of the Ionian Academy stipulated that there should be 28 resident members. Later statutes, dating from 1811, removed the limit, but the number probably never exceeded thirty. The relevant documents are reproduced in the historical sketch of the Academy, published as Appendix II of *Nos anciens à Corfou. Souvenirs de l'aide-major Lamare-Picquot (1807–1814)*, ed. Hubert Pernot, Paris, 1918, pp. 206–37.

11. Dupin, "Sur la régénération de la Grèce, par les progrès des sciences, de l'industrie, de la marine et du commerce", in *Discours et leçons sur l'industrie, le commerce, la marine et sur les sciences appliquées aux arts*, 2 vols., Paris, 1825, vol. 1, pp. 5–22 (16). "Sur la régénération de la Grèce" was an address delivered at the inaugural meeting of the Ionian Academy on 17 July 1808.

12. Pauthier, *Les îles ioniennes pendant l'occupation française, op. cit.*, note 4, pp. 31–2.

13. See the announcement, dated June 1809 (or the first year of the 647th Olympiad, as Dupin liked to call it), reproduced as "Prix olympiadiques", in Dupin, *Discours et leçons, op. cit.*, note 11, vol. 1, pp. 55–62 (61–2).

14. *Ibid.*, vol. 1, p. 60.

15. See, for example, the particularly vivid account of the lawlessness and corruption on Cephalonia in 1810 in Lieut.-Col. Charles Philippe Bosset, *Parga, and the Ionian Islands*, London, 1821,

the time and in retrospect, Dupin laid the blame for this squarely at the door of the Venetian Republic and the servile Ionian nobility who, as vassals of Venice, had maintained a capricious despotic administration, condoned political corruption, and engaged in debilitating blood-feuds, while totally neglecting the economic interests of the citizens.[16] Public policy had been persistently directed at reinforcing the ties with Venice and at weakening such frail bonds as existed between the islands and the mainland of Greece. As a result, even the most elementary form of public education was lacking, and the Greek language, though spoken by some of the peasantry and servants, was neither recognized in official transactions nor used in educated circles.[17]

In publicly denouncing the Venetian past, Dupin was making a political statement whose import could not be missed. He even went so far as to compare the Venetian administration unfavourably with that of the Turks, who elsewhere in Greece had at least allowed the Greek language to be learned and used.[18] There can be little doubt that the denigration of Venice was articulated with one eye on the erratic Ali Pasha, whose presence barely fifty miles away across the Channel of Corfu in Ioannina was a constant threat to the security of the French garrison. But it was primarily directed to the domestic audience, to the mass of the people, and, in that context, it has to be seen as quite deliberately divisive. It was, after all, the ordinary young men of the islands who would benefit from the Academy's ambitious programme of free public lectures that would allow them, as Dupin intimated, to compete with the privately educated scions of the nobility. It was likewise for the mass of the people that the Academy and the French administration combined in schemes to revive the economy by improving harbours and encouraging manufacturing, fishing, and agriculture. As Dupin reminded his hearers at the annual meeting of the Academy in 1809, such schemes for improvement would not have been necessary but for the neglectful stewardship of the Venetians and the local nobles which had left much of Corfu not merely infertile but positively unhealthy.[19]

Few details of the Academy's educational activities have survived.[20] But in

---

pp. 17–18. Bosset was a British officer of French extraction who went to Cephalonia soon after the island seceded from the Septinsular Republic in 1809.

16. See Dupin's comments in his *Discours et leçons, op. cit.*, note 11, vol. 1, pp. 2–3 (written some ten years after the withdrawal of the French), and "Sur l'éducation publique des Grecs" [an address delivered at the annual public meeting of the Ionian Academy on 15 August 1809], *ibid.*, pp. 25–49 (34–7).

17. In fact, the use of Greek seems to have been rare. According to Dupin, *ibid.*, vol. 1, p. 38, the commonest speech on the islands was a form of the Italian patois of Bergamo.

18. Dupin, "Sur l'éducation publique des Grecs", *op. cit.*, note 16, vol. 1, p. 37. However, a footnote added to the version of the address published in 1825 refers to the massacre by the Turks of the teachers and pupils of one of the schools where Greek was taught at Cydonia.

19. *Ibid.*, vol. 1, p. 34 n.

20. Such details as we have of the lectures and teachers are mainly to be found in Dupin, "Prix

the first year, 1808−9, there were lectures in physics, chemistry, natural history, physiology, hygiene, anatomy, and surgery. In the following year, a course in Greek literature was added, and at some later date there was a course in jurisprudence. The teaching seems to have been entrusted to a mixed group of French officers and local men, the majority of them doctors.[21] Among the French contributors, Dupin's leading collaborator was Antoine-Marie Augoyat, an exact contemporary of his at the École Polytechnique and a military engineer who went on to a distinguished career as a teacher at the École de l'Artillerie et du Génie at Metz and as a writer on military matters. The most notable of the Greek teachers was a doctor and writer, Nicolas Mavromati, who gave the course on Greek literature. Mavromati's success clearly owed something to the nature of his subject, but the fact that he had undergone his medical training in Italy and had fled from the mainland of Greece as a refugee from Turkish rule gave him the advantage of a contact with both the older Venetian tradition and the emerging new world of panhellenist aspirations.

At least some of the teaching − in Mavromati's course, for example, and in the lectures on physics and chemistry which Augoyat and Dupin gave jointly − was probably as competent as could be expected in the difficult circumstances which made books and instruments hard to come by. But Dupin's disillusionment soon began to show. His complaints centred, from the start, on the composition of the audiences. It had never been his intention that the officers and doctors who gave the lectures should find themselves discoursing to their colleagues; but that is what happened. The all-important call to the superior youth of Corfu fell on obdurately deaf ears. Dupin complained of this after the first year of teaching[22] and still saw it as the main failing of the Academy when he looked back on its brief history more than a decade later.[23] Throughout the French occupation, the old Greco-Venetian nobility had remained hostile to a force which it saw as a dangerous source of revolutionary ideas. Understandably, the nobles had seen no advantage in abandoning the traditional sequence of private education by Italian teachers followed (where appropriate) by attendance at the universities of Padua or Venice, in order to place their children in an untried system that deliberately disregarded social status.

It is clear that the vitality of the Academy was very dependent on Dupin's personal involvement. In 1811, the President of the Ionian Senate, Emmanuel Theotokis, reported to the French minister of war that the lectures continued to have "a most salutary effect".[24] But they probably began to flag soon after-

et cours de l'Académie Ionienne, annoncés dans la séance publique du 15 août 1809", in *Discours et leçons, op. cit.*, note 11, vol. 1, pp. 50−4.

21. See *ibid.*, vol. 1, pp. 52−4, for a list of the teachers and their courses.
22. *Ibid.*, vol. 1, p. 51.
23. *Ibid.*, vol. 1, pp. 2−3.
24. Pauthier, *Les îles ioniennes pendant l'occupation française, op. cit.*, note 4, p. 29. The com-

wards, as a result of Dupin's departure from the islands, made definitive by a persistent fever which caused him to spend many months in Italy on his journey to France and continued to drain his energies back in Paris in the winter of 1812–13. Another casualty of Dupin's departure and faltering health was the system of *prix olympiadiques*; the prizes, which were due to be distributed for the first time in 1812, were never awarded.

The failures and flagging activity served only to heighten Dupin's sense of disappointment. In the immediate aftermath of the Ionian venture, in 1815, he referred to his experiences with a disenchantment amounting almost to bitterness. Addressing his old friend Augoyat, he recalled how the two of them had seen the Academy descend from a state of early promise into one of lethargy; the only possible conclusion, as he put it, was that their joint efforts in an almost barbarian land had been "useless".[25]

By Dupin's criteria, which gave primacy to the Academy's educational role and to the spirit of emulation that he had hoped to encourage among the youth of Corfu, the word "useless" was no exaggeration. At least for a short time, the Academy survived the departure of the French and the arrival of the British in the spring of 1814. But, with the control of its activities now exclusively in the hands of the Corfiot élite (including members of the great nationalist families of Metaxas and Capo d'Istria), the more extravagant aspects of Dupin's attempts to awaken the "sons of Alcinous" to their ancient heritage were abandoned.[26] Under the British administration, the Academy seems to have assumed a purely symbolic function as a means of bestowing distinction on local citizens and of recognizing, through election to corresponding membership, the islands' cultivated friends and benefactors, nearly all of them in France and Italy.

Despite Dupin's disillusionment, the Academy and its members during the period of French occupation left their mark. Dupin in particular was remembered with affection, and a quarter of a century later, a sense of gratitude was still said to be evident among Greek patriots. It was after encountering this gratitude in his travels in Greece that Jean-Alexandre Buchon dedicated an anthology of Greek historical writing to Dupin, whom he identified as both an old

---

ment appeared in the report given by Theotokis, as head of the delegation from the islands which visited Paris in 1811; see above, note 6.

25. "Dédicace à mon ami, A. Augoyat", in Dupin, *Du rétablissement de l'Académie de Marine*, Paris, 1815. In dedicating his *Développements de géométrie* to Monge in 1813 (see *op. cit.*, note 9, p. vi), Dupin looked back with signs of regret to the years he had been obliged to spend in "contrées presque barbares", an obvious reference to the Ionian Islands.

26. For a reflexion of the Academy's activity and membership after the departure of the French, see the letter of 6 August 1814 from the Academy's secretary to C.P. de Bosset (offering him corresponding membership) and the accompanying list of members, in Bosset, *Parga, and the Ionian Islands, op. cit.*, note 15, pp. 155–9.

friend and a pioneer of Hellenism.[27] There was also a less precise legacy in a continuing tradition of advanced learning on Corfu, which sustained an Ionian University, founded in 1824 by the spectacularly eccentric philhellene, Frederick North, 5th Earl of Guilford,[28] and, from 1836, a Corfu Literary Society.

Perhaps, therefore, Dupin had not really failed but rather, through youth and inexperience, had been led to hope for too much. He had plainly underestimated the impediment presented to his ideals by the bitterly divided society to which he preached. And his notion of a slumbering but proud people awaiting the call to action appears in retrospect romantically naive. That was probably the view also held by the many contemporaries in Greece who shared his panhellenism but who found his constant evocation of lost classical glories irrelevant to the immediate struggle. Such men could hardly have missed the lesson they were intended to draw from the translation and critical edition of Demosthenes' *Olynthics* which Dupin prepared during his first two years in Corfu:[29] Demosthenes' fourteen-year struggle against the foreign tyrant Philip of Macedon was a model for any citizen who aspired to freedom (as Jeremy Bentham, for example, knew very well). But the edition seems to have earned recognition not in Greece, nor as a contribution to the cause of Greek nationalism, but for the vigorous quality of the French translation, which won the admiration of Paul-Louis Courier.[30]

Once Dupin had left Corfu, he quickly re-entered the wider world of European science and scholarship, through a learned edition of Léopold Vaccà Berlinghieri's study of Caesar's victory at Alesia[31] and a series of papers on descriptive geometry, the fruits of nearly ten years of intermittent labour, which he presented to the First Class of the French Institute and which served as the basis for his *Développements de géométrie* (1813).[32] Again, Dupin's adherence to the school of Monge was flaunted, explicitly in a fulsome dedication of the book to his "illustre maître" and, by implication, in the style of the mathematics and in the choice of problems, which focussed, like his earliest work, on the geometry of curved surfaces. Clearly, from the moment of his return to France, Dupin had set his sights on being noticed by the leaders of Parisian mathematics, in

27. *Choix des historiens grecs avec notices biographiques par J.A.C. Buchon*, in the series "Panthéon littéraire", Paris, 1837, pp. vii–viii.

28. For a flavour of Guilford's eccentricities and a brief account of his university, see the article on him in the *DNB*.

29. Dupin, *Essai sur Démosthènes et sur son éloquence, contenant une traduction des Harangues pour Olynthe . . .* , Paris, 1814.

30. *Choix des historiens grecs, op. cit.*, note 27, pp. vii–viii.

31. Léopold Vaccà Berlinghieri, *Examen des opérations et des travaux de César au siège d'Alésia*, Lucca, 1812.

32. For a full reference, see note 9, above. The paper cited in that note was also presented to the First Class of the French Institute, being read there on 12 April 1813.

particular Monge and Carnot. And he succeeded. His work was praised by a committee of the First Class of the French Institute in December 1812,[33] and in November 1814 he was himself elected to corresponding membership of the Académie des Sciences. Four years later, he advanced to full membership.

As Dupin's health improved, he also resumed his service to the Empire, establishing a maritime museum during a posting in Toulon and contributing to the fortification of Lyons during the Hundred Days. After Waterloo, his loyalty to Napoleon caused him no more embarrassment than it did to most other serving officers of the Empire, and he returned yet again to his career as a naval engineer. But his declared closeness to Monge and Carnot (who were the only two members of the Institute to be eliminated from the Académie Royale des Sciences after the restoration) was unmistakably risky, and in the summer of 1815 he provoked active suspicion in governmental circles by his spirited protest against the exile which the Bourbon régime imposed on Carnot.[34] In view of this, it is hardly surprising that his request for permission to visit Britain in order to examine the country's communications and military installations and the state of her economy encountered ten months of delay.[35] But eventually, in 1816, he embarked on the first of the annual tours of Britain that he was to make for the next four years.

Just as Dupin's travels in the Adriatic illustrate very clearly the widening of French cultural horizons under the Empire, so his visits to Britain have to be set in the context of a broader movement which brought Frenchmen flocking to Britain and which, equally, lured the British across the Channel to France. It was a movement which also generated vigorous literary activity. In the words of Richard Chenevix, writing in the *Quarterly Review* in 1820, "The press, in every part of Europe, has teemed of late with publications upon England and France".[36] Many of the British travellers and writers, like the liberal Lady Morgan (to whom I shall return later), were curious about the present state of France; they wanted to see for themselves the effects of the Revolution and the Empire. For others, perhaps more numerous, a main attraction was the oppor-

---

33. Dupin, *Développements de géométrie, op. cit.*, note 9, pp. xiii–xx.

34. By his protest, Dupin certainly ran the risk of having his name added to the (relatively short) list of officers in the army and navy who lost their commissions. A personal approach to at least one minister was accompanied by the preparation of a vehement defence, entitled "Du jugement de M. le lieutenant général Carnot"; but, at Carnot's request, this defence was never published. Four years later, following the death of Monge, Dupin committed an almost equally provocative act by publishing a very favourable *Essai historique sur les services et les travaux scientifiques de Gaspard Monge*, Paris, 1819.

35. Dupin, *Voyages dans la Grande-Bretagne, entrepris relativement aux services publics de la guerre, de la marine, et des ponts et chaussées, en 1816, 1817, 1818, et 1819* [in later volumes . . . *1818, 1819, et 1820*, and . . . *ponts et chaussées depuis 1816*], 6 vols., Paris, 1820–4, vol. 1, p. xx.

36. [Richard Chenevix], "Rubichon – *De l'Angleterre*", *Quarterly Review*, 1820, 23, pp. 174–98 (175).

tunity of renewing contact with a country where the old traditions in religion and society, in particular the traditions of the Middle Ages, seemed to have survived more successfully than was the case in Britain. Here, the artist John Sell Cotman and his patron, the antiquarian Dawson Turner, were typical of the numerous visitors whose interests lay chiefly in medieval architecture.

By contrast, most of the French visitors who crossed the Channel did not come in search of architectural gems or political enlightenment but rather for the wild scenery of northern England, Wales, and Scotland. But even those who came in search of mountains and remote places could not, and they did not, ignore the very different, urban Britain that had been transformed in barely three decades of rapid industrialization. The country which the economist Jean-Baptiste Say had known intimately before the Revolution had little in common with the one that he described in *De l'Angleterre et des Anglais*, following his visit to England in 1814. The proliferation of steam-engines was, for him, one of the clearest signs of the change.

> Everywhere [he wrote] the number of steam-engines has multiplied prodigiously. Thirty years ago, there were only two or three of them in London; now there are thousands. There are hundreds of them in the large manufacturing towns and they are even to be seen in the countryside. Industrial activity can no longer be profitably sustained without the powerful aid they give.[37]

Dupin's reactions, as they are recounted in the six volumes of his *Voyages dans la Grande-Bretagne*,[38] show very clearly that he found both of the contrasting faces of Albion intensely intriguing. In this respect, his response was typical of the generation of young French travellers for whom Britain emerged from the mists of obscurity and rumour at a time when romanticism and industrialization were simultaneously making their mark. It is tempting, but false, to suppose that these two movements were invariably seen as incompatible with each other. For some aristocratically inclined romantics – Chateaubriand and Lamartine, for example – the incompatibility was, of course, beyond question. But there was another, more liberal strand in early romanticism which saw the literary upheaval not merely as reconcilable with industrial development but even as an integral part of a greater movement of renewal. A reading of the periodical *Le globe*, especially in its Saint-Simonian phase after 1830, illustrates the point very clearly: here, liberal politics, romantic sensibility, and support for industry were perfectly contented bedfellows.

In his literary tastes, however, Dupin was in no sense a doctrinaire romantic. Indeed, quite early in the Restoration, he published a passionate defence of Racine and the classical ideal of a well-structured and carefully worked play, in

37. Jean-Baptiste Say, *De l'Angleterre et des Anglais*, Paris and London, 1815, pp. 30–1.
38. For a full reference to the *Voyages*, see above, note 35.

response to the wholesale disparagement of the traditions of French classicism which Lady Morgan had meted out from Dublin in her book *La France* in 1817. [39] Yet the position that Dupin took in this bitter controversy was one with which many romantics would have sympathized. It rested on the argument that in fact Racine had depicted human passion with great clarity; to the modern reader, though, it smacks as much of hurt national pride as of a total hostility to the romantic style which Lady Morgan was espousing. Certainly, when Dupin travelled in the lowlands of Scotland, he was enough of a romantic to make the customary visit to Walter Scott's home at Abbotsford; and he was unreservedly delighted when Scott wrote an "imitation" of a poem of his celebrating the Caledonian Canal as a work in which the awesome grandeur of nature had been matched, through locks and masonry, by the grandeur of a human achievement. [40]

In his account of Britain's new industrial strength, Dupin's few critical comments – for example, on the pitiful condition of certain of the poor and of many veterans of the wars [41] – were far outweighed by the eulogies of a country in which, as he saw it, the classes of society worked in exemplary harmony (an observation published, oddly enough, in 1824, only five years after the Peterloo massacre). [42] According to Dupin, the widespread spirit of association and the freedom from governmental constraints in Britain encouraged a spirit of enterprise and engaged not only the working and commercial classes but also the aristocracy in the common pursuit of industry and trade and in the provision of communications that put those of France to shame. Even the emerging wretchedness of the industrial towns went without comment. Manchester was remarkable for its educational facilities, libraries, and learned societies quite as much as for its manufacturing activity; [43] Liverpool boasted not only libraries and societies but also a handsomely constructed port; [44] in Glasgow, too, edu-

39. Dupin, *Lettre à Mylady Morgan sur Racine et Shakespeare*, Paris, 1818. The work was published anonymously in response to *La France*, which Lady Morgan (Sydney Owenson) published in Paris and London in two volumes in 1817. Lady Morgan, who knew France well as a result of three visits she made there between 1816 and 1830, was an acerbic critic of the Bourbon régime and a champion of the revolutionary tradition. On the conflict between her and Dupin, which put an end to the friendship they had established during Dupin's stay in Dublin, see Marcel Ian Moraud, *Une irlandaise libérale en France sous la Restauration. Lady Morgan 1775–1859*, Paris, 1954, pp. 128–35.

40. Dupin, *Voyages dans la Grande-Bretagne, op. cit.*, note 35, vol. 6, pp. 162–3. A somewhat different version appears in Dupin's *Mémoires sur la marine, op. cit.*, note 9, pp. 66–7, and in the two editions of the (partial) English translation of the work: *Two excursions to the ports of England, Scotland, and Ireland, in 1816, 1817, and 1818*, London, 1819, pp. 98–9, and the shorter *Narratives of two excursions . . .* , London, [1819], p. 96.

41. Dupin, *Voyages dans la Grande-Bretagne, op. cit.*, note 35, vol. 1, p. x.

42. *Ibid.*, vol. 5, pp. xx–xxi.

43. *Ibid.*, vol. 5, p. 168.

44. *Ibid.*, vol. 6, pp. 200–33.

cation, learning, and public buildings had all been handsomely financed as a matter of high priority.[45] And all three towns were lauded for the vast increases in their populations and the prosperity which these increases were assumed to reflect.

Throughout his writings on Britain, Dupin continued to preach the same urgent lesson he had preached on Corfu: that the key to the prosperity and social harmony of a country lay in education and that no education was better suited to achieve this aim than one based on mathematics and science.[46] It is entirely in keeping with his belief in the special status of the sciences that a number of leading English scientists, including Sir Joseph Banks, W.H. Wollaston, and Thomas Young, were among those whose help he respectfully acknowledged.[47] But, for Dupin, the value of the sciences lay not only in the intellectual satisfaction they provided but above all in their usefulness. For his most favoured model, therefore, Dupin looked not to the London of Banks, Wollaston, and Young, but to the very different world of Glasgow, where he made one of the most decisive of all his personal contacts in Britain, with the chemist Andrew Ure.

At the time of Dupin's visit in 1817, Ure had held the chair of natural philosophy at the city's Andersonian Institution for over ten years. In those years, he had earned a high reputation for his evening lectures on chemistry and mechanics and for his contribution to the education of many artisans who had gone on to responsible positions in the industry of the region. The account of Ure's activities and their beneficial effect on local industry which appeared in Dupin's *Mémoires sur la marine* in 1818 reflected an enthusiasm suffused with polemical purpose.[48] The implication was that France had been outdistanced by Britain in the educational facilities provided for artisans and that if the French economy was to reap benefits comparable with those which Dupin had observed in Glasgow, a determined programme of emulation was required.

The forthrightness of Dupin's analysis was enough to ensure it a hearing on both sides of the Channel. In Britain, his comments on British naval establishments provoked the translator of his *Mémoires sur la marine* into adding a tart introduction and sustaining a critical running commentary in a series of footnotes.[49] In these additions, Dupin was accused not only of factual errors and unfounded assertions but also of an arrogant pride in French scientific superiority which ignored and, by implication, disparaged the practical skill and in-

---

45. *Ibid.*, vol. 6, pp. 173–82.

46. For the views expressed on Corfu, see Dupin, "Sur la régénération de la Grèce", *op. cit.*, note 11, vol. 1, pp. 19–20.

47. Dupin, *Voyages dans la Grande-Bretagne, op. cit.*, note 35, vol. 1, p. xv n.

48. Dupin, *Mémoires sur la marine, op. cit.*, note 9, pp. 68–9.

49. Dupin, *Two excursions . . .* , and *Narratives of two excursions . . .* , *op. cit.*, note 40, pp. iii–viii and *passim*.

316

ventiveness of British shipbuilders and engineers. By contrast, the Tory John Barrow, in the *Quarterly Review*, was condescending: he criticized Dupin's account of the commercial and industrial strength of Britain for its naivety and repetitiveness and doubted whether France in any case was capable of emulating British achievements. [50] Most British observers, however, were flattered by Dupin's account. The translator of the last two volumes of his *Voyages* referred in 1825 to his "enlightened and philosophical views" and praised his accuracy. [51] And George Birkbeck saw in Dupin's campaign for popular education a model which it was in the interests of the British to follow (as they had apparently already done in founding the new School of Arts in Edinburgh). [52]

In France, the reception of Dupin's work was, on the whole, less favourable than it was in Britain. As I have already indicated, his standing with the government and the relevant ministries in the immediate aftermath of the collapse of the Empire was not high; and, even though the tide of extreme reaction in the period of the White Terror and the *Chambre introuvable* had been arrested by 1816, his praise for the economic and social order in Britain and, more particularly, his advocacy of popular education were inevitably viewed with suspicion. One consequence of this was the harrassment that occurred in 1820, when two unidentified members of the government tried to censor the *Voyages*, parts of which were judged to be unpatriotic. [53] When Dupin refused to amend his text, the work was removed from naval establishments and all institutions under the control of the Ministry of War, and it remained on what Dupin called the "index" for nearly five years.

By the time this conflict with authority occurred, however, Dupin's experiences in Britain had already left a lasting mark on French technical education, in the form of the highly successful lectures that were launched in Paris at the Conservatoire des Arts et Métiers. This innovation could probably not have come about but for the change in the political climate of France that followed

50. [John Barrow], "Dupin – *Commercial power of England*", *Quarterly Review*, 1823–4, 30, pp. 368–82.

51. See the "Advertisement" to the translation, published as *The commercial power of Great Britain; exhibiting a complete view of the public works of this country . . .* , 2 vols., plus atlas, London, 1825, vol. 1, pp. v–vi.

52. See Birkbeck's Preface to Dupin, *Mathematics practically applied to the useful and fine arts. Adapted to the state of the arts in England, by George Birkbeck*, London, 1827, pp. i–xv.

53. Dupin, "Avant-propos de la seconde édition", in *Voyages*, 2nd ed., Paris, 1825, vol. 1, pp. xv–xvi. A typical passage which might have been regarded as offensive is on pp. 15–16 n. of the fourth volume of the *Voyages*, where Dupin contrasts the rigid discipline in the British navy with the rampant insubordination in the French fleet under the Ancien Régime and the Empire.

The attempt at censorship was made despite the favourable response of a committee of the Académie des Sciences which reported on a manuscript account of Dupin's early visits to British military and naval establishments in March 1818. See *Rapport de M. le Mal Duc de Raguse, sur un ouvrage de M. Charles Dupin . . .* , Paris, 1818; also the text of the report, as reproduced in *Institut*

the elections of October 1818. The immediate consequence of the enhanced support for the left in these elections was the establishment, in December 1818, of the duc de Decazes's ministry, which lasted until February 1820, when the murder of the heir presumptive to the throne, the duc de Berry, heralded a new period of reaction. It was no coincidence that it was in November 1819, during these fourteen months of relatively liberal government, that Dupin's proposals for the lectures at the Conservatoire secured royal approval.

As I have argued elsewhere, Dupin's scheme, which grew in a very obvious way from what he had seen at the Andersonian in Glasgow, was a daring one. [54] By his plan, the Conservatoire, which had served since its creation in 1794 primarily as a museum and educational depository for models of industrial and agricultural machinery, was to be endowed with chairs in mechanics (a chair occupied by Dupin himself), industrial chemistry, and the very sensitive subject of political economy. Moreover, the lectures (unlike those of the École Polytechnique and the other *écoles spéciales*) were to be open to the public, free, and pitched at a level suitable for working men with no formal education (though it must be said that in the 1820s the lectures on industrial chemistry were attended by the mathematician Louis-Benjamin Francoeur and Sadi Carnot, neither of whom fell into this category [55]). The breadth and accessibility of the teaching are both striking, marking as they do an abrupt departure from the older tradition of specialized vocational instruction which J.A.C. Chaptal and the duc de La Rochefoucauld-Liancourt would almost certainly have preferred to develop as the Conservatoire's contribution to technical education.

The new teaching also appeared daring because of the risks of disorder associated, in reactionary minds, with any extension of working-class education that purveyed more than narrowly and immediately useful knowledge. In response, Dupin took the offensive, arguing (as he was to do for the rest of his life) that education would not only foster technical progress but also give dignity to industrial labour and serve thereby to stabilize rather than disrupt society. [56] Educated workmen, Dupin argued, would recognize the debt they owed to God and the king, and the importance of preserving the social order that

de France. Académie des Sciences. Procès-verbaux des séances de l'Académie tenues depuis la fondation de l'Institut jusqu'au mois d'août 1835, 10 vols., Hendaye, 1910–22, vol. 6, pp. 298–301.

54. Robert Fox, "Education for a new age: the Conservatoire des Arts et Métiers, 1815–30", in Artisan to graduate. Essays to commemorate the foundation in 1824 of the Manchester Mechanics' Institution . . . , ed. D.S.L. Cardwell, Manchester, 1974, pp. 23–38, especially pp. 23–32.

55. Robert Fox, "Watt's expansive principle in the work of Sadi Carnot and Nicolas Clément", Notes and records of the Royal Society of London, 1970, 24, pp. 233–53 (236 and 245).

56. See, for example, "Avis aux manufacturiers et aux chefs d'ateliers", published as a prefatory note to an introductory lecture for a new course which Dupin launched at the Conservatoire in November 1824, in Discours et leçons, op. cit., note 11, vol. 2, pp. 153–60 (160). For a later statement of Dupin's view that the greatest threat of working-class disorder lay in ignorance, see Dupin, Harmonies des intérêts industriels et des intérêts sociaux, Parix, 1833.

guaranteed these blessings; ignorance, by contrast, bred disharmony between worker and employer and instability of the kind that was to flare up violently in Lyons and other industrial towns in the 1830s.

By their very success, the lectures at the Conservatoire aggravated Dupin's difficulties with the predominantly reactionary Bourbon governments of the 1820s. It was a matter for concern that by 1824 2,000 young men gathered regularly at Dupin's lectures, and a vigilant watch was maintained on professors and audiences alike.[57] The fact that the students were often restless and that Dupin and his colleagues, Jean-Baptiste Say and Nicolas Clément, were seen to lard their lectures with criticisms of the government only confirmed the worst predictions and made defence of the enterprise more difficult. Defence was also made harder by events in Britain, where the decade of the 1820s turned out to be one of the most turbulent and potentially unstable in modern British history, with social disunity a constant threat and Peterloo a perpetual reminder of the conflict that might resurface at any time. In these circumstances, it is easy to see why Dupin began to resort less readily to a justification in terms of the benefits wrought by working-class education across the Channel. Nevertheless, under a gently exercised surveillance, the lectures survived, and, in a period of quite remarkable expansion after the accession of Charles X, the model was widely adopted in the provinces as well. In one year beginning in 1824, for example, free public lectures for working-men, which until then had been given only at the Conservatoire and in the garrison town of Metz, were established in over fifty towns in France.[58] As Dupin was proud to observe, military and naval engineers and members of the Corps des Ponts et Chaussées and the Corps des Mines, many of them former *polytechniciens*, had taken the lead in this movement, though it is hard to imagine that anything significant would have been achieved but for Dupin's skilful nurturing of influential support in the face of a national administration, under Villèle, that was profoundly unsympathetic to his ideas.[59]

For the rest of his life, Dupin remained a prominent public figure. But, from

57. Fox, "Education for a new age", *op. cit.*1, note 54, pp. 31–2.

58. On this remarkable development, see Dupin, *Géométrie et mécanique des arts et métiers et des beaux-arts*, 3 vols., Brussels, 1825–6, vol. 1, pp. 393–403 (cf. also vol. 3, pp. v–viii); "Extraits du rapport général fait au Ministre de la marine, sur l'enseignement de la géométrie et de la mécanique appliquées aux arts", *Revue encyclopédique*, 1826, 31, pp. 594–608; and "Effets de l'enseignement populaire sur les prospérités de la France", *Revue encyclopédique*, 1827, 33, pp. 40–63 (58–63). For evidence of the encouragement which Dupin's achievements in France gave to Birkbeck in his campaign for the Mechanics' Institutes in Britain, see the Preface to Dupin, *Mathematics practically applied*, cited above, note 52.

59. The most important support which Dupin secured in the mid-1820s was that of the reforming naval minister, the comte de Chabrol. But it seems that both Charles X and the Dauphin also took a genuine interest in his campaign.

the 1820s, as he immersed himself increasingly in his work for the advancement
of popular instruction and industry and in the pursuit of a parliamentary career, he travelled less; and very quickly, the image of the fearless young Turk
with exciting, even radical views, was transformed into that of a versatile but
wordy political orator with a reputation for pliability.[60] By the end of the nineteenth century, Paul Tannery, writing in the *Grande Encyclopédie*,[61] painted
a distinctly unflattering portrait of Dupin as a careerist and unprincipled trim-
mer. The tone of the portrait is palpably unfair, though, composed as it was
at a time when the sin of having rallied to the Second Empire (as Dupin did)
was unpardonable in republican eyes, it is not surprising.

It may well be that Dupin's reputation suffered in Tannery's day and has suffered ever since from the very diversity of his activities and the success he enjoyed in all of them. A steady accumulation of honours and a long life (he lived
on into the Third Republic, dying in 1873 in his eighty-ninth year) have further
compounded the problem of assessment by relegating Dupin's early life to a
rather obscure and distant background. In the years I have treated in this paper,
Dupin was anything but a conformist. As an early enthusiast for Greek nationalism under the First Empire, a perceptive analyst of the industrial and commercial superiority of Britain under the Bourbon Restoration, and a man of the Enlightenment whose views coincided with many of those which came in the 1830s
to be associated with social romanticism, he was in the van of cultural and political change. As I have tried to show, this intellectual adventurousness drew in
important ways on a creative interaction between his observations abroad and
the turbulent economic and social history of his own country. In his travels, Dupin was at once a man of his times and a pacemaker. For the first quarter of
the nineteenth century, many other Frenchmen travelled widely, either by
choice or necessity, and Dupin was by no means alone in the experiences to
which he was exposed in the Ionian Islands and Britain. But I doubt whether
anyone of his generation could match the zest with which he traversed the
length of Europe, still less the sense of engagement that converts his travelogues
into documents of outstanding importance for the historian.

60. See, for example, the critical comment on Dupin's wordiness in "Timon" [Le Vte de Corme-
nin], *Etudes sur les orateurs parlementaires*, Brussels, 1838, pp. 197–200: ". . . il ne peut retenir
le flux de son éloquence dévoyée. Il faut qu'il parle, parle, parle. Le prurit de l'in-quarto le dé-
mange. Il faut qu'il imprime, imprime, imprime" (p. 199).
61. *La grande encyclopédie*, 31 vols., Paris, 1887–1902, vol. 15, p. 81.

VI

320

ACKNOWLEDGEMENTS

The writing of this paper has been greatly helped by the British Academy Readership in the Humanities which I have held since October 1983. I have also benefited from grants towards the cost of my research from the Royal Society of London and the Small Grants Fund of the British Academy.

VII

THE SCIENCE OF FIRE :
J.H. LAMBERT AND STUDY OF HEAT

In their 'Mémoire sur la chaleur' of 1783 Lavoisier and Laplace drew a clear distinction between the two theories of heat that competed for support in their day [1]. On the one hand there was the material theory, or caloric theory as it was soon to be called [2], according to which thermal phenomena were caused by the mere accumulation of a property-bearing 'matter of fire' or 'igneous fluid' *(matière du feu, fluide igné)*, which could combine with substances in the manner of a chemical element. This was the theory which Lavoisier himself had already espoused for more than ten years [3] and which, in various forms, was to hold sway among chemists and physicists alike until the early decades of the nineteenth century. Its rival was the kinetic theory, according to which heat consisted in the vibratory motion of the particles of ordinary matter, the measure of quantity of heat being, in this case, the *vis viva* or *force vive* (mv^2) of the particles. Although it was less widely held than the caloric theory, this theory was not without distinguished and influential supporters – the most notable was the chemist Pierre Joseph Macquer [4] – and it was the kinetic theory, of course, which eventually carried the day, when the new science of thermodynamics was established in the mid-nineteenth century.

---

(1) A.L. LAVOISIER and P.S. LAPLACE, 'Mémoire sur la chaleur', *Mémoires de l'Académie royale des Sciences,* (1780 ; published in 1784), pp. 355-408 (357-8). The paper was read to the Académie, by Lavoisier, on 18 June 1783.

(2) The word *calorique* was first used in print in L.B. GUYTON DE MORVEAU, A.L. LAVOISIER, C.L. BERTHOLLET, and A.F. DE FOURCROY, *Méthode de nomenclature chimique* (Paris, 1787), p. 31, though there is evidence that it was already current in Lavoisier's circle by 1784. On this earlier usage, see Robert FOX, *The caloric theory of gases from Lavoisier to Regnault* (Oxford : Clarendon Press, 1971), p. 6n.

(3) Lavoisier first outlined his views on the matter of fire in an unpublished essay which almost certainly dates from August 1772. Henry Guerlac identifies the essay as Lavoisier's 'Système sur les élémens', which is known to have been handed to and initialled by the secretary of the Académie des Sciences on the 19th of that month. See GUERLAC, 'A lost memoir of Lavoisier', *Isis, 50* (1959), pp. 126-7, and *Lavoisier – the crucial year. The background and origin of his first experiments on combustion in 1772* (Ithaca, N.Y., 1961), pp. 96-7. The essay is reproduced in the latter work, on pp. 215-23. On Lavoisier's first public statement of his views, see below, note 61.

(4) See Pierre Joseph MACQUER, *Dictionnaire de chymie* (2nd edn., 4 vols., Paris, 1778), vol. 2, pp. 168-92. Macquer's pupil, Antoine François de Fourcroy, was another important contemporary who supported the theory ; see his *Elémens d'histoire naturelle et de chimie* (2 vols., Paris, 1786), vol. 1, p. 117. According to W.A. SMEATON, *Fourcroy, chemist and revolutionary, 1755-1802* (Cambridge, 1962), p. 102, the relevant passage in the *Elémens* (which was the second edition of the *Leçons élémentaires d'histoire naturelle et de chimie* of 1782) was written in 1784. By the time the book appeared, in 1786, Fourcroy had been won over to a material theory of the kind adopted by Lavoisier ; see *Elemens*, vol. 1, p. XXIV.

326

Endorsed by nineteenth-century authorities from Haüy and Fischer to Tyndall and Tait[5], the dichotomy between the material and kinetic theories has become firmly enshrined in our historiography, and it is only recently that serious attention has been directed to a cluster of heat theories that bridged the gulf between the material and the kinetic traditions[6]. The theories in this hybrid tradition were kinetic in the sense that they ascribed the sensation of heat to motion ; yet they were also material to the extent that they required the existence of an all-pervading subtle fluid similar to the aethers postulated by Descartes and Newton. For some, it was motion in this aether, or 'fire' as it was commonly called in the eighteenth century, which constituted heat[7] ; for others, the motion of fire merely sustained the vibrations of the particles of ordinary gross matter, and it was these vibrations which caused a body to feel hot[8]. But all who worked

---

(5) René-Just HAUY, *Traité élémentaire de physique* (2 vols., Paris, an XII [1803]), vol. 1, p. 90 ; E.G. FISCHER, *Physique mécanique*, trans. by J.B. Biot (2nd edn., Paris, 1813), pp. 85-6 ; John TYNDALL, *Heat considered as a mode of motion* (London, 1863), pp. 23-5 ; Peter Guthrie TAIT, *Heat* (London, 1884), pp. 21-31. The eighteenth-century literature incorporating the idea of heat as motion in an all-pervading subtle fluid seems to have been quickly forgotten in the early decades of the nineteenth century, but it was still discussed (though not favoured) by Thomas THOMSON in his book *An outline of the sciences of heat and electricity* (London, 1830), p. 5. THOMSON observes that 'the greater number of the French and German chemists of the last century' considered heat to be a 'tremor... of the particles of a subtle, highly elastic and penetrating fluid which is contained among the pores of the hot bodies, or interspersed among their particles'.

(6) A fine pioneering study, useful on the work of Boerhaave and his place in the Cartesian tradition of aether theories, is Hélène METZGER, *Newton, Stahl, Boerhaave et la doctrine chimique* (Paris, 1930), especially pp. 209-28. Important among the more recent literature is Geoffrey R. TALBOT, 'Origins and solutions of some problems in heat in the eighteenth century' (University of Manchester Ph.D. thesis, 1967), in which the Cartesian roots of doctrines of fire are traced back through Boerhaave, the Lemerys, and Homberg. See also the article by Heimann cited in note 16, those by Goldfarb and Brush cited in notes 66 and 67, and Robert E. SCHOFIELD, *Mechanism and materialism. British natural philosophy in an age of reason* (Princeton, N.J. : Princeton University Press, 1970), pp. 182-90.

It is interesting to note the fleeting but significant suggestion in the 'Mémoire sur la chaleur' that the material and kinetic theories were not necessarily incompatible with each other. On p. 358 Lavoisier and Laplace remark, of the two theories they have just distinguished : 'peut-être ont-elles lieu toutes deux à la fois'.

(7) Peter van Musschenbroek was the most widely read exponent of this view. See MUSSCHENBROEK, *The elements of natural philosophy*, trans. by John Colson (2 vols., London, 1744), vol. 2, pp. 3, 19-23, and 46. On p. 46 MUSSCHENBROEK defines 'heat in bodies' as 'A certain quantity of fire in motion in the interstices of the parts, and concealed in the pores of the particles... therefore bodies are so much the hotter, as they contain more fire in motion'. All my references to MUSSCHENBROEK's *Elements* are to the quarto edition ; an octavo was also published.

MACQUER was another major authority who adopted a similar view – at least in this early writings, before being converted to the kinetic theory referred to above (see note 4). Heat is treated as the rapid motion of the particles of a material fire in MACQUER's *Elémens de chimie-théorique* (Paris, 1749), pp. 12-14, and in the first edition of his *Dictionnaire de chymie* (2 vols., Paris, 1766), vol. 1, pp. 498-507.

(8) This, the more common view of the role of fire in causing the sensation of heat, appears in BOERHAAVE, *Elementa chemiae* (2 vols., Leyden, 1732), vol. 1, pp. 142-3 ; and Willem Jakob STORM VAN 'sGRAVESANDE, *Philosophiae Newtonianae institutiones in usus academicos* (Leyden, 1723), p. 190. It is also the view adopted by Euler ; see Euler, 'Dissertatio de igne, in qua ejus natura et proprietates explicantur', in *Pièces qui ont remporté le prix de l'Académie royale des Sciences en MDCCXXXVIII* (Paris, 1739), pp. 1-19. On p. 11 Euler writes : 'Cum enim calor in Motu quodam minimarum particularum corporum consistat, satis perspicuum est Ignem in omnibus corporibus calorem excitare debere'. His conception of heat appears to have remained unchanged to the end of his life ; see, for example, his definition in 'Conjectura circa naturam aeris, pro explicandis phaenomenis in atmosphaera observatis', *Acta Academiae Scientiarum Petropolitanae*, (1779, Part 1 ; published in 1782), 161-87 (175).

in this tradition — Boerhaave, Euler, Musschenbroek, and many others — agreed that an aetherial matter in motion was essential for a satisfactory explanation of thermal phenomena.

In attempting to understand Lambert's version of what I have called the 'science of fire', it is important to recognize not only that theories based on the motion of fire were common currency in the mid-eighteenth century but also that they represented the main stream of speculation on the nature of heat. To the extent that he adopted such a theory, therefore, Lambert was being perfectly conventional, though, as I shall argue, he handled the theory with a boldness and originality that few of his precursors and contemporaries could match.

The element of conventionality in Lambert's work on heat is apparent in the opening pages of the very first of his published papers, the long ' Tentamen de vi caloris' , which appeared in the *Acta helvetica* in 1755[9]. His description of fire particles (*particulae igneae* or *particulae ignis*) as minute and highly ealstic (i.e. mutually repulsive) and of their capacity to expand bodies by separating the particles of ordinary matter until an equilibrium was obtained[10], was a commonplace in the eighteenth century : it could have come from any number of earlier contributors to the doctrine of fire[11]. But what follows is anything but trite. For at this point, in a way that I have come to regard as characteristic of all his work on heat, Lambert introduced a distinctive personal gloss by casting the rest of the discussion in terms of the 'force' (he used the term *vis*) of the fire particles.

The temperature of a body, he asserts, is determined by the total force of all the fire particles contained in a given volume : 'Intensitas caloris est vis particularum in certo spatio. . .'[12]. And this total force depends in turn, first, on the number of fire particles per unit volume (so that compression alone will cause the temperature to rise) and, secondly, on the force contributed by each one of them individually. The whole cast of the argument makes it abundantly clear that the force of a fire particle, though never defined, was regarded by Lambert as a function of the velocity with which it moved[13].

---

(9) J.H. LAMBERT, 'Tentamen de vi caloris, qua corpora dilatat, eiusque dimensione, *Acta helvetica, physico-mathematico-anatomico-botanico-medica*, 2 (1755), 172-242.

(10) Ibid., pp. 172-3.

(11) Cf., for example, MUSSCHENBROEK, *Elements*, op.cit. (note 7), vol. 2, pp. 3 and 19-23. In the 'De vi caloris' LAMBERT follows prevailing opinion in asserting the particulate nature of fire. Some eighteenth-century authorities, however, conceived fire as a continuous fluid. This view was probably favoured by 'sGRAVESANDE ; see his *Philosophiae Newtonianae institutiones*, op. cit. (note 8) and other writings, where there is no mention of the *particles* of fire. In the *Pyrometrie*, published more than twenty years after the 'De vi caloris', LAMBERT was less explicit on the point, though his use, throughout the book, of the term 'fire particles' (*Feuertheilchen*) suggests that his opinion was unchanged. During the colloquium, however, Professor Andreas SPEISER pointed out to me that even the references to 'fire particles' do not in themselves rule out the possibility that LAMBERT thought of fire as a continuous fluid,

(12) LAMBERT, 'De vi caloris', op. cit. (note 9), p. 174.

(13) For LAMBERT, as for virtually all eighteenth-century authorities, it was axiomatic that fire was in constant motion. See, among numerous other sources of the period, BOERHAAVE, *Elementa chemiae*, op. cit. (note 8), vol. 1, pp. 398-401 ; 'sGRAVESANDE, *Philosophiae Newtonianae institutiones*, op.cit. (note 8), pp. 185 and 190 ; EULER, 'De igne' , op . cit. (note 8), *passim ;* MUSSCHENBROEK, *Elements*, op. cit. (note 7), vol. 2, pp. 19-22.

Vague references to the 'force' of fire particles are common in the eighteenth-century literature. Euler, in his 'De igne' , was among those who referred freely to the *vis* of fire particles, without formally defining the term.

Hence, for him, the heating that occured when a solid object was rubbed or struck was a consequence not only of a supposed reduction in its volume – that plausible but rather lame explanation was the one adopted by most calorists, when they did not evade the problem completely[14] – but also of an increase in the 'intestine motion' (*motus intestinus*) of the fire particles[15].

Lambert's explanation of the heats of friction and percussion is a good illustration of the Janus-faced character of his theory of heat. For, while his 'science of fire' has its roots very firmly in a tradition that goes back directly through Euler to Boerhaave and, more distantly, to Newton and Descartes[16], it shows repeatedly the independence and assurance of a mind of almost intemperate originality. With an assertiveness that presumably owed something to his comparative youth – he was in his mid-twenties when the 'De vi caloris' was composed[17] – Lambert presented his views on friction and percussion almost as an aside, worthy of only brief and passing mention. In doing so, he gave no hint that he was departing significantly from what was, in the 1750s, the more normal treatment, as given in the most revered of all the standard texts, Boerhaave's *Elementa chemiae* of 1732. Somewhat in the manner of Lambert, Boerhaave had referred to the way in which friction and percussion would 'excite' fire : 'ignis excitatur' is Boerhaave's phrase[18]. But there the similarity ends. For, in Boerhaave's opinion, as in that of most contemporaries, it was the chief role of the fire to sustain the motion of the particles of ordinary gross matter ; hence it was their motion, not the motion of the fire itself, that caused the sensation of heat.

However much we may glean from Lambert's remarks on heating by friction and percussion, we should not mistake them for the meat of his paper. Most of the seventy pages of the 'De vi caloris' are devoted to material of a quite different character, in particular to a highly speculative mathematical treatment of the relation between the force of fire particles and the volume of bodies undergoing changes in temperature.

---

(14) It is a significant indication of the prevailing beliefs of calorists on this point that they commonly drew an analogy between the heating produced by friction and percussion, and the rise in temperature that accompanies the rapid compression of a gas. See, on this point, FOX, *Caloric theory*, op.cit. (note 2), p. 100.

(15) LAMBERT, 'De vi caloris', op. cit. (note 9), p. 175 : '... si sub angulo acuto ferrum malleo percutiatur, unde non solum motus intestinus particularum ignis, verum & maxima earum compressio oritur, qua simul illarum vis, corporisque calor augetur .

(16) The Cartesian origins of eighteenth-century doctrines of fire are stressed by Metzger and Talbot in the works cited in note 6. The links between the eighteenth-century doctrines and Newton's speculations on the aether are explored in P.M. HEIMANN, 'Nature is a perpetual worker' : Newton's aether and eighteenth-century natural philosophy', *Ambix*, 20 (1973), pp. 1-25 (10-13). Heimann sees the roots of Boerhaave's notion of fire in Newton's aether, as expounded in the 1717 edition of the *Opticks*, and in older chemical doctrines associated with the Helmontian tradition. It appears that, for Boerhaave, fire had a role akin to that of a Newtonian 'active principle'. As Heimann observes, whatever the true source of Boerhaave's ideas may have been, his doctrine of fire had become an integral part of 'Newtonian' natural philosophy by the mid-eighteenth century.

(17) There are several references to the composition of the 'De vi caloris' in Lambert's 'Monatsbuch'. They show that the paper was written in 1753-4. See Karl Bopp (ed.), 'Johann Heinrich Lamberts Monatsbuch', *Abhandlungen der Königlich Bayerischen Akademie der Wissenschaften. Mathematisch-physikalische Klasse, 27*, 6. Abhandlung (1915), pp. 12-14.

(18) BOERHAAVE, *Elementa chemiae*, op.cit. (note 8), vol. 1, pp. 176-81.

In this part of his work, I find it particularly hard to identify Lambert's sources, and it is no less difficult to summarize his rambling and often bizarre argument. The following illustration, drawn from the early part of the paper[19], conveys something of the flavour of his approach, but it gives little impression of the complexities that follow. To them, no summary can do justice.

In my example, Lambert derives an equation expressing the number of fire particles flowing, in a given period of time, between two bodies, of volume A and a, and of different temperatures. He defines the 'relative quantity' (*quantitas relativa*) of the fire particles, Q, as the difference between the total number of fire particles in each body. Suppose, then, that x fire particles pass from the hotter to the cooler body in a time t. After the passage of these particles, the force of those that remain in the hotter body ('relative' to those in the cooler body) will be $V(Q - x)/A$, where V is the force of a single fire particle in the hotter body. (It will be recalled that, for Lambert, the total force of the fire in a body is equal to the force contributed by all the fire particles in a unit volume.) In the same time, the force of the fire in the cooler body will have increased by vx/a, where v is the force of an individual fire particle at the lower temperature. Hence the 'relative' force of the fire will now be

$$\left\{ \frac{V(Q - x)}{A} - \frac{vx}{a} \right\}$$

(compared with its original value of VQ/A). Lambert now assumes that the rate of flow of fire particles between the two bodies, dx/dt, is proportional to this relative force, so that, introducing a constant m :

$$\frac{dx}{dt} = \frac{1}{m} \left(\frac{V(Q - x)}{A} - \frac{vx}{a} \right)$$

$$= \frac{1}{m} \left\{ \frac{aVQ - x(aV + Av)}{Aa} \right\} .$$

Hence

$$t = \left\{ \frac{maA}{aV + Av} \right\} \log \left\{ \frac{aVQ}{aVQ - (aV + Av)x} \right\}$$

and, when an equilibrium of temperature has been attained,

$$x = \frac{aVQ}{(aV + Av)}$$

As Lambert goes on to argue[20], a further assumption allows this result to be modified very simply to express changes in volume. For, if the volume occupied by each particle in the hotter body is S, and the corresponding quantity for the cooler body is s, it follows, according to Lambert, that the changes in volume resulting from the passage of x fire particles will be xS for the hotter body, and xs for the cooler. The insertion of appropriate conditions will then permit testing for specific cases. Lambert

---

(19) LAMBERT, 'De vi caloris', op.cit. (note 9), vol. 1, pp. 176-81.

(20) *Ibid.*, pp. 181 and 187-8.

had observed, for example, that the readings on an alcohol thermometer suddenly exposed to open sunlight and then left in the sun for two hours rose in accordance with his theoretical predictions.

It is hardly surprising that no one appears to have shown an interest in Lambert's speculations. In the arguments I have summarized and in the very detailed mathematical arguments that make up the rest of the paper, Lambert departed so far from traditional views on fire that few can have regarded his work as anything but eccentric. Perhaps Lambert himself came to this conclusion in due course, for despite occasional references in his later writings, he never built on the theory of the 'De vi caloris'.

But, for all its extravagance, the paper has its roots unmistakably in the tradition of speculation on the nature and properties of fire that I have already identified as standard in the mid-eighteenth century. This is not to say, of course, that individual influences can be readily distinguished. Indeed, I think it inconceivable that these influences could ever be unravelled with certainty. Of all the sources, the likeliest, at first sight, is Euler, with whom Lambert conducted an amicable scientific correspondence from 1758 to 1772[21]. But in that correspondence, there is only a fleeting reference to Lambert's work on heat[22], and, moreover, Lambert's notion of fire departs significantly from Euler's[23]. In his emphases and terminology, Lambert probably comes closest to Musschenbroek, among the eighteenth-century authorities, and the two men did meet, at Leyden in 1757 or 1758 (though the encounter seems to have been notable not so much for an instant *rapport* as for Musschenbroek's disquieting failure to recognize Lambert[24]). Conceivably, despite Musschenbroek's lapse, the encounter was a significant one, but it came some four years after the 'De vi caloris' was written, and, in any case, I am convinced that personal encounters were far less important for a self-taught polymath like Lambert than a command of the published literature that was probably unrivalled among his peers in the study of heat. In large measure, Lambert owed this command to his uncommon facility with languages — he evidently read German, French, Latin, and English with ease — but it also reflects an internationalism that would have made the scientific traditions of any one school or any one nation unduly restricting. Lambert, in fact, was subject to flow of ideas and influences that knew no national boundaries, and it is for this reason above all that the precise identification of his sources is such a daunting task.

It was, naturally enough, in later life that the attitudes I have described bore full fruit. And this brings me to the second of Lambert's two main works on the theory of heat and fire, the *Pyrometrie*[25]. As it happens, the dating of the various parts of this book is difficult. Entries in Lambert's 'Monatsbuch' show that he first contemplaced writing such a work as early as 1756, and over the next twenty years the project was

(21) Karl Bopp (ed.), 'Leonhard Eulers und Johann Heinrich Lamberts Briefwechsel', *Abhandlungen der Preussischen Akademie der Wissenschaften (phys.-math. Kl)*, 2. Abhandlung (1924), pp. 7-45.

(22) *Ibid.*, p. 15. In a letter dated 4 April 1760, Lambert refers to his plans for a work on ' Pyrometrie'.

(23) See note 8.

(24) [J.H.S. FORMEY], 'Eloge de M. Lambert', *Histoire de l'Académie royale des Sciences et Belles-Lettres*, pp. 79-80 ; published in the *Nouveaux mémoires de l'Académie royale des Sciences et Belles-Lettres [de Berlin]*, (1778 ; published in 1780), pp. 72-90 (79-80).

(25) J.H. LAMBERT, *Pyrometrie oder vom Maasse des Feuers und der Wärme* (Berlin, 1779).

kept fitfully alive and various sections were drafted[26]. Even when his attention turned to photometry, the work on heat and fire was never forgotten. In the Preface to the *Photometria* Lambert stated that his study of light was to be completed by a work referred to as *Pyrometria* : this work, in fact, was the true goal (*scopus*) of his optical researches[27]. It is quite clear, then, that for Lambert the measurement of light and that of heat were but two aspects of the same enterprise – a view entirely consistent with the prevailing opinion (to which Lambert apparently inclined) that fire in motion was the cause of the phenomena of light as well as those of heat. But while the *Photometria* was published by 1760, the *Pyrometrie* was only completed during a final burst of activity in the spring of 1777, just a few months before the author's death[28]. And even then the text required the attention of his long-standing friend, Johann Gustav Karsten, as editor, before it finally appeared, posthumously, in 1779[29]. Despite this history of

---

(26) For evidence of this continuing interest, see the 'Monatsbuch', reproduced in Bopp, op.cit, (note 17), pp. 18, 27, and 33. In an entry dated August 1756, Lambert writes : 'Incepi principia physica de calore a § 1 ad § 189 [apparently a reference to the first 189 sections of the *Pyrometrie*]. Scripsi Schema universae Theoriae caloris, Pyrologiam et Pyrometriam complectens'. In September 1756, referring again to early sections of the *Pyrometrie*, he writes : 'Scripsi Hypotheses atque positiones de Igne et Luce (§ 1 . . . 20) quibus prima principia Physica de igne et luce conduntur'. In September 1756, he refers to 'Experimenta pyrometrica', and between March and May 1777 he was evidently engaged in preparing the manuscript of the *Pyrometrie* for publication. By the middle of May 1777, it seems, the work was completed. The very last entry in the 'Monatsbuch' reads : 'Mai. Die Pyrom. geendigt, den 16ten,

Lambert's correspondence also contains several references to his work on heat and fire. The references are summarized in Max STECK, *Bibliographia Lambertiana. Ein Führer durch das gedruckte und ungedruckte Schrifttum und den wissenschaftlichen Briefwechsel von Johann Heinrich Lambert 1728-1777* (new edn., Hildesheim : Verlag Dr. H.A. Gerstenberg, 1970), p. 15.

Perhaps the most striking evidence of Lambert's sustained interest in heat and fire is to be found in his manuscripts preserved in the Library of the University of Bâle. For an inventory, see *Die Handschriftliche Nachlass von Johann Heinrich Lambert (1728-1777). Standorts-Katalog auf Grund eines Manuskriptes von Max Steck herausgegeben von der Universitätsbibliotek Basel* (Bâle : Offentliche Bibliotek der Universität, 1977), especially the entries on pp. 5-6, 10-12, 65-75, 82-3, 87-9, and 94. The notes on heat and fire, which deserve a far more detailed study than I have been able to undertake, are chiefly concerned with thermometry. Lambert clearly had a special interest in the use of the thermometer in meteorology, and he made meteorological observations and scrutinized the observations of others from about 1750.

(27) J.H. LAMBERT, *Photometria sive de mensura et gradibus luminis, colorum et umbrae* (Augsburg, 1760), 'Praefatio' (unpaginated). Lambert writes in the Preface : '. . . hoc unum adiungam, Pyrometriam, quam curatius evoluendam suscepi, scopum huius libri primarium fuisse. Quanam vero ratione mensura luminis ad mensuram caloris & ignis quicquam faciat, & quis inter utramque sit nexus, hoc in ipso Pyrometriae opere, Deo adiuvante, ob oculos ponetur'.

During the colloquium, Monsieur Henri Pfeiffer drew my attention to the draft of a final chapter of the *Pyrometrie* on the measurement of the intensity of colour. This draft chapter, which was withdrawn from the book at Lambert's request, is preserved in the Library of the University of Bâle.

(28) See the evidence in the 'Monatsbuch', referred to above, note 26.

(29) The *Pyrometrie* has a nine-page Foreward by Karsten, outlining (among other things) the background to the composition of the book and its posthumous publication. It also has an essay by Johann August Eberhard on the philosophical context of Lambert's work.

A letter from Lambert to Karsten, dated 4 March 1777, shows that Karsten was kept informed of Lambert's progress with the writing of the *Pyrometrie* in the last few months of his life. See Johann Bernoulli, III (ed.), *Joh. Heinrich Lamberts deutscher gelehrter Briefwechsel* (5 vols., Berlin, 1782-7), vol. 4, pp. 336-7.

piecemeal composition, it seems reasonable to suppose that the *Pyrometrie* represents Lambert's mature position, and it is as the culmination of his work on heat that I shall regard it in this paper.

For the range of its references alone, the *Pyrometrie* is a dazzling performance : in it, the greatest names of eighteenth-century natural philosophy jostle with a host of others that must have been virtually unknown even in the 1770s. So if it had been offered as no more than a *summum* of the science of fire as it was understood in the mid-eighteenth century, the *Pyrometrie* would have been notable. But Lambert's readiness to take a stand on obscure points of theory and his determination to advance rather than merely compile knowledge are as apparent in this book as in the 'De vi caloris'. The special quality that emerges in the *Pyrometrie*, however, is Lambert's enthusiasm for experiment[30]. Of course, enthusiasm cannot be equated with competence, and it was suggested by John Leslie, who knew the *Photometria* well (though probably not the *Pyrometrie*), that Lambert's insistence on making all his own apparatus was an impediment to his achievements as an experimenter[31]. Leslie may well have been right to criticize the optical work, but a comparison of the long and rich section of the *Pyrometrie* in which Lambert reviewed the various thermometers and temperature scales that were available in his own day with similar studies in the works of Martine and De Luc[32] leaves no doubt in my mind of the excellence of Lambert's contribution. The patient scrutiny of existing information – much of it culled from little-known readings made in the seventeenth century – augmented by careful observations of his own on mercury, alcohol, and air thermometers was the foundation of the finest guide we possess to an anarchic aspect of eighteenth-century experimental philosophy. In all, nineteen scales were compared, the verdict which Lambert gave being resoundingly in favour of the air thermometer as a standard instrument. When air expanded at a constant pressure, according to Lambert, it did so in proportion to the quantities of heat added ; it showed, as he put it, the 'actual degrees of heat'[33]. Spirit of wine, by contrast, expanded most irregu-

(30) Although the *Pyrometrie* provides the best published evidence of the quality of Lambert's experimental work, it should not be thought that he neglected experiment in his earlier writings. Even in the highly speculative 'De vi caloris', he laboriously compares his theoretical predictions (in particular concerning the readings on thermometers exposed to the sun and immersed in cooling water) with observations of his own ; see 'De vi caloris', op. cit. (note 9), pp. 187-93, 207-8, and 220-1.

(31) John LESLIE, *An experimental inquiry into the nature and propagation of heat* (London, 1804), p. 405. Speaking of the *Photometria*, Leslie deplores the 'lame and inaccurate' experimental evidence on which, all too often, Lambert's conclusions rested. Cf. a similar criticism expressed in the unsigned 'Encomium', which prefaces Lambert, *The system of the world*, trans. by James Jacque (London 1800), p. XXVII.

(32) For the work of George Martine, see 'Some observations and reflections concerning the construction and graduation of thermometers' (dated 1738) and 'The comparison of different thermometers with one another' (dated 1740), both published in MARTINE's *Essays medical and philosophical* (London, 1740), pp. 175-230. (These essays also appeared in the three subsequent editions of the work, published in Edinburgh in 1772, 1780, and 1792, as *Essays and observations on the construction and graduation of thermometers.)* Martine's essays are accompanied by a plate displaying the equivalent points on fifteen different temperature scales ; it is similar in style to Plate VIII of the *Pyrometrie*, where nineteen scales are compared.

Jean André DE LUC published his work on thermometry in his *Recherches sur les modifications de l'atmosphère* (2 vols., Geneva, 1772), vol. 1, pp. 331-93.

(33) LAMBERT, *Pyrometrie*, op. cit. (note 25), p. 78.

larly, and even the mercury thermometer, so strongly favoured by De Luc[34], deviated significantly from the air scale.

Quite apart from the guidance they give on the choice of a scale of temperature, Lambert's observations on the expansion of air are notable for their bearing on a major theoretical issue. In experiments of unprecedented accuracy Lambert showed that air expanded by 370/1 000 of its original volume between the melting point of ice and the boiling point of water (points which he marked respectively as 1 000 and 1 370 on his own scale)[35]. Extrapolation of this observation yielded what we should now term an absolute zero of temperature at $-270°C$, and Lambert was convinced not only that an absolute zero did exist but also that extrapolation from the data on the expansion of air was permissible in order to locate it. For him, the point of 'absolute cold' *(absolute Kälte)* was that at which, in the absence of any fire particles whatsoever, the volume of air would be reduced to zero, or at least to the negligible volume occupied by the air particles themselves[36].

In seeking to locate the point of absolute cold at all, Lambert was taking a characteristically bold, though not original position, for the very existence of such a point was viewed with suspicion until the last twenty years or so of the eighteenth century. Boerhaave had been adamant that although the 'limits of cold' could be defined in principle, they could not be located[37], and in 1765 the writer of the official history of the Académie des Sciences in Paris endorsed the opinion of Jean-Jacques Dortous de Mairan in asserting that the zero was 'a chimera that exists nowhere in nature'[38]. Yet amid this scepticism there existed a very different thread of speculation which has its origins in one of those disregarded authors whose work Lambert resurrected : Guillaume Amontons. It was Amontons, in 1703, who gave the earliest formal definition of an absolute zero : his 'extreme cold' *(extrême froid)* was the hypothetical point at which air exerted no pressure[39]. To recast this in the more speculative terms which Amontons

---

(34) DE LUC, op. cit. (note 32), vol. 1, pp. 285-330.

(35) LAMBERT, *Pyrometrie*, op. cit. (note 25), p. 47.

(36) *Ibid*, p. 29 : 'In der absoluten Kälte fällt die Luft so dicht zusammen, bis sich ihre Theilchen durchaus berühren, oder bis sie, so zu reden, wasserdicht wird'.

(37) BOERHAAVE, *Elementa chemiae*, op. cit. (note 8), vol. 1, pp. 151 and 189. For Boerhaave, the limits of cold lay, in principle, at the point where no fire was present (p. 151) or where fire was perfectly at rest (p. 189). But he stressed that such a condition was, in his view, quite unrealizable. The search for an upper limit of temperature was likewise vain.

(38) *Histoire de l'Académie royale des Sciences*, (1765 ; published in 1768), p. 10. For the views of Dortous de Mairan, see *Mémoires de l'Académie royale des Sciences* for 1765 (invariably bound with the *Histoire*), pp. 204-9. Dortous de Mairan was severely critical of Amontons's definition of the absolute zero : even if such a point existed, it certainly could not be located. As late as 1818 Gay-Lussac could still assert that there was probably no limit to the degree of cold that might be attained by the rapid (i.e. quasi-adiabatic) expansion of air. If this was in fact the case, he observed, 'la détermination du zéro absolu de chaleur doit paraître une question tout-à-fait chimérique'. See Joseph-Louis Gay-Lussac, 'Sur le froid produit par la dilatation des gaz', *Annales de chimie et de physique*, 9 (1818), 305-10 (310).

(39) Guillaume AMONTONS, 'Le thermomètre réduit à une mesure fixe et certaine...', *Mémoires de l'Académie royale des Sciences*, (1703 ; published in 1705), pp. 50-6 (52). In fact, the existence of the zero had already been hinted at in a paper by Amontons published on pp. 161-80 of the *Mémoires* for 1702 ; see p. 171. The latter paper and one published in the *Mémoires* for 1699 (pp. 112-26) are the best sources for Amontons's experiments on the pressure of air.

used and which must have struck a responsive chord in Lambert's mind, it was also the point at which the air contained no particles of fire, for it was these particles which, according to Amontons, gave air its elasticity[40].

In fact, Amontons did not locate the point of extreme cold himself, though, as George Martine showed nearly forty years later, he possessed data which pointed to a zero at $-239.5°C$[41]. In his discussion, Lambert departed from Amontons in defining his zero in terms of the volume of air rather than its pressure, but he gave full credit to both Amontons and Martine. As it transpired, the warm endorsement of Amontons's work had little effect, and when interest in the absolute zero revived in the 1780s, the methods of calculation were founded on completely different (and totally erroneous) principles, laid down by William Irvine, a pupil of Joseph Black[42]. It was only in the early years of the nineteenth century that methods of locating the zero by reference to the expansion of air gained ground and eventually, from about 1820, became standard[43].

Lambert's flair for seizing on and developing earlier work, irrespective of its standing in his own day, is apparent enough in his discussion of the absolute zero. But in this respect, the case of Johann Christian Arnold, who became professor of physics at the University of Erlangen in 1759, is even more striking. For I suspect that Arnold's investigation of what we now recognize as an example of adiabatic heating and cooling was retrieved, by the *Pyrometrie*, not from marginality but from total oblivion.

In his inaugural dissertation as professor in 1759, Arnold reported the changes in temperature which he had observed when air was allowed to enter, or was abstracted from, the receiver of an air pump[44]. Such changes had first been noted, though not explained, only four years earlier, by William Cullen, in Glasgow[45]; but until they were referred to by Arnold, Cullen's observations had apparently passed unnoticed. No doubt

---

(40) Amontons's discussion of the nature and properties of fire extends only to its role as the cause of the elasticity of air. In 1702, in his 'Discours sur quelques proprietez de l'air, et le moyen d'en connoître la température dans tous les climats de la terre', *Mémoires de l'Académie royale des Sciences*, (1702 ; published in 1704), pp. 161-80 (173), Amontons describes how air particles are put in motion by 'les parties du feu'. Then, in a paper dating from the following year, he states that the elasticity of air results entirely from 'le mouvement des particules ignées dans lequel [sic] il [i.e. l'air] nage' ; see 'Que les nouvelles expériences. . .', *Mémoires de l'Académie royale des Sciences*, (1703 ; published in 1705), 101-8 (102).

(41) MARTINE, *Essays medical and philosophical*, op. cit. (note 32), pp. 291-2.

(42) On the theory of heat proposed by William Irvine and the method of calculating the absolute zero based on it, see FOX, *Caloric theory*, op. cit. (note 2), pp. 25-8, and 'Dalton's caloric theory', in D.S.L. CARDWELL (ed.), *John Dalton and the progress of science* (Manchester : Manchester University Press, 1968), pp. 187-202 (188-91). The 'Irvinist' method of calculation was used by Jean Hyacinthe de Magellan, Adair Crawford, and John Dalton, among others.

(43) See FOX, *Caloric theory*, op. cit. (note 2), pp. 61-2 and 147. The method was endorsed in the work of Alessandro Volta (1793) and J.T. Mayer (1809). In answer to the prize competition of 1812 on the specific heats of gases, set by the Institut de France, Nicolas Clément and Charles Bernard Desormes used the same method, but also a version of the 'Irvinist' method (see above, note 42), in order to locate the absolute zero at $-266.66°C$.

(44) Johann Christian ARNOLD, *De thermometri sub campana antliae pneumaticae suspensi variationibus* (Erlangen, 1759).

(45) William CULLEN, 'Of the cold produced by evaporating fluids and of some other means of producing cold', *Essays and observations, physical and literary. Read before a society in Edinburgh, and published by them* (Edinburgh, 1756), vol. 2, pp. 145-56. The paper is dated 1 May 1755.

misled by the context of Cullen's remarks — they appeared in a paper entitled 'Of the cold produced by evaporating fluids...' — Arnold associated the cooling vaguely with evaporation (probably from the surface of the thermometer bulb) ; and he attributed the heating to friction between the inrushing air and the walls of the entrance channel. In short, he completely misconstrued the nature of the effect, and it was left for Lambert to recognize, at least two years (and probably very much longer) before anyone else, that a previously unknown thermal property of air had been discovered.

Lambert records, in the *Pyrometrie,* that Arnold demonstrated the experiments with the air pump to him in 1761[46]. At that time, he had not yet seen Arnold's dissertation, so that he could claim complete originality for his own explanation, as given in the *Pyrometrie*[47]. There seems no reason to doubt Lambert's account : he was quite clearly the first person to infer from the experiments with the evacuated receiver that the rapid compression and expansion of air causes changes in its temperature.

Lambert's interest in the phenomenon was undoubtedly heightened by the ease with which it could be incorporated into his views on fire. For it was a natural corollary of these views that when the receiver of an air pump was evacuated, fire particles as well as air were abstracted. Therefore, if the density of the air was reduced by, say, one half, and if no fire entered from the surroundings, the density of the fire particles in the receiver would also be reduced in the same proportion. It followed that the temperature — and here Lambert clearly had in mind absolute temperature — would also be reduced to half of its initial value. Indeed, if the process was carried still further and all the fire particles were abstracted, it would be possible to attain the point of absolute cold merely by pumping. At least, it would be possible in principle ; in reality, as Lambert knew, fire particles were constantly entering the reciver to replace those which had been withdrawn. Heating, by contrast, was caused by the compression of fire particles : a reduction in the volume of air by one half, without heat loss, would double both the density of the fire and hence the (absolute) temperature of the air. The effect of this doubling of the temperature, added to that of an isothermal decrease in volume, would be to increase the pressure of the air fourfold, but, as in the case of evacuation, heat exchange with the surroundings vitiated any possibility that the prediction might be tested.

Having achieved such a precocious understanding of the nature of adiabatic phenomena, Lambert might have been expected to treat the rise in temperature accompanying the opening of an evacuated receiver (correctly) as yet another example of heating by compression. In fact, despite a clear description of the effect, he gave no explanation at all. The reason for his failure to make what may now appear an obvious application of his insight is, I suspect, that he simply did not see compression as the cause of the heating in the receiver. I think it far more likely that he associated the rise in temperature with his belief that even a vacuum contained fire particles. Now this belief may sound innocuous enough, even obvious, but it does imply, as Lambert pointed out, that temperature changes will occur if the volume of the Torricellian vacuum in a barometer tube

(46) LAMBERT, *Pyrometrie,* op.cit. (note 25), p. 266.

(47) *Ibid.,* pp. 266-9.

is altered rapidly[48]. It also implies — and here I come back to the experiment with the evacuated receiver — that the entry of air, with its accompanying fire particles, into a vacuum will have the effect of increasing the density of the fire particles, so that heating will result[49].

Despite the paucity of the literature on adiabatic phenomena that existed in 1779, Lambert's perceptive discussion shared the fate of the *Pyrometrie* as a whole. Apart from references in the writings of Saussure and Pictet[50], it provoked no public comment, favourable or unfavourable. Why was this so ? And why did Lambert's work even fail to surface in the 1780s, when adiabatic phenomena began to attract attention in Britain, or after 1800, when a knowledge of them at last reached France, through Pictet ? It is no explanation, of course, to dwell on what might have been, but I cannot resist prefacing my answer with a passing mention of two missed opportunities which, if seized, might have brought both adiabatic phenomena and Lambert together to the very centre of late-eighteenth-century physics.

The opportunities arose from two classic problems that engaged the attention not only of Lambert but also of a succession of the most distinguished of eighteenth-century physicists. The problems were, first, the explanation of cold at high altitudes and, second, the existence of a discrepancy of some ten per cent between the experimental value for the velocity of sound and the value predicted by Newton's theory[51]. Within some thirty years of Lambert's death, correct solutions to both of these problems, based on adiabatic phenomena, had won general acceptance, with experimental evidence no better than that which was available to Lambert in the 1770s. Hence my reference to missed opportunities.

---

(48) *Ibid.*, pp. 268-9. In a paper read to the First Class of the Institut de France on 15 September 1806, Gay-Lussac described his attempt to test just such a prediction. He had failed to observe any change in temperature when the volume of the Torricellian vacuum above the column of a mercury barometer was altered by brusquely tipping the tube. See Gay-Lussac, 'Premier essai pour déterminer les variations de température qu'éprouvent les gaz en changeant de densité et considérations sur leur capacité pour le calorique', *Mémoires de physique et de chimie de la Société d'Arcueil, 1* (1807), 180-203. It seems certain that Gay-Lussac was ignorant of Lambert's discussion of the matter, as were John Dalton and Clément and Desormes, who also speculated on the heat capacity of the void in the early nineteenth century. Eighteenth- and nineteenth-century views on the capacity of the void are discussed in P. Costabel, 'Le "calorique du vide" de Clément et Desormes (1812-1819)', *Archives internationales d'histoire des sciences*, 21e année (1968), pp. 3-14, and Fox, *Caloric theory*, op.cit. (note 2), pp. 48, 51-3, 130, 142-8, and 152.

(49) This was the explanation ascribed to Lambert by Marc Auguste Pictet in his *Essais de physique* (Geneva, 1790), pp. 19-20. Despite the absence of an explanation in the *Pyrometrie*, Pictet's reconstruction of Lambert's view seems entirely plausible.

(50) H.B. de SAUSSURE, *Essais sur l'hygrométrie* (Neuchâtel : Samuel Fauche, 1783), p. 232 (my reference is to the quarto edition, Pp. xxiv + 367). For a reference to Pictet's discussion, see above, note 49.

(51) For studies of eighteenth-century interest in these problems, see (on the cold at high altitude) W.E. Knowles MIDDLETON, *A history of the theories of rain and other forms of precipitation* (London : Oldbourne, 1965), pp. 103-6, and (on the velocity of sound) Bernard S. FINN, ' Laplace and the speed of sound', *Isis, 55* (1964), 7-19 ; FOX, *Caloric theory*, op.cit. (note 2), pp. 81-6 ; and Part II of C.A. Truesdell's introduction to *Leonhardi Euleri opera omnia* (Lausanne, 1955), 2nd ser. vol. 13, pp. XIX-LXXII.

Lambert's own explanation of the cold at high altitudes, as expounded in a paper in 1772 and repeated in the *Pyrometrie*[52], turned on the supposed lightness of the particles of fire. He regarded it as axiomatic that their specific gravity was less than that of air, so that they would rise with an ever-increasing velocity, impeded only by the physical obstacle presented by the air itself. As the upward velocity increased, so the density of the fire particles, and hence also the temperature, would fall. Since there was no independent evidence that the density of the fire particles did decrease with altitude, the argument had an element of circularity which a complicated mathematical treatment of the upward force on the fire particles did nothing to obscure, incorporating as it did a number of dubious assumptions. But it was not the weakness of the argument which prevented Lambert's solution from winning support : rather, his discussion (like the whole of the *Pyrometrie*) simply suffered from neglect. Saussure was the only author to mention it in print, and even he did so critically[53]. By the time Erasmus Darwin correctly ascribed the cold at high altitudes to the expansion of warm air as it rises, in a paper to the Royal Society of London in 1787[54], Lambert's work was forgotten ; and it was Darwin's theory which, in a curiously serene way, gradually became standard between 1787 and the 1820s.

And so to the second of my missed opportunities in the application of adiabatic phenomena. The discrepancy between the theoretical and experimental values for the velocity of sound was one of the great scandals of eighteenth-century physics, with Euler, Johann Bernoulli (II), Daniel Bernoulli, d'Alembert, and Lagrange all involved more or less closely in the quest for a solution. In ascribing the discrepancy to Newton's false assumption that the pressure of air (P) in a sound wave was proportional to its density ρ – i.e. effectively, that isothermal conditions were maintained – Lambert anticipated the correct solution which Laplace proposed in 1802, through his protégé Biot[55]. But whereas Laplace explained the departure from Boyle's Law by suggesting that the compression and rarefaction of air in a sound wave took place adiabatically, Lambert invoked the presence of 'foreign particles' (*particules étrangères*), such as those of water vapour, in the air[56]. The effect of these particles, he argued in a paper dating from 1768, was to increase the density of the air without contributing to the pressure, so that P in atmospheric air was, in reality, by no means proportional to ρ. As in his speculations on the cold air on mountains, there was a strong element of cir-

(52) LAMBERT, 'Sur la densité de l'air', *Nouveaux mémoires de l'Académie royale des Sciences et Belles-Lettres [de Berlin]*, (1772 ; published in 1774), pp. 103-40 (114-15). Lambert's explanation of the cold at high altitudes is summarized, very briefly, in the *Pyrometrie*, op.cit. (note 25), p. 232.

(53) SAUSSURE, *Voyages dans les Alpes, précédés d'un essai sur l'histoire naturelle des environs de Genève* (4 vols., Neuchâtel, 1779-96), vol. 2 (1786), p. 351.

(54) Erasmus DARWIN, 'Frigorific experiments on the mechanical expansion of air', *Philosophical transactions of the Royal Society of London, 78* (1788), pp. 43-52. The paper was read on 13 December 1787.

(55) Jean-Baptiste BIOT, 'Sur la théorie du son', *Journal de physique, 55* (1802), 173-82.

(56) LAMBERT, 'Sur la vitesse du son', *Histoire de l'Académie royale des Sciences et Belles-Lettres [de Berlin]*, (1768 ; published in 1770), pp. 70-80. See also his 'Sur la densité de l'air', op.cit. (note 52), pp. 131-8.

338

cularity, with Lambert reaching the unconfirmed conclusion that the ratio between the density of atmospheric air and that of pure air (i.e. air free from 'foreign particles') must be as 37:25 if the error in Newton's theory was to be removed.

I end this somewhat unhistorical digression on missed opportunities as I began it, apologetically. It may remind us how close Lambert came to making two contributions to physics which would certainly have earned him at least posthumous recognition. But it leaves us without an explanation, or even a clear assessment, of the degree of neglect to which his work on heat was subject in his own lifetime. And so I return to what I regard as major problems. First, was Lambert's science of fire totally without influence ? And, secondly, if it was without influence, why was this so ?

Necessarily, evidence of neglect is negative evidence, but in this case it is at least of striking uniformity. My own examination of the literature of the eighteenth and early nineteenth centuries suggests that Lambert's writings on heat and fire had few repercussions, and that they remained little-read until historians closer to our own day began their search for the usung heroes of science. Of course, Lambert's work was known among physicists working in what may loosely be termed the German tradition. It seems to have been one source of inspiration for the work on the expansion of air by the German cleric, Johann Friedrich Luz[57]. And Benjamin Thompson, Count Rumford, who moved in the scientific circles of Bavaria in the 1780s and 1790s, also knew Lambert's work and even provided Fourier with a translation of parts of the *Pyrometrie*[58]. But the signs of influence are both few in number and of modest importance. And their insignificance is underlined by the apparent absence of indebtedness to Lambert on the part of some of the leading representatives of the German tradition in physics : I think here particularly of Johann Tobias Mayer (the younger), who wrote voluminously on heat in the early years of the nineteenth century.

Yet Lambert's work on heat was not entirely without admirers. The most enthusiastic by far were in the school of Swiss physicists, centred on Saussure, whose chief interests were meteorology and hygrometry. Now such recognition is in no way to be disparaged, for the school was distinguished, and, despite points of disagreement, Saussure's admiration for the 'great geometer', as he called Lambert, was warm. In the second volume of his *Voyages dans les Alpes,* published nine years after Lambert's death, Saussure pleaded for the translation of the *Pyrometrie* as a way of making it accessible to those who did not read German.

---

(57) Luz's observations on the expansion of air are referred to in his *Vollständige und auf Erfahrung gegründete Beschreibung* (Frankfurt and Leipzig, 1784), pp. 413-14.

(58) See Louis Charbonneau's contribution to the colloquium : 'Lambert et la physique mathématique au XIXe siècle', note 37. Monsieur Charbonneau has discovered translations, which Rumford undertook and sent to Fourier, of paragraphs 327, 331-5, and 452-99 of the *Pyrometrie.* The translations are in the Bibliothèque Nationale, cabinet des manuscripts, fonds français, 22525, ff. 170-2, and 22529, f. 50. I am grateful to Monsieur Charbonneau for drawing my attention to these papers, of which I was previously unaware.

L'homme de lettres qui traduiroit cet ouvrage [he wrote] rendroit un grand service aux physiciens qui n'entendent pas la langue dans laquelle il est écrit[59]

As it turned out, the plea was in vain. But its very appearance in the *Voyages* points to the most obvious reason for Lambert's restricted influence. As Saussure evidently recognized, the *Pyrometrie* had a severely limited readership. Through choice or ignorance, few of Lambert's contemporaries in France and Britain read works in German. And even those who could read the language would not easily have come across the book : in France and Britain it was, and is, a great rarity[60]. The practical impediments to the circulation of Lambert's only other major work on heat – the 'De vi caloris' – were arguably less formidable, but even this paper was handicapped by appearing, in Latin, in the regrettably little-known *Acta helvetica.*

So I suggest that the inaccessibility of Lambert's writings on heat, compounded perhaps by the way in which knowledge of them was handed down through a somewhat localized research tradition – namely the type of meteorology pursued by Lambert's Swiss admirers – goes a long way towards explaining their lack of influence. Yet this is not the whole story. For there remains the possibility that the *Pyrometrie* might have survived the short-term neglect, to be resurrected long after its author's death as a depository of anticipations of later developments, of theories vindicated, quite independently, by the work of others. But the book was denied even this accolade, for reasons that are obvious enough. We need look no further than to the timing of its publication, which made its theoretical aspects obsolete almost from the moment they appeared.

It was the misfortune of the *Pyrometrie* to appear just two years after Lavoisier had begun expounding the doctrine which developed, over the next decade, into the full-blown caloric theory[61]. The effect of Lavoisier's successful advocacy of caloric was two-fold. First, it gave unprecedented authority to the view that the mere accumulation of a subtle fluid was sufficient to explain thermal phenomena. This is not to say, incidentally, that such a view originated with Lavoisier. An increase in the 'density' of fire had commonly been thought to be one cause of a rise in temperature, and, by the time he wrote the *Pyrometrie,* Lambert himself placed far more emphasis on the number of fire particles

(59) SAUSSURE, *Voyages,* op. cit. (note 53), vol. 2, p. 350. Pierre Prévost was another member of the Swiss school who acknowledged, albeit perfunctorily, the importance of Lambert's work in heat. In several places he follows Saussure in crediting Lambert with the discovery that the thermal effects of radiant heat are unrelated to the luminosity of the scource. See PREVOST, *Recherches physico-mécaniques sur la chaleur* (Geneva and Paris, 1792), p. 10 ; 'Quelques remarques sur la chaleur, et sur l'action des corps qui l'interceptent', *Philosophical transactions of the Royal Society of London, 92* (1802), pp. 403-47 (439) ; and *Du calorique rayonnant* (Paris and Geneva, 1809), pp. 3 and 87. For Lambert's work on radiant heat, see *Pyrometrie,* op. cit. (note 25), pp. 210-13.

(60) In Britain, for example, I know of only two copies, in the British Library and the library of the Royal Institution of Great Britain, London.

(61) Lavoisier's first public statement of his views on the 'matter of fire' appears in his paper 'De la combinaison de la matière du feu avec les fluides évaporables, et de la formation des fluides élastiques aëriformes', *Mémoires de l'Académie royale des Sciences,* (1777 ; published in 1780), pp. 420-32. He read the paper to the Académie on 5 September 1777.

in a given space, irrespective of their motion, than he had done in the 'De vi caloris'[62]. The second effect of the rise of Lavoisier's caloric theory was to orient the study of heat decisively towards chemistry and away from many of the areas of physics that interested Lambert. As the French chemists of the 1780s knew well, the great strength of caloric was that it gave a plausible explanation of heats of chemical reaction – a problem which Lambert and his immediate precursors and admirers did not touch upon. And if Lavoisier's theory was weak in areas where Lambert's was strong – for example, in the explanation of the heat of friction, which raised notorious problems for calorists long before Rumford made a rather obvious point in his cannon-boring paper in 1798[63] – such flaws, which had no bearing on chemistry, were easily banished to the periphery.

In short, it would be wrong to see the doctrines of fire of the kind that Lambert espoused foundering under the weight of Kuhnian-style anomalies, to be replaced by a new theory that was, comparatively, anomaly-free. For both theories had grave weaknesses. All that happened – and it happened to the detriment of Lambert's reputation – is that a shift in the orientation of the study of heat obscured the strengths of the fire-particle tradition, while pointing up the merits of caloric.

I would argue, therefore, that it is far from obvious that doctrines of fire had outlived their usefulness by the time they were abandoned about 1780 : confronted with a skilfully promoted and eye-catching rival, they just ceased to be taken seriously. It is a mark of their undimmed vitality in the 1770s that in 1775, on the very eve of the so-called chemical revolution, Joseph Black at last overcame his innate caution and sense of the intractability of the nature of heat sufficiently to give tentative endorsement to the fire-particle theory. Of all theories, he declared in a lecture at Edinburgh, that of the 'German philosophers', according to which hotness was caused by the 'tremulous motion' of a 'subtle matter', was 'most agreeable to Chymical experiments'[64]. Even more positive,

---

(62) Lambert's explanation of the changes in temperature associated with the rapid expansion and compression of air illustrates this point.

The wider context of the ousting of the traditional doctrines of fire by the caloric theory deserves far more attention than I have been able to give it in this paper. The primary literature on the subject is voluminous and largely unexplored. However, my reading leaves me in no doubt that an emphasis on the mere accumulation of 'fire' as the cause of a body's hotness began to rival the older notions in the 1770s ; see, in particular, my comments on the work of William Cleghorn, who took this view, in *Caloric theory*, op. cit. (note 2), pp. 8-9 and 12. Hence I see Lambert and even Lavoisier as endorsing rather than inaugurating new departures in the study of heat. As I argue in *Caloric theory*, pp. 14-19, the new attitudes probably owed a great deal to developments which had occurred, since the 1740s, in theories based on the fluids of electricity and magnetism.

(63) Benjamin THOMPSON (Count Rumford), 'An inquiry concerning the source of heat which is excited by friction', *Philosophical transactions of the Royal Society of London, 88* (1798), 80-102.

(64) The comment appears in a set of notes taken by an unidentified student at Black's lectures on chemistry at Edinburgh. The notes are preserved in Edinburgh University Library, as MS.Dc.3.11. In his *Lectures on the elements of chemistry*, ed. John Robison (2 vols., Edinburgh, 1803), vol. 1, p. 34, Black cautiously supports Cleghorn's theory and so turns his back on the theories involving the motion of fire. This probably represents Black's true opinion towards the end of his life, although Robison was a notoriously bad editor who was not above adjusting Black's views in order to present them in the best possible light. On Robison's editorial shortcomings, see Donald S.L. CARDWELL, *From Watt to Clausius. The rise of thermodynamics in the early industrial age* (London : Heinemann, 1971), pp. 41-2.

if less authoritative evidence of the high standing of the theory in the last years of the eighteenth century can be found in the third edition of the *Encyclopaedia Britannica*, published in 1797, where the author of the unsigned article 'Chemistry' confidently defines heat as 'a certain violent action of the elementary fire'[65]. And I leave until last the most celebrated of all latter-day supporters of the theory, Count Rumford. As Stephen Goldfarb has shown, when Rumford declared, in his famous cannon-boring paper of 1798, that heat was motion, he was referring not to the motion of the particles of gross matter, as has generally been supposed, but to motion in an all-pervading aether[66].

It would appear, therefore, that the tradition of theorizing in which I have placed Lambert did not perish completely with the rise of Lavoisierian chemistry. Indeed, Stephen Brush has argued that theories of heat involving the motion of an all-pervading aether underwent a modest revival in the 1830s and 1840s[67]. But by then the roots of such theories in the eighteenth century were quite forgotten, and in any case the revival was short-lived. With the emergence of thermodynamics in the 1850s, the tradition sank without trace, and with it went a not undistinguished literature, including the *Pyrometrie*.

In conclusion, I find it hard to ascribe a significant, distinctive influence to Lambert in the study of heat. It was his misfortune to contribute to the elaboration of a theory which Lavoisier condemned to obscurity in that first paper on the 'matter of fire' which he read to the Académie des Sciences on 5 September 1777, just three weeks before Lambert's death[68]. Of course, the *Pyrometrie* contained much that did not stand or fall by the existence of fire particles : potentially, the experimental work on thermometry was of lasting value (though, here too, developments in the later eighteenth century, in particular the general adoption of the Celsius and Fahrenheit scales, to the effective exclusion of all others, soon made Lambert's contribution redundant). But in science, bathwater and babies tend to go out together, and this, I believe, was the case with Lambert.

If I appear to be relegating Lambert to the periphery, therefore, I do so because of his slender influence on the future course of the study of heat, not for any intrinsic weaknesses in his work. Hence my assessment in no way impugns his prescience : it is

---

(65) *Encyclopaedia Britannica*, vol. 4 (1797), p. 388. Cf. a similar view in the article 'Heat', ibid., vol. 8, pp. 350-2.

(66) Stephen J. GOLDFARB, 'Rumford's theory of heat : a reassessment', *The British journal for the history of science, 10* (1977), pp. 25-36. Rumford's famous but enigmatic comment that heat is 'motion' appears on p. 99 of the paper cited in note 63.

Whereas Goldfarb attaches special importance to Rumford's reading of Boerhaave as the catalyst for his views on heat, I am more inclined to place Rumford in the continuing tradition of speculation about fire which I have outlined in this paper. Indeed, Rumford may even have been one of the very few heat theorists to be influenced by Lambert ; see above, note 58, on the translation of passages from the *Pyrometrie* which he prepared for Fourier.

(67) Stephen G. BRUSH, 'The wave theory of heat : a forgotten stage in the transition from the caloric theory to thermodynamics', *The British journal for the history of science, 5* (1970-1), pp. 145-67.

(68) See above, note 61.

historiographically not a very sophisticated point, but Lambert repeatedly broached ideas, even discoveries,which were later attributed to others. Similarly my comments do not deny Lambert an honoured place in the broad movement towards an experimentally based mathematical theory of heat — a movement that gained great force in his lifetime and matured in the nineteenth century in the work of Fourier, among others. And, finally, I do not underestimate Lambert's experimental dexterity and the boldness of his theorizing in an area of physics that was prone to vacillation. These are qualities which, for me, mark Lambert as a physicist of exceptional gifts and originality. They amply endorse John Leslie's high opinion of his 'ardent and fertile genius'[69].

(69) LESLIE, Experimental inquiry, op. cit. (note 31), p. 405.

VIII

THE BACKGROUND TO THE DISCOVERY OF DULONG AND PETIT'S LAW*

1. Introduction

THE years immediately after the final downfall of Napoleon Bonaparte could easily have been years of anti-climax in French science. In 1815, after two decades of undoubted greatness, the time, I feel, was ripe for decline. And decline might well have occurred if the traditions and the style of science as practised in France in the period of Napoleon's rule had been carried on unchanged by the disciples of the two great men who had dominated work in the physical sciences for so many years. These men, of course, were the chemist Claude Louis Berthollet and the mathematician and physicist Pierre Simon Laplace.

As it happened, science in France did not proceed in the same way after 1815 as it had done before. In fact, the period 1815-20 was one of great and significant intellectual change, change which, at least temporarily, seems to have averted any process of decline. Of course, there were those who resisted change and who remained loyal to the teachings of Berthollet and Laplace on such important questions as the nature of heat and the atomic theory. But the fact remains that there was also at this time a small, though highly influential, group of gifted young men who felt the need to break with a number of the traditional doctrines which they themselves had assimilated in their early scientific education. Of these men, two of the most prominent were Pierre Louis Dulong and Alexis Thérèse Petit.

In view of these comments, it is hardly necessary to state that I shall be concerned in this paper to do more than merely elucidate the process by which Dulong and Petit discovered their famous law in 1819. Certainly this process is of considerable interest and it is duly examined in section 2. In later sections, however, I shall try to show that the work of Dulong and Petit has a far wider historical significance than that associated simply with the discovery of their law. In particular, I shall suggest that their work can throw light both on the general state of French science at what I believe to have been a crucial period (1815-20) and, more specifically, on two matters of great concern to historians of early nineteenth-century science. These matters are:

(i) The fortunes of the atomic theory in France at a time when the influence of Berthollet was beginning to be less strongly felt.

(ii) The way in which the generally held opinion that the phenomena

* An abbreviated version of this paper was read at the summer meeting of the Society in Oxford on 2 July 1966. The author is indebted to Dr. J. R. Ravetz for some valuable discussions during the early stages of the preparation of this paper.

of heat were to be attributed to a highly elastic and weightless fluid, known as caloric,[1] came to decline, giving way eventually to what we may term the "vibrational" theory, according to which heat consisted in the vibrations of the particles of ordinary ponderable matter.

2. The origins of the law

A good deal of uncertainty surrounds the way in which Dulong and Petit's law was discovered. If we are to believe an account given by the great French chemist Jean-Baptiste Dumas in 1881, it would seem that the discovery owed almost as much to François Arago as it did to the two men who normally take the credit for it. Dumas wrote:

"Le lundi 5 avril 1819, date mémorable, Petit . . . montrait, en confidence, à son beau-frère Arago,[2] un chiffon de papier, sur lequel se trouvaient inscrits les rapports selon lesquels les corps simples se combinent, et les quantités de chaleur exigées par chacun d'eux pour s'échauffer d'une manière égale sous le même poids. Au premier aspect, c'était le désordre; mais, en multipliant pour chacun de ces corps les deux chiffres l'un par l'autre, tous les produits se trouvaient égaux. Une heure après, l'illustre secrétaire perpétuel,[3] convaincu que Dulong, toujours hésitant, pourrait s'opposer à la divulgation de cette belle loi, en entretenait ses confrères, par une indiscrétion calculée.[4] Huit jours plus tard, les deux collaborateurs l'énonçaient devant l'Académie elle-même . . ."[5]

And so it was, according to Dumas, that on 12 April 1819 Dulong appeared before the *Académie des Sciences* in Paris to read the short joint paper[6] in the course of which he and Petit announced their conclusion that "Les atomes de tous les corps simples ont exactement la même capacité pour la chaleur".[7]

Dumas's recollection, written as it was over 60 years after the event, is inevitably suspect, the more so since Dumas, who was 18 at the time and working in Geneva, would not have been present.[8] Doubt concerning its accuracy also arises from the fact that no other reference to the crucial

[1] The term "calorique" for the fluid of heat was first introduced in L. B. Guyton de Morveau, A. L. Lavoisier, C. L. Berthollet and A. F. de Fourcroy, *Méthode de Nomenclature Chimique* (Paris, 1787), 31.
[2] Petit married Arago's sister in November 1814.
[3] In fact Arago became permanent secretary of the *Académie des Sciences* only in 1830.
[4] For evidence of Dulong and Petit's fear of plagiarism see Berzelius's letter to Alexandre Marcet, 27 April 1819, in *Jac. Berzelius Bref*, ed. H. G. Söderbaum (Uppsala, 1912-32), i, Part 3, 193. This is cited hereafter as Berzelius, *Bref*.
[5] J. B. A. Dumas, "Éloge historique de Henri-Victor Regnault", *Mém. Acad. Sci.*, xlii (1883), p. xlviii.
[6] A. T. Petit and P. L. Dulong, "Sur quelques points importants de la théorie de la chaleur", *Ann. Chim. Phys.*, x (1819), 395-413. Dulong, who seems to have overcome any hesitation he may have felt by 12 April, is identified as the reader in Académie des Sciences. *Procès-verbaux des Séances de l'Académie tenues depuis la fondation de l'Institut jusqu'au mois d'août 1835* (Hendaye, 1910-22), vi, 437. This is cited hereafter as *Procès-verbaux*.
[7] Petit and Dulong, *op. cit.* (6), 405.
[8] All biographers of Dumas state that he came to Paris for the first time in 1823, some three years after Petit's death. This strongly suggests that Dumas's account of the discovery of Dulong and Petit's law originated with Arago.

encounter between Petit and Arago has come to light. Indeed, the only additional evidence, that of Dulong and Petit themselves, gives a somewhat different view of the matter, for their paper of 12 April makes no mention either of the unexpectedness of the discovery or of any debt to Arago. The first paragraph implies rather that they had been working all along in the belief that certain natural laws would appear in their simplest form once they could be related to the properties of individual atoms and not, for example, of equal weights or volumes.[9] Written as an introduction to one of the most striking pieces of evidence which had yet been obtained for the truth of the atomic theory, the story was a convincing one; but it was all too easy to write in this vein after the discovery had been made and it is clear that the accounts of neither Dumas nor Dulong and Petit are to be trusted without further examination. On the important question whether the law emerged by chance or whether it was the natural consequence of a planned research programme,[10] they unfortunately provide conflicting evidence and it is with this problem particularly in mind that we turn now to examine the work of Dulong and Petit in the years immediately preceding their discovery.

In their earlier writings it would be difficult to find any ground for believing that Dulong and Petit's interest in specific heats began in the way which they themselves suggested. They appear to have been concerned not with the investigation of the properties of matter on the atomic scale but with some of the familiar problems of contemporary physics. The traditional nature of their early interests is emphasized above all by the fact that the initial incentive for their work on specific heats almost certainly came from an academic prize competition. The competition in question was announced by the First Class of the French Institute[11] on 9 January 1815.[12] Competitors were asked to measure the expansion of mercury in a thermometer between 0°C and 200°C and then to determine the rate at which a body cooled both in a vacuum and in certain specified gases (air, hydrogen and carbon dioxide) at different temperatures and pressures. It seems likely that the subject was set

[9] Such an opinion would not have been entirely original. In Thomas Thomson's *Annals of Philosophy*, ix (1817), 428, J. B. Emmett, a minor English writer, had already proposed that specific heats should be measured not for equal weights or volumes but for weights in proportion to the atomic (or molecular) weights of the substances concerned, but there is no evidence that Dulong and Petit knew of this. They certainly did know, however, of a result somewhat similar to their own law which had been derived by John Dalton in 1808, but any possibility that Dalton was a precursor is easily ruled out (see below, section 7).

[10] The latter is the view taken by Charles Laurens in the most detailed of the existing biographies of Dulong. See J. Girardin and C. Laurens, *Dulong de Rouen. Sa Vie et ses Ouvrages* (Rouen, 1854), 67.

[11] Despite the first restoration of the Bourbon monarchy in 1814, it was not until March 1816, after the second restoration, that the title *Académie Royale des Sciences* was restored.

[12] The subject appears on page 2 of a pamphlet announcing the subjects for prizes to be awarded by the First Class of the Institute in 1816 and 1817. The pamphlet was printed for the public meeting held on 9 January 1815. An announcement also appeared in J. N. P. Hachette, *Correspondance sur l'École Polytechnique*, iii (issue for May 1815), 250.

4

primarily to answer problems raised by the competition of 1811 in which Joseph Fourier had won the prize for his theoretical treatment of heat conduction in solids.[13] Although it was fundamental to his theory that the rate of flow of heat from a point inside a solid bar to another either inside or outside the bar was proportional to the difference in temperature between the points, an assumption which he based tenuously on Newton's law of cooling, Fourier had taken good care to emphasize that his results could be modified easily, should Newton's law ever be proved inexact.[14] In fact, it had long been thought that the law was only approximate for any but the smallest temperature differences,[15] so that a determination of its precise form became an obvious subject for research once Fourier's work had drawn new attention to the problem.

That it was the resulting prize competition which brought about the collaboration between Dulong and Petit seems in little doubt, although the point cannot be proved conclusively. At the time the subject was announced, in January 1815, Dulong and Petit would have found the prospect of victory in the competition extremely attractive. Although they had already achieved some distinction, both men were young, being 29 and 23 years old respectively, and neither of them was a member of the Institute (a fact which made them eligible to compete). Of the two Petit[16] had the more distinguished academic record. After completing the entrance requirements for the *École Polytechnique* at the remarkably early age of $10\frac{1}{2}$, he came to the notice of the mathematician and engineer J. N. P. Hachette, who invited him to a school in Paris run by teachers from the *École*. Here Petit filled in the time before reaching the statutory age for entry (16). In 1809 he graduated from the *École* itself with high honours, being placed "*hors de ligne*", with the next student in the year designated "first". In the years which followed, his merit did not go un-

[13] Fourier's paper appeared (in two parts) only in 1824 and 1826, as "Théorie du mouvement de la chaleur dans les corps solides", *Mém. Acad. Sci.*, iv (1819-20), 185-555, and v (1821-22), 153-245. For comments on the remarkable delay in publication, which seems to have been the result of criticisms made by the influential judges for the competition (Laplace, Lagrange, Legendre, Malus and Haüy), see G. Darboux in *Oeuvres de Fourier* (Paris, 1888), i, pp. vi-viii, and D. F. J. Arago, "Éloge historique de Joseph Fourier", *Mém. Acad. Sci.*, xiv (1838), cxii-cxiii.

[14] Fourier, *Mém. Acad. Sci.*, iv (1819-20), 202-203. The assumption that Newton's law could be extended from the case of a hot body losing heat in a cooler surrounding medium to the case of heat flow between two points in a solid is, of course, unfounded. Fourier's cursory and quite inadequate attempt to justify the assumption appears on pp. 200-201 of the first part of his paper as published in the Academy's *Mémoires*.

[15] Although much of the evidence against the law appears inconclusive to modern eyes. See G. Martine, *Essays Medical and Philosophical* (London, 1740), 233-247; J. C. P. Erxleben, *Novi Commentarii Societatis Regiae Scientarum Gottingensis, commentationes physicae et mathematicae classis*, viii (1777), 74-95; J. Dalton, *A New System of Chemical Philosophy* (Manchester, 1808), i (Part 1), 12 and 108-123. The work with the greatest influence in France, however, was probably that of François Delaroche, published in the *Journal de Physique*, lxxv (1812), 201-228. Although Delaroche's paper was read to the Institute on 3 June 1811, it was not published until September 1812 and so was in all likelihood unknown to Fourier, who was working in Grenoble, at the time he submitted his entry on 28 September 1811.

[16] For biographical details see J. B. Biot, "Notice historique sur M. Petit", *Ann. Chim. Phys.*, xvi (1821), 327-335. This biographical sketch also appeared in *Journal de Physique*, xcii (1821), 241-248.

noticed and by the beginning of 1815 he was already teaching physics at the *École Polytechnique* as a replacement for his former professor, J. H. Hassenfratz, who had been "invited to resign" by the Minister of the Interior in 1814, ostensibly for reasons of health and age, although it is hard to believe that memories of Hassenfratz's earlier Jacobin associations did not play at least some part in the matter.[17] Unhappily the full promise of what would surely have been a most brilliant career was never realized, for in June 1820, when he was not yet thirty, Petit died from tuberculosis, a grave loss to French science in general and to his close friend Dulong in particular.

Dulong's rise to prominence had been less spectacular. His student days at the *École Polytechnique* had been dogged by ill health and within a year of his admission in 1801 he was granted a month's leave of absence on three occasions, before leaving without completing the course in October 1802.[18] He then practised medicine in Paris for some years, to the serious detriment of his finances, and it was not until he came to the notice of Berthollet about 1810 and was invited to work in the latter's celebrated private laboratory at Arcueil,[19] near Paris, that he began to make his mark as a chemist, notably by his discovery of nitrogen trichloride[20] and by some important work on metal oxalates.[21] By 1815 he held teaching posts at both the *École Normale* in Paris and the veterinary school at Alfort, but he would probably still have regarded the select group which made up Berthollet's circle at Arcueil as the chief focus for his intellectual activities, although regular meetings of the circle had ceased by this date.

Since Petit does not appear to have worked at Arcueil and since Dulong, at least until 1820, held no post at the *École Polytechnique* other than that of examiner for the graduates going on into the public services, it is not certain how the two men first came into contact, although it is probably significant that they numbered Arago among their mutual friends. The first evidence of their association is the joint paper on expansion and temperature measurement which they presented before the

---

[17] See *Biographie Nouvelle des Contemporains*, ed. A. V. Arnault, A. Jay, E. Jouy and J. Norvins (Paris, 1823), ix, 57 (article "J. H. Hassenfratz"). The reasons for Hassenfratz's resignation which have been cited here are given by the Director of Studies at the *École* in a MS. note dated 20 October 1815. Another note, in the report of the *Conseil de Perfectionnement* of the *École* and dated 30 October 1815, states that Petit took over from Hassenfratz on 1 September 1814. Both of these items are in the archives of the *École Polytechnique* and were consulted on my behalf by the librarian, Monsieur A. Moreau.

[18] From information on Dulong's record card at the *École Polytechnique*. For other biographical details see Girardin and Laurens, *op. cit.* (10).

[19] For a detailed account of the society which met at Arcueil and of the work which was performed there see M. P. Crosland, *The Society of Arcueil* (London, 1967).

[20] P. L. Dulong, "Mémoire sur une nouvelle substance détonnante", *Mémoires de Physique et de Chimie de la Société d'Arcueil*, iii (1817), 48-63.

[21] See pp. cxcviii-cc of Georges Cuvier's report on the work of the First Class for the period 1813-15 (physical sciences) in *Mémoires de la Classe des Sciences Mathématiques et Physiques de l'Institut de France* (1813, 1814, 1815).

6

Institute on 29 May 1815,[22] and it is above all the nature of the problems
tackled in this paper which indicates that they were working with the
Institute's prize competition in mind. Since the paper also suggests how
they were led to concern themselves with specific heat measurements, it
is of special interest for our purpose. In it Dulong and Petit pointed out
that a determination of the precise form of the law of cooling could not
be made without some clear definition of temperature or, better still,
the establishment of a rational temperature scale, and although, not
surprisingly, they made little progress towards the ideal of rationality in
temperature measurement, they at least did something to clarify the
problem. They recognized that it was not enough merely to relate the
expansion of the thermometric substance to the quantities of heat which
it absorbed, for even if a substance were found which expanded regularly
with the addition of equal amounts of heat, it could still not be assumed
that these equal increases in volume corresponded to the same temperature increment at all points on the scale. Only if the specific heat of the
substance were shown to be independent of temperature would this
conclusion be justified. The status which Dulong and Petit would have
given to such determinations of specific heat made *before* the establishment
of a rational temperature scale is far from clear, but it was the experimental and not the theoretical difficulties which they cited as the greatest
obstacle in their work.[23] The measurement of specific heats at high temperatures in particular was seen as presenting very serious problems. No
details of any experiments on specific heat were given, however, and their
paper consisted of little more than a careful comparison of the expansive
properties of mercury, glass, air and certain metals. It hardly justified
their strong preference for the gas thermometer as an ultimate standard,
a preference which they based on Joseph Louis Gay-Lussac's well-known
demonstration that all gases expanded to the same extent for any given
rise in temperature.[24]

Owing to a number of unspecified interruptions,[25] possibly connected
with the political unrest of 1815[26] or with Petit's promotion to the post of
full professor (*professeur titulaire*) of physics at the *École Polytechnique* in
October of that year,[27] the experiments were discontinued and in July
1816 Dulong and Petit published their paper in its original form, having
apparently given up any intention they might have had of entering for

[22] P. L. Dulong and A. T. Petit, "Recherches sur les lois de dilatation des solides, des
liquides et des fluides élastiques, et sur la mesure exacte des températures", *Ann. Chim. Phys.*, ii
(1816), 240-263. Dulong was the reader (see *Procès-verbaux*, v, 514).
[23] Dulong and Petit, *op. cit.* (22), 241-242.
[24] *Ibid.*, 243 and 263. For Gay-Lussac's work see *Ann. Chim.*, xliii (1802), 137-175.
[25] Dulong and Petit, *op. cit.* (22), 263n.
[26] On the effect of the unrest see G. Pinet, *Histoire de l'École Polytechnique* (Paris, 1887),
74-102 and Berthollet's letter to Berzelius, 27 August 1815, in Berzelius, *Bref*, i, Part 1, 54-55.
[27] A. Fourcy, *Histoire de l'École Polytechnique* (Paris, 1828), 331.

the competition.[28] Fortunately the judges announced in March 1817 that no entry of sufficient merit had been received[29] and that the same subject would be set again for the following year.[30] This time Dulong and Petit produced a masterpiece,[31] which won the prize and which was to be acclaimed as a model of experimental method by such varied authorities as Auguste Comte, Siméon Denis Poisson, Gabriel Lamé and William Whewell[32] (among many others). Again they began with the problem of thermometry and again they came down in favour of the air thermometer, convinced that increments in temperature measured on this instrument, or indeed on any gas thermometer, were increments in the true temperature. This was no more than they had said in 1815, of course, but they had now made great progress with the experimental determination of specific heats, acquiring data which were still important, even after their choice of a temperature scale had been made, in the task of interpreting observed changes in the temperature of a cooling body in terms of quantity of heat lost. Using a method of mixtures, they had succeeded in measuring the specific heats of several solids (including five metallic elements) and of mercury over a number of different temperature ranges up to 350°C. So accurate were the results obtained that the Dulong and Petit law could easily have been deduced from them,[33] but the very fact that no such conclusion was drawn demonstrates almost conclusively that Dulong and Petit had no interest in the specific heats of individual atoms at this stage and that, even in 1818, they still regarded their work on specific heats as doing little more than provide a tool which could be applied in other more established problems such as cooling.

Roughly a year separated the announcement of Dulong and Petit's

[28] The decision to publish at this stage was almost certainly connected with Dulong's (unsuccessful) candidature for a place in the physics section of the First Class. See Crosland, *op. cit.* (19), 167.

[29] A decision which was fully justified, if we are to judge by the two insubstantial entries which have survived in the archives of the *Académie des Sciences*.

[30] *Procès-verbaux*, vi, 164 (17 March 1817) and *Ann. Chim. Phys.*, iv (1817), 302-303. In fact the wording of the subject was slightly modified for the new competition. It was now specified that the observations of the mercury-in-glass thermometer should be made "comparativement à la marche du thermomètre à air" and between 20°C and 200°C rather than between 0°C and 200°C. Otherwise the wording was identical.

[31] P. L. Dulong and A. T. Petit, "Recherches sur la mesure des températures et sur les lois de la communication de la chaleur", *Ann. Chim. Phys.*, vii (1818), 113-154, 224-264 and 337-367. That the importance and high quality of the work were immediately recognized in England as well as in France is apparent from the fact that Thomson's *Annals of Philosophy*, xiii (1819), contained not only a full translation but also a lengthy, and generally favourable, criticism by Thomson himself (on pp. x-xiv, xvi-xviii and xxi-xxiii).

[32] A. Comte, *Cours de Philosophie Positive* (Paris, 1835), ii, 534; S. D. Poisson, *Théorie Mathématique de la Chaleur* (Paris, 1835), 6; G. Lamé, *Cours de Physique* (Paris, 1836), i, pp. i-ii; W. Whewell, *A History of the Inductive Sciences* (London, 1837), ii, 485.

[33] If Dulong and Petit had used Berzelius's most recent values for atomic weights, as given for example in Thomson's *Annals of Philosophy*, iii (1814), 362-363, they would have found that the atomic heats of no fewer than five of the seven metals whose specific heats they had determined were very nearly constant. An equally striking result would have been obtained if they had used John Dalton's figures given in the latter's *New System* (Manchester, 1810), i (Part 2), 546.

victory in the competition in March 1818[34] and the discovery of their law. In this period they used a refined method of cooling to extend their determinations of the specific heats of solids, but there is still nothing to suggest that they were looking for a relationship of the type which emerged in April 1819, so that essentially, if not in matters of detail, Dumas's account would appear to be substantiated. We may not accept that Petit collated specific heats and atomic weights in the way that Dumas described or that he met Arago, but of the unexpectedness of the discovery there can be little doubt.

CHALEURS SPÉCIFIQUES (1).		POIDS RELATIFS des atomes (2).	PRODUITS du poids de chaque atome par la capacité correspondante.
Bismuth,	0,0288	13,30	0,3830
Plomb,	0,0293	12,95	0,3794
Or,	0,0298	12,43	0,3704
Platine,	0,0314	11,16	0,3740
Etain,	0,0514	7,35	0,3779
Argent,	0,0557	6,75	0,3759
Zinc,	0,0927	4,03	0,3736
Tellure,	0,0912	4,03	0,3675
Cuivre,	0,0949	3,957	0,3755
Nickel,	0,1035	3,69	0,3819
Fer,	0,1100	3,392	0,3731
Cobalt,	0,1498	2,46	0,3685
Soufre,	0,1880	2,011	0,3780

Fig. 1. Dulong and Petit's table of results from Ann. Chim. Phys., x (1819), 403. The scales used in columns 1 and 2 are such that the specific heat of water and the atomic weight of oxygen are unity.

The boldness and simplicity of Dulong and Petit's statement of their law deserves comment in view of the evidence from which it was derived. In arriving at it, Dulong and Petit used the highly reputable figures for atomic weight recently established by the Swedish chemist Jöns Jacob Berzelius,[35] but they found it necessary to modify Berzelius's values in the case of no fewer than eleven of the thirteen elements for which they gave results. Berzelius's atomic weights in eight cases were halved and for three other elements different simple fractions of his

---

[34] *Procès-verbaux*, vi, 292 (2 March 1818).
[35] These are most easily referred to in J. J. Berzelius, *Essai sur les Proportions Chimiques* (Paris, 1819), a translation of part of his *Lärbok i Kemien* (Stockholm, 1818), iii.

figures were used. Only for sulphur and platinum was modification thought to be unnecessary.[36] That Dulong and Petit could feel justified in making changes of this sort at all reflects the unreliability of existing determinations of atomic weight based on the combining weights in chemical reactions, an unreliability which resulted above all from the uncertainty surrounding the numbers of different elementary atoms which combined to form any given "compound atom" (or molecule).[37]

3. *Objections to the caloric theory*

From the circumstances which led to the discovery we proceed now to consider the significance of Dulong and Petit's work in the broader context of early nineteenth-century science. It will be argued that the paper of 1819 contained far more than the announcement of a now familiar law and that the attacks on established opinion which it delivered were seen as being no less important, both by Dulong and Petit themselves and by their contemporaries. The attacks were directed on two major fronts—the nature of heat and the atomic theory. As far as heat was concerned, Dulong and Petit rejected the caloric theory in all its forms, although it was only with respect to what may be termed the "Irvinist" version of the theory[38] that their evidence was really convincing. The principles on which this version was based originated with the Scottish chemist William Irvine, a talented pupil of Joseph Black who lectured at Glasgow University from 1766 until his death in 1787. Although Irvine seems to have been reluctant to commit himself, at least openly, on the question of the nature of heat,[39] he had maintained that in unit mass of any substance at a given temperature the quantity of heat (whether it was

---

[36] In their choice of the elements whose atomic weights should be left unchanged, Dulong and Petit were probably influenced by Berzelius, who had long been convinced that the molecular formulae of compounds containing oxygen were analogous to those in which the oxygen was replaced by sulphur. Thus the quantities of oxygen and sulphur in pairs of analogous compounds yielded atomic weights for those elements which were thought not to be subject to the usual uncertainty concerning the numbers of different atoms in compounds. On this see especially Berzelius's letter to Berthollet in *Ann. Chim.*, lxxvii (1811), 68. Although I have assumed here that Berzelius's value for the atomic weight of platinum was not modified for the purposes of the 1819 paper, it should be pointed out that the precise figure used by Dulong and Petit is in some doubt. In their table of results (reproduced opposite) the weight for platinum appears as 11.16, whereas in his *Essai* Berzelius gives it as 12.15 (the atomic weight of oxygen on this scale being unity). Since the product 11.16×0.0314 is in fact 0.3504 and not 0.3740, as stated in the last column of the table, it is clear that the figures given for platinum are not to be relied upon and that some error, possibly involving a misprint of the atomic weight of platinum, was made. For a discussion of some possible explanations of this curious mistake see J. W. Van Spronsen, "The history and prehistory of the law of Dulong and Petit as applied to the determination of atomic weights", *Chymia*, xii (1967), 165. The point remains, however, that a simple fraction of Berzelius's atomic weight for platinum was certainly not used.

[37] On the importance of Dulong and Petit's law in removing this doubt see note 86.

[38] For a detailed account of the "Irvinist" doctrines, with full references to the work of the leading "Irvinists" and an illustration (from John Dalton's *New System of Chemical Philosophy*) of the cylinders of caloric referred to below, see my paper, entitled "Dalton's caloric theory", in D. S. L. Cardwell (ed.), *John Dalton and the progress of science* (Manchester 1968), 187-201.

[39] *Essays, chiefly on Chemical Subjects by the late William Irvine, M.D., F.R.S.Ed., and by his son William Irvine, M.D.* (London, 1805), 71.

a fluid or a vibration) was proportional to the specific heat of the substance. The idea is conveniently illustrated by the analogy of a cylinder of uniform cross-section filled with a liquid, where the liquid in the cylinder represents the heat in a body and where the height of the surface of the liquid above the bottom of the cylinder indicates the (absolute) temperature. It can be seen that any decrease in the cross-section of the cylinder (i.e. in specific heat) would cause the level of the liquid (or temperature) to rise, unless heat could escape readily to the surroundings, and it was to sudden changes of this type that the Irvinists, among the most notable of whom were the Irish-born physician Adair Crawford, and John Dalton, attributed heats of chemical reaction and the latent heats of vaporization and liquefaction. That the Irvinist view was extremely vulnerable to experimental refutation is obvious, since just one instance of an exothermic reaction in which the total heat capacity of the reactants was not greater than that of the products was sufficient to make it quite untenable. Hitherto the difficulty and unreliability of experiments designed to determine specific heats had provided a measure of protection against critics, but Dulong and Petit were now convinced, apparently as a result of experiments of their own, that heats of reaction and changes in specific heat were not related and that "in most cases" the release of heat in a chemical reaction was not accompanied by an over-all decrease in heat capacity. Their line of attack, we should note, was not original, but the introduction of new experimental evidence made the argument still more effective than it had been, for example, in the hands of François Delaroche and Jacques Étienne Bérard in 1812.[40]

The evidence against the other main version of caloric theory, a version associated principally with the names of Antoine Lavoisier and Laplace,[41] was less decisive. In this case no relationship between heats of reaction and specific heat was expected and it was assumed that temperature was determined not by the *total* quantity of heat in a body but only by a part of it, which was termed its "sensible" heat. The rest of the heat content was made up of "latent" heat,[42] which was incapable

[40] F. Delaroche and J. É. Bérard, *Ann. Chim.*, lxxxv (1813), 169-176. This paper by Delaroche and Bérard won the Institute's prize competition in physics for 1812.

[41] The main sources for the opinions of Lavoisier and Laplace are their "Mémoire sur la chaleur", *Mém. Acad. Sci.* (1780), 355-480, and their "3e mémoire . . . contenant les expériences faites sur la chaleur, pendant l'hiver de 1783 à 1784", in Lavoisier's *Mémoires de Chimie* (Paris, 1805), 121-147. The latter paper was written in 1793. In the former paper (see especially p. 358) Lavoisier and Laplace were careful not to commit themselves on the question of the nature of heat, although Laplace later became a convinced calorist. In Britain, Joseph Black was usually cited as the leading exponent of the view that bodies contained both latent and sensible heat and in his *Lectures on the Elements of Chemistry*, ed. J. Robison (Edinburgh, 1803), i, 193-197, he appears as a critic of the Irvinist doctrine that changes in specific heat could *cause* the emission or absorption of heat. But, despite his undoubted priority with regard to much in the opinions of Lavoisier and Laplace on heat, Black is of less importance here, concerned as we are with an episode in French science.

[42] "Sensible" and "latent" were just two of a number of terms in use at this time to denote the states of caloric. In the second of the papers cited in note 41, for example, Lavoisier and Laplace used the adjectives "interposé" and "combiné".

of affecting a thermometer or, indeed, of being detected in any way at all, although it was generally thought to bear some unknown relationship to the volume of a body. The result of a decrease in volume, for example, was supposed to be a decrease in the quantity of latent heat which the body could contain and a consequent conversion of some heat from the latent to the sensible state, which was manifested as a rise in temperature. Applying this principle in 1793 to the case of combustion,[43] Lavoisier and Laplace had shown how it led to the conclusion that the quantity of heat evolved in any such reaction would be largest when the resulting oxide was a solid, since oxygen would then have undergone the greatest possible decrease in volume. This view was not wholly borne out in their experiments, but to account for the discrepancies it was necessary simply to assume that in the process of combustion oxygen was capable of retaining varying quantities of heat according to the nature of the oxide which was formed. The solution was an arbitrary one without any independent justification and it was above all at this arbitrariness that Dulong and Petit directed their attack, although they had little more evidence than was available to, and used by, Lavoisier and Laplace themselves. So here, as in their criticisms of the Irvinist theory, their case was hardly novel and the conviction with which they stated it almost certainly owed a good deal to the fact that a quite different and (to them) more acceptable explanation of chemical heat was now to hand. The origin of this, although the fact was never fully acknowledged,[44] was very probably Berzelius.

4. *The electrochemical theory*

Berzelius was in France from 23 August 1818 until 10 July 1819. At Berthollet's instigation, he and Dulong worked together on the composition of water in the laboratory at Arcueil[45] and there the two men struck up an extremely close friendship ended only by Dulong's death in 1838. By the time they first met, in September 1818, Berzelius had been known for some years as a critic of the caloric theory of chemical heat. In 1808 he had adopted the "Lavoisier and Laplace" version of the theory[46] but by 1811 he had rejected it, principally on the ground that he had observed no decrease in volume during the exothermic reaction between sulphur and copper,[47] and he had substituted the view, suggested

---

[43] Lavoisier, *Mémoires de Chimie*, 138-147.
[44] Dulong and Petit made only one brief reference to Berzelius in their paper, on p. 412. Berzelius seems to have taken no part in the experimental work (see Berzelius's letter to Alexandre Marcet, 27 April 1819, in Berzelius, *Bref*, i, Part 3, 193).
[45] The result was a joint paper by the two men in the *Ann. Chim. Phys.*, xv (1820), 386-395.
[46] In J. R. Partington, *A History of Chemistry* (London, 1964), iv, 168, it is stated that Berzelius adopted Lavoisier's theory of chemical heat in volume i of the first edition of his *Lärbok i Kemien* (Stockholm, 1808), but a copy of the latter work has not yet been traced.
[47] L. W. Gilbert's *Annalen der Physik*, xxxvii (1811), 278-280, and *Annales de Chimie*, lxxix (1811), 249-251. *Cf.* the similar criticism made independently by Augustin Fresnel in a letter to his brother Léonor, 5 July 1814, in *Oeuvres Complètes d'Augustin Fresnel* (3 vols., Paris, 1866-70), ii, 820-821.

earlier by Humphry Davy,[48] that the heat was electrical in origin. It was already well known that an electrical discharge could give rise to heat even though no chemical reaction occurred and it was therefore a short step from Davy's demonstration that a piece of graphite was heated, even in the absence of oxygen, by being placed across the poles of a cell,[49] to the view that it was an electrical discharge between the oppositely charged reactants entering into combination which caused the emission of heat. Berzelius's opinion was unchanged in 1818 when he published an extended account of his electrochemical theory and with it a lengthy criticism of both the main versions of the caloric theory of heats of reaction,[50] a criticism which he based on arguments remarkably similar to those adopted by Dulong and Petit in the following year. It should be emphasized that Berzelius's opinions on chemical heat could be adopted without necessarily abandoning belief in the fluid nature of heat. Thus, even after he had expressed his support for the electrochemical theory, Berzelius was still able to write (in 1812) in a manner which strongly implied his conviction that heat was a fluid,[51] although some far from complimentary notes on Davy's *Elements of Chemical Philosophy*, which he communicated to the author soon after the book appeared in 1812, suggest that he might have preferred not to commit himself on the matter. Referring to Davy's exposition of a somewhat unusual form of the vibrational theory,[52] he wrote:

"Je suis intimement convaincu qu'il n'existe aucune hypothèse qui explique suffisament les phénomènes du calorique. Donc, plutôt que de tromper nos lecteurs par des apparences imparfaites, avouons notre ignorance à cet égard . . ."[53]

The blow dealt to caloric by the growth of the electrochemical theory, although indirect, was nevertheless a serious one. Certainly the hypothesis of the electrical origin of heats of reaction did not in itself either exclude the fluid theory or imply the vibrational nature of heat, but one of caloric's greatest strengths had been precisely that it provided such a convincing explanation of chemical heat. Once an alternative explanation was accepted, then the whole question of the nature of heat would become a completely open one again, with the fluid and vibrational theories fighting, so to speak, on equal terms. It was in 1819 that this alternative became really well known in France, initially through the paper by Dulong and Petit and a little later, in June, by the publication of Berzelius's own views on the matter in a readily available form in his *Essai sur les*

48 *Phil. Trans.*, xcvii (1807), 42-44.
49 *Ibid.*, xcix (1809), 71-72.
50 Berzelius, *Essai*, 58-73. This is a translation, with only minor modifications, of pp. 49-63 of vol. iii of the *Lärbok*. See note 35 for bibliographical details.
51 J. S. C. Schweigger's *Neues Journal für Chemie und Physik*, vi (1812), 139-141. These comments appeared in French in *Ann. Chim.*, lxxxvi (1813), 168-171.
52 Described in H. Davy, *Elements of Chemical Philosophy* (London, 1812), 95-96.
53 Berzelius to Davy, n.d., in Berzelius, *Bref*, i, Part 2, 41.

Proportions Chimiques.[54] Just how important the *Essai* was seen to be in this respect, as well as in the support which it gave to the atomic theory, is brought out well in Georges Cuvier's report on the work of the *Académie des Sciences* for 1819, which contains a lengthy account of the book and in which special reference is made to Berzelius's comments on chemical heat.[55]

5. *Support for the vibrational theory*

Dulong and Petit expressed their acceptance of the electrical origin of heats of reaction in their paper in April 1819. They were careful to present their belief as no more than a highly probable conjecture, to the point of admitting that the more traditional theories of chemical heat might still be applicable in accounting for a small part of the over-all effect. Yet private correspondence, as we might expect, tells us rather more. That Dulong at least had unreservedly abandoned the caloric theory by January 1820 is evident from the following passage in a letter which he wrote in that month to Berzelius:

Nous avions déjà porté un coup funeste à la théorie chimique de la chaleur dans le mémoire que nous avons lu à l'Institut pendant votre séjour à Paris.—De nouvelles expériences me portent à regarder comme une vérité incontestable que tous les phénomènes qui n'ont point de rapport avec la chaleur rayonnante ne sont que le résultat des mouvements vibratoires des molécules matérielles elles-mêmes. Le calorique rayonnant se propagerait, d'après cette manière de voir, par les vibrations du même fluide qui, avec une plus grande vitesse, produit sur nous la sensation de la lumière. Ainsi la pile voltaïque ne développerait le phénomène du feu qu'en excitant par le courant électrique les vibrations des particules matérielles. MM. Clément et Desormes ont publié un fait qui vient à l'appui de mon opinion dont je n'ai au reste parlé encore à personne. C'est qu'un même poids de vapeur d'eau, prise sous une force élastique quelconque et avec la température qui convient à cette force élastique, contient toujours la même quantité de chaleur. M. Despretz a trouvé qu'il en était de même pour les autres liquides.[56] Or, je puis prouver qu en faisant varier subitement le volume d'un gaz ou d'une vapeur, on produit par là des changements de température incomparablement plus grands que ceux qui résulteraient des quantités de chaleur développées ou absorbées, s'il n'y avait pas de chaleur *engendrée* par le mouvement. Rumford avait déjà employé à peu près le même mode de raisonnement pour soutenir son opinion, mais il avait pris pour sujet de ses observations les corps solides, ce qui rendait ses arguments beaucoup plus faciles à attaquer.[57]

The nature of the evidence which had now not only confirmed Dulong in his rejection of the caloric theory but also made him an unhestitating

[54] Berzelius, *Essai*, 68-73. On the date of publication see Berzelius's letter to Alexandre Marcet, 1 June 1819, in Berzelius, *Bref*, i, Part 3, 195. I am indebted to Dr. C. A. Russell for the latter reference.

[55] *Mém. Acad. Sci.*, iv (1819-20), pp. lxxxi-lxciii.

[56] On the work of Despretz, see note 67.

[57] Dulong to Berzelius, 15 January 1820, in Berzelius, *Bref*, ii, Part 1, 13-14.

advocate of the vibrational theory of heat is of considerable interest. The word "engendrée",[58] which Dulong underlined, strongly suggests that he saw the temperature changes which were known to accompany the rapid compression of a gas as being due, at least in part, to a conversion process, although there is no reason to believe that he even considered the possibility of a conservation principle (in this case relating heat and work). Unfortunately he nowhere enlarged on the conjectures contained in this letter, so that his views would have been known to few contemporaries.[59] Precisely how he applied the work of Clément and Desormes was thus never made clear and we can now do no more than attempt a reconstruction of the argument.

The crucial piece of new evidence which Dulong used had been provided in August 1819 by Nicolas Clément and Charles Bernard Desormes,[60] who at the time were associated in running a factory for the manufacture of alum at Verberie (Oise). Although they could hardly be said to have been at the centre of the Parisian scientific community, Clément and Desormes had produced a number of papers, usually jointly, which had attracted attention and praise and they were therefore not strangers to most academicians[61] when, on 16 and 23 August, a paper by them on the theory of heat engines (*"machines à feu"*) was read before the *Académie des Sciences*.[62] In the paper Clément and Desormes described experiments with an ice calorimeter in which they had shown, to their own satisfaction at least,[63] that the quantity of heat in unit mass of any given saturated vapour was always the same, being independent both of its temperature and pressure.[64] For anyone unfamiliar with interconvertibility of heat and work it was an obvious consequence of this that the mere compression of saturated vapour under adiabatic conditions (i.e. under conditions such that there was no heat exchange with the surroundings) would cause no change in the initial state of saturation.

[58] *Cf.* the term "excited" used by Rumford to express a similar idea in *Phil. Trans.* (1798), 88 and 99.
[59] The originality of Dulong's interpretation is apparent only in the light of contemporary opinion, which was almost entirely committed to a "calorist" explanation of the heat produced in the rapid compression of a gas. On the latter see T. S. Kuhn, "The caloric theory of adiabatic compression", *Isis*, xlix (1958), 132-140.
[60] The *Dictionnaire de Biographie Française* (Paris, 1933—in progress), contains convenient biographical sketches of Clément and Desormes. Clément appears as Clément-Desormes, the name which he adopted after his marriage to Desormes's daughter.
[61] Both men, moreover, had been candidates for admission to the *Académie des Sciences* on more than one occasion and as recently as 5 July 1819 Desormes had finally secured election as a corresponding member in the chemistry section (see *Procès-verbaux*, vi, 466).
[62] The paper was never published, but an extract appeared in the *Bulletin des Sciences par la Société Philomathique de Paris* for August 1819, pp. 115-118.
[63] On the discrediting of their results see below, and also notes 67 and 68.
[64] On earlier versions of this "law" proposed by James Watt and J. N. P. Hachette, see Black, *op. cit.* (41), i, 190, and "Tableau de M. Clément-Desormes, relatif à la théorie générale de la puissance mécanique de la vapeur (extrait)", *Nouveau Bulletin des Sciences par la Société Philomathique de Paris* (1826), 53. The extract in the *Bulletin* was prepared by Hachette, who signed the article.

Neither condensation nor any departure from conditions of complete saturation would be expected to occur. Therefore, the vapour could be treated as a perfect gas and the volume of, say, unit mass of a saturated vapour at any temperature could be determined by applying Boyle's law in conjunction with Gay-Lussac's figure for the expansion co-efficient of gases[65] and with the widely-used tables relating saturated vapour pressure and temperature. From an argument conducted in this way[66] it followed, for example, that the adiabatic compression of a mass of steam initially at 60° C (and at the corresponding saturated vapour pressure of approximately 14·5 cm of mercury) to one-fifth of its original volume would require that the temperature should rise by roughly 40° C if the steam were to remain saturated and if no condensation were to occur. Hence, on the basis of Clément and Desormes's "law" and on the assumption that the heat content of the vapour remained constant, it could be argued that a rise in temperature of roughly 40° C should be observed. In the light of a rapidly growing body of experimental evidence concerning the temperature changes accompanying the rapid (if not truly adiabatic) compression of air, Dulong was almost certainly convinced that this figure was far too small and that the discrepancy was due not to any error in Clément and Desormes's "law" but rather to the fact that the steam did *not* remain saturated and that the "law" consequently became inapplicable. Since he accepted Clément and Desormes's "law", Dulong would undoubtedly have associated this departure from a state of saturation with an increase in the heat content of the steam and he may well have been seeking to account for this increase when, as in the letter of January 1820 quoted above, he concluded that heat could actually be generated by the movement of a piston.

It is unfortunate that in the absence of any further comment, even from Berzelius, we can do no more than speculate on the effect that Dulong's argument and the support which it gave to the vibrational theory might have had on his contemporaries. That it would have been seen as an important contribution to the debate concerning the nature of heat can hardly be doubted, for Dulong had done more than point to the inadequacies of the caloric theory, many of which were already familiar enough. Like Rumford some twenty years earlier, he had also tackled the far more difficult problem of establishing the vibrational theory. In 1820 Dulong's case might well have appeared more plausible than Rumford's, but its ultimate effectiveness was limited by the nature of the evidence on which it was based. In particular, Clément and

---

[65] Given in Gay-Lussac, *op. cit.* (24).

[66] For an argument similar to that outlined here, in which the data for saturated vapour pressures are taken from J. B. Biot's *Traité de Physique Expérimentale et Mathématique* (Paris, 1816), see N. L. S. Carnot, *Réflexions sur la Puissance Motrice du Feu* (Paris, 1824), 67-68n.

Desormes's "law" was untrue and it was known to be so by 1827.[67] Indeed, the fact that Dulong himself conducted an experimental refutation of the "law" about this time[68] may even provide a much-needed clue in understanding why he did not pursue his championing of the vibrational theory.

6. *Support for the atomic theory*

The support which Dulong and Petit's law gave to the chemical atomic theory dealt a blow at established opinion which, in 1819 at least, was no less effective (and certainly more original) than their views on heat. The existence of a completely unexpected relationship between the familiar observable quantity specific heat and the individual atoms of matter was important evidence, especially in France, where even by 1819 the theory was still far from being generally accepted.[69] During the first decade and a half of the nineteenth century Berthollet's system, which excluded the concept of atoms of the Daltonian type and which taught that elements combined in varying and not in fixed proportions by weight, had dominated French chemistry, so that it was only to be expected that Dulong, as a young chemist and more particularly as one who had been privileged to belong to Berthollet's circle at Arcueil, should have adopted the traditional views in his early work. In fact his first paper,[70] which he read to the Institute in July 1811, gives every indication that at this time he wholeheartedly accepted Berthollet's theory with regard to chemical reactions and over the next five years his work proceeded without any evidence of a change of opinion on his part or, indeed, of any concern with the great theoretical issues involved in the debate over the truth of the atomic theory. In 1816, however, it quite suddenly became clear that Dulong had been deeply impressed by the theory, for it was in that year that he prepared an extract from William Prout's papers on the structure of atoms, which had recently appeared in

[67] In his *Traité Élémentaire de Physique* (1st edn., Paris, 1825), 100, C. M. Despretz stated that he had shown experimentally that Clément and Desormes's "law" could be applied to a number of vapours other than steam, but in the second edition of the work (Paris, 1827), pp. 113-114, he wrote that he had now disproved the law. In the pamphlet *Résumé des Travaux de Physique de M. Despretz* (Paris, n.d. but, according to internal evidence, probably 1828), 3-4, Despretz described how he had announced his revised conclusions before the *Société Philomathique*.

[68] C. M. Despretz, *Traité Élémentaire de Physique* (4th edn., Paris, 1836), 212 (and note) and Lamé, *op. cit.* (32), i, 487. The former reference suggests that Dulong's experiments were performed at about the same time as Despretz's, since Despretz states that Dulong had supported him in his criticisms of the "law" during a discussion with Clément which had taken place at the *Société Philomathique* "some years" earlier. Dulong's experiments may well date from the period 1824-1829 when he was engaged in a great deal of work on vapour pressures and related matters in connection with a commission set up by the *Académie*.

[69] See M. P. Crosland, "The first reception of Dalton's atomic theory in France", in Cardwell, *op. cit.* (38), 274-287.

[70] P. L. Dulong, "Recherches sur la décomposition mutuelle des sels insolubles et des sels solubles", *Ann. Chim.*, lxxxii (1812), 273-308.

England, and inserted it in the *Annales de Chimie et de Physique* for April,[71] despite the scepticism of all the other members of the distinguished editorial board.[72] It was in that year also that Dulong described before the Institute a classic example of what we should now know as the law of multiple proportions in the course of a paper on the oxides of phosphorus.[73] In the paper he discussed atomic weights freely and throughout he seems to have had no hesitation in interpreting his experiments in terms of the atomic theory, a fact which Berthollet and L. J. Thenard, the Institute's referees, duly noted in a generally favourable report.[74] It is unfortunate indeed that the text of a paper by Berthollet describing his reservations towards Dulong's conclusions, with special reference to the atomic theory,[75] was never printed. Clearly the theory was still the object of much scepticism[76] and the precise circumstances which led Dulong to appear as an atomist are therefore all the more difficult to ascertain. It is possible that he had now been finally persuaded by his work on the composition of the oxides of phosphorus, and it may be also that he had been impressed by the fact that almost simultaneously Berzelius cited his own work on the subject as strong evidence in favour of the physical reality of atoms.[77] As an editor of the *Annales de Chimie et de Physique* Dulong certainly knew of Berzelius's work and his esteem for the Swedish chemist was such[78] that he would not have dismissed it lightly.[79]

But, whatever the motives which first caused Dulong to dissent from the attitudes towards the atomic theory which were so established at Arcueil, we may be certain that he would have required little further

[71] *Ann. Chim. Phys.*, i (1816), 411-416. Prout's papers had appeared in November 1815 and February 1816, in Thomson's *Annals of Philosophy*, vi (1815), 321-330 and vii (1816), 111-113.

[72] See Dulong's letter to Berzelius, 8 January 1822, in Berzelius, *Bref*, ii, Part 1, 36-37. Among the other members of the board were Berthollet, Chaptal, Gay-Lussac, Thenard, Chevreul, Biot and Arago.

[73] P. L. Dulong, "Mémoire sur les combinaisons du phosphore avec l'oxigène", *Mémoires de Physique et de Chimie de la Société d'Arcueil*, iii (1817), 405-452. The paper was read on 1 and 15 July 1816.

[74] *Procès-verbaux*, vi, 101-103 (21 October 1816).

[75] *Ibid.*, vi, 103. Berthollet's paper was read immediately after the report on Dulong's work. Crosland states (*op. cit.* (69)), that the paper is now missing from the archives of the *Académie des Sciences*.

[76] But even in 1816 there is evidence that opinions were beginning to change. For example, although the main text of the first edition of L. J. Thenard's *Traité de Chimie Élémentaire, Théorique et Pratique* (Paris, 1813-16) contained no mention of the atomic theory, a translation by Collet-Descotils of W. H. Wollaston's celebrated paper on 1814 on chemical equivalents was included as an appendix, on pp. 247-254, to the last (fourth) volume, published early in 1816 (on this date see *Procès-verbaux*, vi, 33 (4 March 1816)). By 1817, when the first volume of the second edition appeared, Thenard revealed a very sympathetic attitude to Dalton's theory (see *Traité* (2nd edn.), i, 21-23).

[77] *Ann. Chim. Phys.*, ii (1816), 320-339. The first part of this lengthy paper had appeared in June 1816 in the same issue of the *Annales* as a shortened version of Dulong's paper on the subject.

[78] See, for example, *Ann. Chim. Phys.*, ii (1816), 174n.

[79] This is evident also in a letter of 5 August 1816 from Dulong to A. M. Ampère quoted in P. Lemay and R. E. Oesper, "Pierre Dulong, his life and work", *Chymia*, i (1948), 175.

18

persuasion from Berzelius in 1818 and 1819, although his adoption of the electrochemical theory may well date from this time. Any doubts he might still have had were to be finally dispelled in 1819, of course, though not only by the Dulong and Petit law. In November 1819 Berzelius, now back in Sweden, sent word to Dulong of Eilhard Mitscherlich's discovery of isomorphism,[80] which Dulong immediately saw as yet another proof of the existence of atoms. After welcoming the news, he wrote to Berzelius:

"Je suis convaincu, nonobstant les objections de M. de Laplace[81] et de quelques autres, que cette théorie [i.e. the atomic theory] est la conception la plus importante du siècle et que d'ici à une vingtaine d'années elle fera prendre à toutes les parties des sciences physiques une extension incalculable."[82]

7. Later influence

As it happened, Dulong was over optimistic in his prediction, for the establishment of the atomic theory proved a far more difficult task than he could ever have imagined. Dulong and Petit's law seems to have had little effect on the most vigorous critics of the theory, many of whom were to be found not in Berthollet's school but in England, and in later years it could not prevent a tide of scepticism towards atoms which became particularly strong in France from the 1830's[83] and which was manifested throughout the rest of the century in positivistic beliefs of varying strength.[84] However, in the years immediately following the discovery and among those who had at least some sympathy for the concept of atoms it did exert a considerable influence, both as an additional argument in favour of the atomic theory in general[85] and in the practical task of determining atomic weights.[86] It is ironical that John Dalton, whose views Dulong

---

[80] See Berzelius's letter to Dulong, 5 November 1819, in Berzelius, Bref, ii Part 1, 10-11.

[81] It is interesting that Dulong should have chosen to mention Laplace in this context, when Berthollet was generally recognized as the leading opponent of the atomic theory in France. It seems likely that a genuine personal attachment to Berthollet (on which see Dulong's letter to Berzelius, 20 November 1823, in Berzelius, Bref, ii, Part 1, 45) mitigated Dulong's criticisms.

[82] Dulong to Berzelius, 15 January 1820, in Berzelius, Bref, ii, Part 1, 12.

[83] See, for example, G. Buchdahl, "Sources of scepticism in atomic theory", The British Journal for the Philosophy of Science, x (1959), 120-134.

[84] On scepticism in England, with special reference to Davy, see W. H. Brock and D. M. Knight, "The atomic debates: memorable and interesting evenings in the life of the Chemical Society", Isis, lvi (1965), 5-25. For convenient statements of the influential views of P. E. M. Berthelot, see his La Synthèse Chimique (2nd edn., Paris, 1876), 154-171, and also Comptes Rendus hebdomadaires des Séances de l'Académie des Sciences, lxxxiv (1877), 1189-1195.

[85] In his Leçons sur la Philosophie Chimique (Paris, 1837), 271, J. B. A. Dumas cited the discovery of Dulong and Petit's law as the "beau moment" of the atomic theory. For a similar view see, for example, A. Wurtz, Leçons de Philosophie Chimique (Paris, 1864), 24.

[86] The approximate nature of the law, which was soon widely recognized, greatly restricted its usefulness in this respect, but the law did allow a choice to be made in cases where there was uncertainty concerning the number of atoms of various elements making up a "compound atom". Berzelius, for example, had halved many of his atomic weights by 1826 in the light of the law and also of Mitscherlich's work. See J. C. Poggendorf's Annalen der Physik und Chemie, vii (1826), 414.

and Petit did so much to promote,[87] appeared as one of the few critics of their work. In some comments published in 1827 in the long-delayed second volume of his *New System of Chemical Philosophy*[88] Dalton cast serious doubt on the truth of the law, criticizing equally Dulong and Petit's determinations of specific heat and the values which they adopted for atomic weight. He also attacked Dulong and Petit's conviction, which was not entirely unjustified, that the law might be found to apply for elements in the gaseous state as well as for solids,[89] although he took the opportunity of pointing out the similarity between the form which the law would have when extended in this way and a view which he had put forward some twenty years earlier. Then, writing in the first volume of his *New System*,[90] Dalton had argued that under similar conditions of temperature and pressure the quantities of caloric attached to the ultimate particles (whether atoms or molecules) of all gases were the same, irrespective of the nature of the gas concerned. As an Irvinist, he had been able to deduce from this that the specific heat of any gas must be inversely proportional to its atomic or molecular weight and so had arrived at a result which, superficially at least and naturally enough in Dalton's eyes, resembled the Dulong and Petit law. Any claim that Dalton anticipated the discovery of 1819[91] can be easily discounted, however. Even Dalton himself admitted that his own conclusions applied equally to the ultimate particles of both elements and compounds, whereas Dulong and Petit had stated their law for elements only, in the belief that a different though as yet undetermined law existed for compounds.[92] But of still greater weight in dismissing any Daltonian claims to priority is the highly suspect nature of Dalton's reasoning. In the first place, his views on the quantity of caloric attached to gas particles were based on the most tenuous evidence and anyway were quite irrelevant to men such as Dulong and Petit who were unsympathetic to the caloric theory. Moreover, the Irvinist version of the theory, which was essential in his prediction of specific heats, had been explicitly rejected by Dulong and Petit and it was therefore the height of absurdity that Dalton should

---

[87] Despite their disagreement with many of his opinions, Dulong and Petit seem to have held Dalton in genuine esteem. A courteously inscribed copy of their prize-winning memoir of 1818, which they sent to Dalton, is now in the Burndy Library, Norwalk, Conn. I wish to thank Mrs. A. Matthysse, Librarian of the Burndy Library, for providing me with a description of this copy of the memoir.

[88] Dalton, *New System* (Manchester, 1827), ii, 280-281 and 293-297.

[89] Petit and Dulong, *op. cit.* (6), 406-407. Dulong and Petit almost certainly believed, with Berzelius (see Berzelius, *Essai*, 52-55), that equal volumes of all elementary gases at a given temperature and pressure contained the same number of atoms and they would therefore have expected by their law that the volume specific heats of such gases would be identical. This prediction was very nearly borne out by the experimental results given in Delaroche and Bérard, *op. cit.* (40), 157.

[90] Dalton, *op. cit.* (15), 66-75. For a fuller account of Dalton's view see Fox, *op. cit.* (38).

[91] A claim made in W. C. Henry, *Memoirs of the Life and Scientific Researches of John Dalton* (London, 1854), 68, and in the *Dictionary of National Biography*, article "John Dalton".

[92] Petit and Dulong, *op. cit.* (6), 407-408.

have included among his criticisms one to the effect that the Frenchmen had uncritically adopted the typically Irvinist belief that at any given temperature the quantity of heat in unit mass of any body was proportional to its specific heat. We should be correct in concluding from the weakness of such arguments that the Irvinist cause, with Dalton as its chief surviving representative, was now all but lost.

Since this paper is concerned primarily with the immediate influence of Dulong and Petit's work, any lengthy discussion of the fortunes of the atomic theory later in the nineteenth century would clearly be out of place. For the same reason we must pass briefly over the subsequent history of the caloric theory, though with the suggestion that the part played by the views of Berzelius, Dulong and Petit in effecting the downfall of the theory has not received the attention it deserves. If these men had done no more than point to the grave weaknesses in the prevailing theory of chemical heat, their contribution would have been of some significance, but, as we have seen, they did far more than this. Following a lead given by earlier writers, notably Davy, they had formulated a really plausible alternative, one which, unlike its rivals founded in the caloric theory, neither necessitated nor even implied that heat was a fluid. Their views may have done little positively to promote the vibrational theory, a task which was extremely difficult at this time, as Dulong seems to have found, but they did remove much of the *raison d'être* of caloric and it is in this that their importance lies. The fact that the period from 1820 until the advent of energy conservation about 1850 was characterized by a widespread agnosticism on the question of the nature of heat[93] can hardly be seen in isolation from the weakening of the position of caloric which Berzelius, Dulong and Petit helped to bring about. It is interesting to note in passing that in this period of agnosticism Dulong and Petit's law was at the centre of a minor, though lively, research tradition which encouraged the experimental study of heat, and in particular of specific heats, while divorcing it from the traditional disputes between the various fluid and vibrational theories.[94] When we note that probably the greatest incentive for work in this tradition was provided by the search for a law, analogous to that of Dulong and Petit, which could be applied to compounds, it is hardly necessary to state that the problems raised in studies of this type demanded neither knowledge of nor concern with the nature of heat.

[93] This view of the period 1820-1850 clearly requires more detailed documentation than is possible here, but an illustration of the sort of changes which had taken place since the early years of the century is provided in Thomas Thomson's *An Outline of the Sciences of Heat and Electricity* (London, 1830), 335, where Thomson put the then widely-held opinion that the question of the nature of heat was unsolvable, whereas in his *A System of Chemistry* (Edinburgh, 1802), i, 259, he had considered the question as settled (in favour of caloric).

[94] A convenient account of this later work, in which F. E. Neumann, Victor Regnault and Hermann Kopp figured particularly prominently, appears in I. Freund, *The Study of Chemical Composition* (Cambridge, 1904), 361-384. Dulong himself participated in the work after Petit's death, but with little success.

8. *Conclusion*

There can be no denying that chance played an important part in the events described in this paper, and the fact has been duly emphasized. We have seen, for example, that Dulong and Petit discovered their law not in the course of a programme of planned research but unexpectedly, as the result of work on purely traditional problems. We have seen also that their views on the origin of heats of reaction may well have owed a great deal to the fact that Berzelius happened to visit Paris in 1818 and 1819. But for chance, we may conclude, the law would not have been discovered and Dulong and Petit might never have felt able to break with the caloric theory, despite their growing suspicions of its falsity.

Yet I believe that chance events such as these do not constitute the whole story. Above all, they do not seem to account satisfactorily for the extreme confidence with which Dulong and Petit expressed their opinions on the nature of heat and the atomic theory after their discovery of 1819, for the evidence available to them was, as we have seen, somewhat less than conclusive. However, the conviction with which they expressed their views may well become accountable if it is explained not simply as the result of the sudden acquisition of fresh evidence but also as a manifestation of the new critical spirit in French physical science to which reference has been made in the introduction to this paper. The work of Dulong and Petit themselves provides some of the best evidence that this new spirit was abroad in the period 1815-20, for examples of the sort of changes which were taking place are provided by Dulong's sudden appearance in 1816 as a champion of the atomic theory after his previous commitment to Berthollet's chemistry and by the way in which he and Petit felt able to criticize the caloric theory in 1819, although as recently as November 1814 Petit had been expounding the properties of the fluid caloric to students at the *École Polytechnique* without even a mention of the vibrational theory.[95] But this is not the only evidence of change. We should recall that the period 1815-20 also saw the young Augustin Fresnel's first attempts to establish his wave theory of light against the prevailing corpuscular theory advocated by Laplace and by his two closest disciples, Jean-Baptiste Biot and Poisson.[96] Since Fresnel numbered among his earliest converts Arago, Petit and Dulong[97] and since he was

---

[95] The evidence for this is found in a set of notes taken by Auguste Comte at Petit's lectures on physics at the *École Polytechnique* in the winter of 1814-1815. A copy of the notes was kindly supplied to the author by Monsieur D. Cantemir, archivist of the Maison d'Auguste Comte, Paris.

[96] For an account of Fresnel's work see Émile Verdet's introduction to Fresnel, *op. cit.* (47), i, pp. ix-xcix.

[97] Details of this and a number of other developments in the period 1815-1820 will be given in a joint paper which the author is preparing in collaboration with Dr. J. R. Ravetz. However, evidence of Dulong's support for Fresnel will be found in the letter of 15 January 1820, from Dulong to Berzelius, quoted in section 5.

himself critical of the traditional versions of the caloric theory,[98] the views which these men adopted on the important questions of the nature of heat, the nature of light and the atomic theory have a marked unifying theme. Clearly the case for seeing the work of Arago, Petit, Dulong and Fresnel as forming anything resembling a concerted movement would need to be made at far greater length than is possible here, but it is hoped that enough has been said at least to suggest, if not to establish, a broader context in which Dulong and Petit's work might meaningfully be interpreted.

[98] See his "Complément au mémoire sur la diffraction", dated 10 November 1815, in Fresnel, *op. cit.* (47), i, 59-60; also his letters of 5 July 1814 and 11 July 1814 to Léonor Fresnel, *ibid.*, ii, 820-822 and 827-829.

IX

The Rise and Fall of Laplacian Physics

By means of these assumptions, the phenomena of expansion, heat, and vibrational motion in gases are explained in terms of attractive and repulsive forces which act only over insensible distances (distances imperceptibles). *In my theory of capillary action I related the effects of capillarity to such forces. All terrestrial phenomena depend on forces of this kind, just as celestial phenomena depend on universal gravitation. It seems to me that the study of these forces should now be the chief goal of mathematical philosophy. I even believe that it would be useful to introduce such a study in proofs in mechanics, laying aside abstract considerations of flexible or inflexible lines without mass and of perfectly hard bodies. A number of trials have shown me that by coming closer to nature in this way one could make these proofs no less simple and far more lucid than by the methods used hitherto.*[1]

1. INTRODUCTION

The period from Napoleon Bonaparte's assumption of power as First Consul in 1799 until his final overthrow in 1815 is generally recognized to have been one of the most glorious in the whole history of French science. It was a period when France led her European rivals in the quantity and in the quality of her scientific contributions, especially in the physical sciences. Great names, such as those of Laplace, Berthollet, Biot, Poisson,

*Department of History, University of Lancaster, Bailrigg, Lancaster, England.

Abbreviated versions of this paper have been read at meetings of the Northern Seminar in the History of Science in Manchester on 7 May 1969 and of the British Society for the History of Science in London on 23 November 1970, and parts of it have since been discussed at seminars in Oxford and Cambridge. It is a pleasure to express my thanks for the generous help and criticisms I have received from J.R. Ravetz during the preparation of the paper for publication. I am also grateful to E. Frankel, P.M. Heimann, and an anonymous referee for some valuable comments.

[1]P.S. Laplace, *Traité de mécanique céleste*, 5 vols. (Paris, 1799–1825), *5*, 99. Although the title page of the fifth volume bears the date 1825, the six books that made up the volume were published and dated separately. Book XII, in which this passage appears, is dated April 1823. For an earlier and slightly different statement see *Connaissance des tems . . . pour l'an 1824* (Paris, 1822), p. 323; also *Journal de physique*, *94* (1822), 90. The translations throughout the paper are my own.

Gay-Lussac, Thenard, and Malus, abounded, and there were some remarkable successes, of which the most celebrated is perhaps the discovery and
study of the polarization of light. It is not surprising, therefore, that both
the "declinists" of Britain in 1830, such as Charles Babbage and David
Brewster,[2] and those who complained no less bitterly about the state of
French science in the 1860's and 1870's, notably Louis Pasteur and
Adolphe Wurtz,[3] looked back to the years of Napoleon's rule as a truly
golden age for science.

It is not difficult to identify at least some of the conditions that allowed
French science under Napoleon to become a byword for excellence. As
Maurice Crosland has shown, the supply of able graduates from the École
Polytechnique, public recognition and encouragement given even by
Napoleon himself, the attractive career possibilities that were available to
young men trained in science, and the select research school centered on
Berthollet's country house at Arcueil, just outside Paris, all played their
part.[4] And equally, thanks above all to the work of L. Pearce Williams and
Roger Hahn, we know something of the weaknesses of French science in
the Consulate and First Empire. For example, although weaknesses were
rarely acknowledged at the time, at least in public, Williams has identified
grave deficiencies in education, especially at the elementary level,[5] and
Hahn has pointed to the harm that was being done as the First Class of the
Institute (the revolutionary successor of the Academy of Sciences) became
increasingly a manifestation of Napoleon's *Kulturpolitik*.[6]

It is the purpose of this paper to take further the investigation of both
the strengths and the weaknesses of Napoleonic science, with special

[2] In Britain admiration for Napoleonic science was naturally greatest some years
after 1815 and was particularly strong about 1830. Babbage wrote glowingly of
French achievements in his *Reflections on the Decline of Science in England* (London, 1830), especially pp. 25–27 and 30–36, maintaining that under Napoleon "the
triumphs of France were as eminent in Science as they were splendid in arms" (*ibid.,*
p. 26). An equally favorable view was given by David Brewster in his unsigned "Decline of Science in England," *Quarterly Review, 43* (1830), 313–317, although
Brewster stressed that the enlightened policy toward science, which had been so
characteristic of the First Empire, had continued under Louis XVIII and Charles X.

[3] See, for example, Pasteur's contribution of 1871 to the Lyons newspaper *Salut
public,* quoted in R. Vallery-Radot, *La vie de Pasteur* (Paris, 1900), pp. 278–279, and
C.A. Wurtz, *Les hautes études pratiques dans les universités allemandes* (Paris, 1870),
pp. 5–6.

[4] M.P. Crosland, *The Society of Arcueil. A View of French Science at the time of
Napoleon I* (London, 1967).

[5] L.P. Williams, "Science, Education and Napoleon I," *Isis, 47* (1956), 369–382.

[6] R. Hahn, *The Anatomy of a Scientific Institution. The Paris Academy of Sciences,
1666–1803* (Berkeley, Los Angeles, and London, 1971), pp. 310–312.

reference to physical science. In Section 2, I shall argue that the course and content of much of French physics and physical chemistry under Napoleon was determined by the zeal of the mathematician and physicist Pierre Simon Laplace and the chemist Claude Louis Berthollet for a program of research which they jointly sought to pursue. The program was seen, at least by them, as a natural culmination of eighteenth-century work in the Newtonian tradition, so that although it is described here as Laplacian, it was in reality not entirely the creation of Laplace himself, or indeed of Berthollet; it was Laplacian only to the extent that Laplace gave it a number of its characteristic features, stated it explicitly, and was its most brilliant exponent from the time he began to formulate it, probably in the 1790's, until his death in 1827.

In the years of its greatest success, from 1805 to 1815, the program both raised problems and laid down the general principles for solutions; and, by doing so, it gave French physical science a most uncommon unity of style and purpose. It also stimulated much good work; yet, as I hope to show, it owed its dominant position not merely to its merits, considerable though these appeared to be at the time, but equally to the effectiveness with which Laplace and Berthollet were able to control the scientific establishment of France in teaching as well as research. Once this control was lost (a process that I discuss in Section 3), the program became vulnerable, and, beset by the challenges of new discoveries and theories in heat, optics, electricity, magnetism, and chemistry and by a new generation of younger scientists who felt no allegiance to Laplace and Berthollet, it was abandoned, quite suddenly, between 1815 and 1825. By examining the downfall of the Laplacian program and its attendant doctrines, I hope to demonstrate the precariousness of what I see as the leading research tradition in physical science in Napoleonic France and to suggest that the successes of the period owed far more than has previously been recognized to Laplace's personal commitment to his program and to his ability to engage other men of exceptional ability in the same enterprise.

2. LAPLACIAN PHYSICS

Laplacian physics was a style of physics that depended on and was embraced by what J. T. Merz, with an acknowledgment to Maxwell, first called the astronomical view of nature.[7] It was a physics that sought to

[7]J.T. Merz, *A History of European Thought in the Nineteenth Century,* 4 vols. (Edinburgh and London, 1896–1914), *1*, 347–348.

account for all phenomena, on the terrestrial and, more particularly, the molecular scale as well as on the celestial scale, in terms of central forces between particles which, although treated by analogy with Newtonian forces of gravitation, could be either attractive or repulsive. Since attempts to "explain" gravitation had been generally abandoned long before, forces of this character could readily be accepted as "mechanical" and hence in need of no further explanation.[8] The forces were conceived as being exerted by and upon imponderable as well as ordinary ponderable matter; indeed, an essential and highly characteristic element in Laplacian physics was the system of imponderable fluids of heat, light, electricity, and magnetism. In accordance with beliefs that had come to be widely accepted by the end of the eighteenth century, each fluid was thought to consist of particles which were mutually repulsive but which in all cases were attracted by ponderable matter.[9] In the hands of the Laplacians, models of such fluids, founded on the assumption that the forces between imponderable and ponderable matter were effective only over "insensibly small" distances, were capable of being translated into systems of differential equations whose approximate solutions could "save" the phenomena already known and even predict new ones. And it was in the attempt to refine and quantify a theory of the imponderables which had hitherto been vague and qualitative that Laplacian physics found its main problems and had its most notable achievements.

The imponderable fluids, like the rest of Laplacian physics, did not originate with Laplace himself. As far as the basic model for their structure is concerned, they have their roots in Newton's speculations on the subtle electrical spirit in the General Scholium of 1713,[10] in his speculations on the ether which appeared in the second and subsequent editions

---

[8] However, the charge against the imponderables that they merely allowed the explanation of action at distance to be postponed, since their own properties presupposed the existence of such action, was raised in the late eighteenth and early nineteenth centuries, notably by Lavoisier and Davy; see R. Fox, *The Caloric Theory of Gases from Lavoisier to Regnault* (Oxford, 1971), pp. 17 and 118.

[9] Those who adopted the two-fluid theories of electricity and magnetism (see below, p. 93) made the additional assumption that there was attraction between the vitreous and resinous electrical fluids and between the austral and boreal magnetic fluids.

[10] Newton, *Philosophiae Naturalis Principia Mathematica*, 2nd ed. (Cambridge, 1713), p. 484. On the interpolation of the words "electric and elastic" to describe the spirit in Andrew Motte's translation of 1729, see A. Koyré and I.B. Cohen, "Newton's 'Electric & Elastic Spirit'," *Isis, 57* (1960), 337.

of the *Opticks*,[11] and even more clearly perhaps in what was generally recognized through the eighteenth century as the Newtonian view of gas structure—the view that the particles of gases were stationary and that repulsive forces between these particles accounted for gas pressure and the other characteristic gaseous properties.[12] This Newtonian model had been applied with increasing frequency in discussions of the properties of imponderable fluids since the 1740's, when Franklin used it in his widely read speculations on the nature of the electric fluid.[13] In fact, by about 1780 Franklin's belief that electrostatic phenomena could be explained in terms of the supposed repulsion between the particles of the electric fluid and the supposed attraction between the fluid and the particles of ordinary ponderable matter had become standard doctrine; and, as a result of the work of such men as Aepinus, Priestley, and Lavoisier, the same was true of the other "Newtonian" imponderables which had emerged by then, the fluids of magnetism, heat (or fire), and light.[14] There were divergences of opinion, of course, notably between the supporters of the "one-fluid" theories of electricity and magnetism, associated principally with the names of Franklin and Aepinus, and those like Coulomb who favored the "two-fluid" theories, in which a vitreous and a resinous electrical fluid and an austral and a boreal magnetic fluid were postulated.[15] But such differences did nothing to make the fundamental concept of the imponderable elastic fluid any less acceptable. By 1780 the

[11] Newton, *Opticks*, 2nd ed. (London, 1718), pp. 322-328 (Queries 17-24). In the *Principia* the subtle spirit was simply described as "electric and elastic" and no further details of its supposed structure were given. In Query 21 of the *Opticks*, however, it was suggested that the postulated ether might consist of mutually repulsive particles. Although the structure of Newton's ether was similar to that of the later imponderables, its function, primarily as the basis for an explanation of gravitation and optical phenomena, was quite different; see J.E. McGuire, "Force, Active Principles, and Newton's Invisible Realm," *Ambix*, 15 (1968), 154-208.

[12] First stated (as no more than a mathematical hypothesis) in the first edition of the *Principia* (London, 1687), pp. 301-303 (Book II, Proposition xxiii).

[13] Most easily consulted in I.B. Cohen, ed., *Benjamin Franklin's Experiments* (Cambridge, Mass., 1941), especially pp. 213-215.

[14] For accounts of the rise of the imponderables between the 1740's and the 1780's see I.B. Cohen, *Franklin and Newton* (Philadelphia, 1956), pp. 365-554; R.E. Schofield, *Mechanism and Materialism* (Princeton, 1970), pp. 157-190; Fox, *op. cit.* (note 8), pp. 6-20; and J.E. McGuire and P.M. Heimann, "Newtonian Forces and Lockean Powers: Concepts of Matter in Eighteenth-Century Thought," *Historical Studies in the Physical Sciences*, 3 (1971), 233-306.

[15] Also the Newtonians such as Gowin Knight, P.D. Leslie, Cadwallader Colden, Bryan Higgins, James Hutton, and Adam Walker, who worked in the predominantly British tradition discussed by Heimann and McGuire (see their paper cited in note

theory of the imponderables was recognized to be imperfect to the extent
that it was still almost entirely qualitative, yet it was coherent, simple,
and, above all, it had the merit of possessing what appeared to be a
thoroughly Newtonian pedigree.

So for a Frenchman such as Laplace, whose interest in the experimental
aspects of physical science, especially the study of heat, grew rapidly
about 1780,[16] it was to be expected that the imponderable fluids would
provide an entirely acceptable explanation of the phenomena of physics.
This was a time when in a strongly Newtonian France few qualms were felt
about accepting the possibility of forces acting at a distance and when in
any case the physics of the eighteenth-century Newtonian tradition was
seen as having had so many notable successes. Moreover, it was just at this
time that the writings of Laplace's friend Lavoisier were beginning to
attract favorable attention to the fluid or caloric theory of heat[17] and
to give it the form it was to have for the next seventy years; and it was
only shortly afterwards that Coulomb produced some of his most important work in electricity and magnetism,[18] work which did much to win

---

14), held views on subtle fluids that differed appreciably from those of Franklin,
Aepinus, Lavoisier, and Coulomb.

[16]The stimulus for this new interest came from Lavoisier, according to Laplace's
letter to Lagrange, 21 August 1783, in Oeuvres de Lagrange, 14 vols. (Paris, 1867-
1892), 14, 124. In a letter to Lagrange of 11 February 1784 (ibid., 14, 130), however, he made no show of reluctance, writing, with reference to Haüy's recently
published Essai d'une théorie sur la structure des cristaux, which he and Daubenton
had read on behalf of the Academy of Sciences: "It contains an interesting application of mathematics to nature, and the hope that we may be able to extend the
realm of geometry cannot be put too strongly. It is with this in view that I have
devoted a little of my time to physics, and I am not without hope that I shall be
able to grasp some other physical problems sufficiently well to be able to apply the
methods of analysis to them."

[17]Notably in his papers in the Mémoires . . . de l'Académie Royale des Sciences for
1777 (pp. 420-432 and 592-600), and the joint "Mémoire sur la chaleur," written
with Laplace, which appeared on pp. 355-408 of the Mémoires for 1780 (although
it was not read until June 1783 and was published only in 1784). It is important to
note that Lavoisier was always cautious on the question of the nature of heat, and in
the joint paper with Laplace the view that heat consists in the motion of the particles of ordinary ponderable matter was stated to be no less plausible than the fluid
theory. However, especially in his Traité élémentaire de chimie, 2 vols. (Paris, 1789),
1, 4-27, he was easily taken for a convinced calorist.

[18]The seven papers describing this work, which all date from the period 1785-
1789, are most easily consulted in Collection de mémoires relatifs à la physique,
publiés par la Société Française de Physique. Tome 1. Mémoires de Coulomb (Paris,
1884), pp. 107-318. For a study of Coulomb's work in electricity and magnetism
see C.S. Gillmor, Coulomb and the Evolution of Physics and Engineering in Eighteenth-Century France (Princeton, 1971), pp. 175-221.

support for the two-fluid theories that Coulomb himself favored.[19] But Laplace, of course, did more than merely accept the legacy of Newtonian physics as this was normally understood towards the end of the eighteenth century. It was Laplace's great achievement to build on the Newtonian tradition, to restate many of its principles in a mathematical form, to take up its outstanding problems, especially with regard to the short-range forces that were thought to operate on the molecular scale, and thereby to create a physics which, although Newtonian in origin, was unmistakably and characteristically Laplacian.

Signs of what was later to emerge as the true Laplacian program can be detected as early as 1796. In the first edition of the *Exposition du système du monde*, published in that year, Laplace stated that not only optical refraction and capillary action but also the cohesion of solids, their crystalline properties, and even chemical reactions were the result of an attractive force exerted by the ultimate particles (*molécules*) of matter, and he looked forward to the day when the law governing the force would be understood[20] and when, as he put it, "we shall be able to raise the physics of terrestrial bodies to the state of perfection to which celestial physics has been brought by the discovery of universal gravitation."[21] In Laplace's view there was good reason to believe that the molecular forces might themselves be gravitational in nature, even though they did not obey the simple inverse-square law, a complication that resulted from the effect, on the molecular scale, of the shape of the individual molecules.

But in all this he was saying no more than what so many eighteenth-century Newtonians, with an eye on the Queries, had already accepted as standard doctrine. Throughout the eighteenth century, molecular forces, usually assumed to be negligible except over a very short range, had been invoked as a standard element of Newtonian physics in treatments of optical refraction, capillary action, surface tension, and crystal structure,[22]

---

[19]*Collection de mémoires . . . Coulomb* (note 18), pp. 250–252.

[20]P.S. Laplace, *Exposition du système du monde*, 2 vols. (Paris, an VII [1796]), *2*, 196–198.

[21]*Ibid., 2*, 198.

[22]The most important source for this acceptance was, of course, Query 31 of the *Opticks;* see Newton, *Opticks*, 4th ed. (London, 1730), pp. 350–382. For references to the work of Clairaut and Buffon, the leading exponents of the "molecular forces" tradition in eighteenth-century France, see notes 24, 25, and 26. Views on crystal structure are discussed in H. Metzger, *La genèse de la science des cristaux* (Paris, 1918), pp. 165–170, and J.G. Burke, *Origins of the Science of Crystals* (Berkeley and Los Angeles, 1966), pp. 34–35 and 78. Although Laplace did not discuss the matter in 1796, he almost certainly shared the belief of many eighteenth-century New-tonians that the molecular forces were operative only over a very short range. For

and they had been invoked also in a continuing tradition of work on chemical affinities.[23] Although molecular forces had been equally acceptable to British and to French Newtonians, it was probably the French, in particular Alexis Claude Clairaut and Georges Louis Leclerc de Buffon, who (next to Newton himself) exerted the greatest influence on Laplace, and it is by examining their work that we can see most clearly the continuity between Laplace and his Newtonian precursors in the eighteenth century. Clairaut, for example, had ascribed "the roundness of drops of fluid, the elevation and depression of liquids in capillary tubes, the bending of rays of light, etc." to gravitational forces that became large at very small (molecular) distances,[24] and Buffon had discussed similar forces, though in his case with special reference to the forces of chemical affinity.[25] Although both men were proud to call themselves Newtonians and stated explicitly that the molecular forces they postulated were of the same nature as those operating between celestial bodies, they disagreed fundamentally over the relationship between force and distance on the molecular scale. The debate in which this disagreement became apparent arose because of a discrepancy between the predicted and observed periods of the apogee of the moon.[26] To remove the discrepancy, Clairaut suggested, in 1747, that the force law should contain a term inversely proportional to the fourth power of the distance, $1/r^4$, in addition to the usual term proportional to $1/r^2$.[27] This additional term, he argued, would

earlier expressions of this belief see, for example, John Keill, "Epistola ad Cl. virum Gulielmum Cockburn, Medicinae Doctorem. In qua leges attractionis aliaque physices principia traduntur," *Philosophical Transactions, 26* (1708–1709), 97–110, especially p. 100; F. Hauksbee, *Physico-Mechanical Experiments on Various Subjects* (London, 1709), pp. 157–160; and, for a contemporary French statement, by Coulomb, *Collection de mémoires relatifs à la physique . . . Mémoires de Coulomb* (note 18), pp. 125–126. For Newton's earliest comment on the matter, see his *Optice* (London, 1706), pp. 322–348 (Query 23), especially p. 335.

[23] For references to this tradition see note 32.

[24] A.C. Clairaut, "Du système du monde dans les principes de la gravitation universelle," *Mémoires . . . de l'Académie Royale des Sciences* for 1745 (published 1749), p. 338. The paper was read on 15 November 1747. Cf. his later comment on p. 547 of the same volume.

[25] G.L.L. de Buffon, "De la nature. Seconde vue" (1765), in his *Histoire naturelle, générale et particulière,* 44 vols. (Paris, 1749–1804), *13,* xii–xv.

[26] The debate is best followed in the papers by the two men that were published in the *Mémoires . . . de l'Académie Royale des Sciences* for 1745, pp. 329–364, 493–501, 529–548, 551–552, and 577–587. For accounts of the debate see P. Brunet, *La vie et l'oeuvre de Clairaut (1713–1765)* (Paris, 1952), pp. 82–88, and A.W. Thackray, *Atoms and Powers. An Essay on Newtonian Matter-Theory and the Development of Chemistry* (Cambridge, Mass., and London, 1970), pp. 157–160.

[27] Clairaut, *op. cit.* (note 24), pp. 337–339.

at once remove the anomaly in question and be consistent with the existence of molecular forces that were very large at insensible distances. Concerned at what he saw as a loss of elegance and simplicity and hence as a threat to the Newtonian system, Buffon upheld the familiar $1/r^2$ law, though some years later he admitted that such a law would be modified at short range by the shape of the particles of matter.[28] The details of the debate do not concern us here, and it is sufficient to note that it ended abruptly in 1749 when Clairaut found that, after all, he could reconcile the motion of the apogee with a simple inverse-square law.[29] However, he was careful not to concede victory to Buffon with regard to the short-range forces, and the form of the law obeyed by such forces remained undiscovered at the end of the century, having attracted little further attention.

Laplace's comment on the effect of the shape of individual particles on the force law suggests that he was especially indebted to Buffon, but even though he was reticent about his precursors, there can be little doubt that he owed quite as much to Clairaut, who had treated refraction and capillary action in a mathematical way of which he would surely have approved.[30] There is, then, evidence of continuity between Laplace and at least one kind of eighteenth-century Newtonianism which had a strong following in France, and the degree of continuity is so great that it could be argued that Laplace merely brought together a group of ideas which previously had been rather disparate. However, his statement of the range of phenomena that could be treated in terms of molecular forces, even as given in 1796, was certainly more comprehensive than any made previously, and it did direct renewed attention to the unification of terrestrial and celestial physics which, in his view, would result from further research. Above all, it looked forward to future work, in the manner of a research program.

It was not until the publication of the fourth volume of the *Traité de mécanique céleste* in 1805 that Laplace's brief and rather formal state-

[28] Buffon, *loc. cit.* (note 25).

[29] "Avertissement de M. Clairaut ... sur le système du monde, dans les principes de l'attraction," *Mémoires ... de l'Académie Royale des Sciences* for 1745, pp. 577–578. This is dated 17 May 1749.

[30] See Clairaut, "Sur les explications cartésienne et newtonienne de la réfraction de la lumière," *Mémoires ... de l'Académie Royale des Sciences* for 1739 (published 1739), pp. 259–275, and (on capillary action) his *Théorie de la figure de la terre, tirée des principes de l'hydrostatique* (Paris, 1743), pp. 105–128. In both of these discussions he used short-range molecular forces. For a reference to one of Laplace's very few comments on his precursors, in this case to Clairaut, see note 40.

ment of 1796 was transformed into the basis for a truly Laplacian style of
science. It seems crucial to our understanding of this transformation that
by 1805 Laplace had gained a close and highly influential ally in
Berthollet, his intimate friend since the early 1780's, his next-door
neighbor at the village of Arcueil from 1806, and a man who saw chemistry in precisely those Newtonian terms that Laplace sought to apply
more particularly in physics. In the *Recherches sur les lois de l'affinité* of
1801 and more fully in the *Essai de statique chimique* of 1803, Berthollet
had expounded his view that chemical affinity was the result of attractive
forces between the particles of matter. Indeed, he had gone so far as to
begin the *Statique chimique* by declaring:

> The forces that bring about chemical phenomena all derive from the
> mutual attraction between the molecules of bodies. The name affinity
> has been given to this attraction so as to distinguish it from astronom-
> ical attraction.
>
> It is probable that both are one and the same property.[31]

Of course, in putting forward this idea, Berthollet was making no claim to
originality. The idea was taken straight from the Newtonian tradition in
the chemistry of affinites as this had come down through the English
Newtonians such as the Keills and Hauksbee, and in France through
Buffon and the French chemists Macquer, Guyton de Morveau, and
Lavoisier, among others.[32] But in his lengthy writings on affinity
Berthollet, like Laplace, did far more than reiterate a conventional view.
Perhaps he did not succeed in answering many (or any) of the outstanding
problems of the eighteenth century, for neither he nor his disciples car-
ried through the systematic determination of chemical affinities that good
Newtonian chemists saw as their goal. And his recognition of the dif-
ficulty of the tasks ahead may well have spread more despondency than
encouragement. But by his rigorous, critical restatement of the Newtonian

---

[31] C.L. Berthollet, *Essai de statique chimique,* 2 vols. (Paris, an XI [1803]), *1*, 1.

[32] For accounts of this tradition, which was virtually unaffected by the innovations
of Lavoisier, see M.P. Crosland, "The Development of Chemistry in the Eighteenth
Century," *Studies on Voltaire and the Eighteenth Century, 24* (1963), 369–441,
especially pp. 382–390; A.W. Thackray, "Quantified Chemistry—the Newtonian
Dream," in D.S.L. Cardwell, ed., *John Dalton & the Progress of Science* (Manchester,
1968), pp. 92–108; and Thackray, *op. cit.* (note 26), pp. 199–233. The basic belief
of eighteenth-century Newtonian chemistry was that chemical phenomena could be
explained in terms of short-range forces which it was the goal of the chemist to
quantify and systematize; hence the keen eighteenth-century interest in tables of
affinity.

principles he succeeded in creating a coherent system and, what is even more important for our purpose, in laying down a program for future work, where before there had been a jumble of rather vague beliefs.

So when, between 1805 and 1807, Laplace published his theoretical studies of the refraction of light and capillary action in the fourth volume of the *Mécanique céleste*[33] and when he based these studies on the assumption that there were short-range attractive forces both between the particles of ponderable matter and between the particles of ponderable matter and those of light, he was using an approach which, in his own eyes and in those of his contemporaries, had the sanction not only of a strong and much admired eighteenth-century Newtonian tradition but also of the most eminent French chemist of the day. Berthollet's may indeed have been one of the last of all attempts to realize the "Newtonian dream" of a quantified chemistry based on the measurement of the forces between the particles of matter; it was also one of the most distinguished of all such attempts, and the fact that it was soon to be made obsolete by the new approach to chemistry initiated by the Daltonian atomic theory in no way diminishes its importance in helping to set the course and objectives of French chemistry in the Napoleonic period. Nor should its imminent rejection lead us to underestimate the influence that it must have exerted on Laplace in bringing him to publish his first detailed study of the molecular forces operating in physics just two years after the analogous forces of chemistry had been treated in the *Statique chimique*. Nor, indeed, should its influence on others be discounted, for it was no coincidence that the Abbé René Just Haüy developed his model for acid-alkali reactions in terms of short-range intermolecular forces in the three years following the publication of the *Statique chimique*.[34] We may be certain, of course, that by the period 1803–1806 Haüy was being influenced quite as much by Laplace as by Berthollet. But the precise circumstances of Haüy's work need not concern us; we need only note it as an important early product of the revival of interest in molecular forces that Berthollet and Laplace jointly fostered.

[33] Laplace, *Mécanique céleste* (note 1), *4* (1805), 231–281 (on refraction). The two supplements (separately paginated) contained his theory of capillary action. On the date of their publication see J.B. Biot, *Journal de physique, 65* (1807), 88; also *Académie des Sciences. Procès-verbaux des séances de l'Académie tenues depuis la fondation de l'Institut jusqu'au mois d'août 1835,* 10 vols. (Hendaye, 1910–1922), *3*, 344 (28 April 1806) and 553 (6 July 1807).

[34] See S.H. Mauskopf, "Haüy's Model of Chemical Equivalence: Daltonian Doubts Exhumed," *Ambix, 17* (1970), 182–191.

There can be little doubt that by 1805 Laplace already had a clear conception of his program for physics and for physical science as a whole.
Certainly he had not yet set down his program formally, as he was to do
rather sketchily some three years later[35] and again in 1823 in the definitive form of the passage quoted at the beginning of this paper; but the
1805 volume of the *Mécanique céleste* and its two supplements, published
in 1806 and 1807, indicate clearly enough that Laplace had formulated
the basic idea of the reduction of all physical phenomena to a system of
densely distributed particles exerting attractive and repulsive forces on one
another at a distance (albeit at a very short distance). In this earliest work
on his program Laplace gave lengthy mathematical treatments of optical
refraction and capillary action, basing both treatments on the supposed
existence of short-range attractive forces of the type that had been first
postulated by Newton and discussed so often through the eighteenth
century.[36] In the case of capillary action these forces were assumed to
exist between the particles of ordinary ponderable matter; in the case of
refraction the attraction was between the particles of ordinary matter and
those of the imponderable light.

Laplace's choice of optical refraction and capillary action as the subjects
for his first *sorties* into the realm of molecular physics reveals clearly his
caution at this stage in his work. For these were both manifestations of
action at a distance on the molecular scale which had been of special

[35] This statement of his program appears in a note added to his "Mémoire sur les
mouvemens de la lumière dans les milieux diaphanes," *Mémoires de la Classe des
Sciences Mathématiques et Physiques de l'Institut de France,* 10 (1809), 300–342.
It dates presumably from between January 1808, when the main paper was read, and
August 1810, when the volume was published. In the note (p. 329) he wrote, with
reference to short-range molecular forces:

In general, all the attractive and repulsive forces in nature can be reduced,
ultimately, to forces of this kind exerted by one molecule on another. Thus, in
my *Theory of Capillary Action,* I have shown that the attractions and repulsions
between small objects floating on a liquid, and generally all capillary phenomena,
depend on intermolecular attractions which are negligible except at insensible
distances. Similarly an attempt has been made to reduce the phenomena of
electricity and magnetism to intermolecular action. The behavior of elastic bodies
also may be treated in the same way.

Later (p. 338), after discussing the relevance of short-range forces to the study of heat
flow, he stated the purposes of his note as follows: "I have sought to establish that
the phenomena of nature can be reduced ultimately to action *ad distans* between
molecules and that the theory of these phenomena must be based on a study of such
action."

[36] For references to this earlier work see notes 22, 24–26, 29, 30, and 32.

interest to eighteenth-century Newtonians, as well as to Newton himself.[37] Moreover, both phenomena had raised problems that remained unsolved even in 1805. The detailed mathematical treatment of refraction in particular had proved difficult, and the study of the molecular forces that caused refraction, although of such obvious interest to Newtonians, had scarcely begun.[38] Likewise, even Clairaut's treatment of the theory of capillary action,[39] which for Laplace was the only discussion worthy of serious consideration,[40] was brief and incomplete. By contrast, Laplace's treatments of both refraction and capillarity were lengthy and, to all appearances, comprehensive. Not surprisingly, the problem of discovering the law relating molecular force and distance, which remained in much the same state it was in after the inconclusive confrontation between Clairaut and Buffon nearly sixty years earlier, was one that the *Mécanique céleste* and its supplements did not solve, but Laplace's demonstration that the form of the law was unimportant at least made the situation appear less scandalous, and to that extent it was a minor triumph.

However, in Laplace's mind there was obviously far more to be achieved by a study of short-range forces than the mere tying up of loose ends, and between 1805 and 1807 he appears to have immersed himself totally in the problems of molecular physics. In this period he brought his writings on capillary action to the notice of the Institute on no fewer than four occasions[41] and engaged others in experiments designed to confirm and enlarge on his theoretical work. For example, Haüy, the Parisian engineer Jean Louis Trémery, and Joseph Louis Gay-Lussac, then a young *protégé* of Berthollet, were all asked to undertake experiments on capillary action,[42] while Jean Baptiste Biot, perhaps Laplace's closest disciple at this

[37]Newton, *Opticks* (note 22), 4th ed., pp. 324-327, 345-349, and 367-371 (Queries 19-21, 29, and 31).

[38]Although the existence of the molecular force that caused refraction was not in doubt for Newtonians, they could say little more than that it diminished rapidly with distance. This was true even of a quite detailed mathematical treatment like Clairaut's (cited in note 30).

[39]See note 30 for reference.

[40]See pp. 2-3 of the first supplement on capillary action in Laplace, *op. cit.* (note 33). Hauksbee was the only other authority mentioned.

[41]*Procès-verbaux, 3,* 293 (2 nivôse an 14; 23 December 1805); *3,* 344 (28 April 1806); *3,* 431 (29 September 1806); *3,* 553 (6 July 1807).

[42]Laplace, "Extrait d'un mémoire sur la théorie des tubes capillaires," *Journal de physique, 62* (1806), 120-128, and the first of the two supplements on capillary action that were added to the tenth book of the *Mécanique céleste* (vol. 4), especially pp. 52-55.

time,[43] was chosen to undertake the experimental investigation of refraction in gases which was proposed to the Institute by Laplace. The paper that resulted from this last investigation, read in March 1806,[44] is an important one in the history of the Laplacian program, since Biot and his young collaborator François Arago were convinced that their observations on the bending of light rays had yielded an accurate measure, indeed the first accurate measure, of forces on the molecular scale which had hitherto almost defied quantification. If, as I believe, Laplace had already conceived his program by this time, the work of Biot and Arago must have marked an advance of outstanding significance in his eyes; and its significance was certainly not lost on Biot and Arago themselves. Adopting what appears to have been standard Laplacian doctrine—that the short-range forces which caused the refraction of light were also the forces of chemical affinity—they maintained that their work, quite apart from its more obvious consequences in the field of optics, might also help to solve the great problems of chemistry, by which, of course, they meant Berthollet's chemistry of affinities and molecular forces. And it was in fact with some justification that they proposed the study of optical refraction as a more promising tool for the investigation of the forces that governed the course of chemical reactions than the direct observation of the reactions themselves, in which the complexities were quite daunting.[45] A table of affinities for light, they believed reasonably enough, would be easier to establish than a table of affinities for, say, oxygen, such as Berthollet had proposed.[46]

Laplace's great strength in the Napoleonic period was, of course, the

---

[43] However, later there emerged certain issues on which Biot and Laplace were apparently not in full agreement. For example, in two contributions to the *Mercure de France* in 1809 Biot took the opportunity of pointing to the dangers of accepting the physical reality of the various imponderable fluids. Such fluids, he wrote, were no more than "a convenient hypothesis to which [true natural philosophers (*physiciens*)] are careful not to attach any idea of reality"; see Biot, *Mélanges scientifiques et littéraires*, 3 vols. (Paris, 1858), 2, 97–116, especially pp. 102–103 and 113–116, and (for the quotation) p. 114. There is further evidence of at least a temporary weakening of the bond between the two men in Laplace's decision to vote for Fourier rather than Biot in the election for the post of permanent secretary of the Academy of Sciences in 1822; see note 143.

[44] Biot and Arago, "Mémoire sur les affinités des corps pour la lumière, et particulièrement sur les forces réfringentes des différens gaz," *Mémoires de la Classe des Sciences Mathématiques et Physiques de l'Institut National de France*, 7 (1806), 301–387. Read 24 March 1806.

[45] *Ibid.*, pp. 327–330.

[46] Berthollet, *op. cit.* (note 31), 2, 3–6.

influence that he was able to wield in the scientific community, an influence that was matched in France only by that of Berthollet, whose zeal for Newtonian science was in any case no less than his own. Laplace naturally did not hesitate to use his position to promote his beliefs, and direct patronage which he and Berthollet dispensed to promising young graduates of the École Polytechnique—Biot, Arago, Gay-Lussac, and Siméon Denis Poisson among them—was only one of his methods.[47] Laplace also wielded to extraordinarily good effect the system of prize competitions organized by the First Class of the Institute. Hence it is not surprising to find him serving on, and presumably dominating, the five-man committee which in December 1807 proposed a mathematical study of double refraction as the subject for the prize for mathematics to be awarded some two years later.[48]

Clearly the intention in setting the subject was that Laplace's treatment of refraction, given in the fourth book of the *Mécanique céleste,* should be extended to embrace double refraction as well. The prospect of success in the enterprise was all the more attractive from the Laplacian point of view because the explanation of double refraction had presented such notorious problems since the discovery of the phenomenon in crystals of Iceland spar in 1669.[49] Even Huygens himself had admitted that his wave theory was inadequate to explain all the observations associated with double refraction;[50] in particular, the behavior of the ordinary and extraordinary rays when passed through a second crystal defied explanation by anyone who, like Huygens, postulated a longitudinal wave motion, and it was only when Fresnel introduced the concept of transverse waves in 1821 and so made an understanding of polarization possible that this grave weakness in the wave theory was removed. Although Huygens' failure was seized upon in the *Opticks,*[51] Newton's own explanation, in terms of the "sides" of rays of light, was vague and difficult to reconcile with a corpuscular theory, and even by the end of the eighteenth century it had

[47] For an excellent study of the patronage system of Arcueil see Crosland, *op. cit.* (note 4).

[48] *Procès-verbaux,* *3,* 632-633 (21 December 1807). The other members of the committee were Lagrange, Legendre, Lacroix, and Lazare Carnot.

[49] For accounts of earlier theories of double refraction see A.I. Sabra, *Theories of Light from Descartes to Newton* (London, 1967), pp. 221-229, and V. Ronchi, *The Nature of Light. An Historical Survey,* trans. V. Barocas (London, 1970), pp. 153, 188-190, and 205-206.

[50] C. Huygens, *Traité de la lumière* (Leiden, 1690), pp. 88-91.

[51] Newton, *Opticks* (note 22), 4th ed., pp. 332-336 and 336-339 (Queries 25, 26, and 28).

been little improved. So to Laplace the theory of double refraction presented the enticing challenge of a long-notorious anomaly; it was, moreover, an anomaly that was likely to become increasingly troublesome,
since, for some years in France, Haüy had been using double refraction as a
valuable exploratory tool for the investigation of crystal structure, and his
work was still proceeding.[52]

Predictably, it was a confirmed supporter of the corpuscular theory of
light and one of the most brilliant of Laplace's disciples at Arcueil,
Étienne Malus, who was awarded the prize of 3000 francs on 1 January
1810. Indeed, it is probable that the subject for the competition was conceived by Laplace not simply in the hope of a glorious victory for the
corpuscular theory but with Malus specifically in mind. For by 1807,
despite some ten years of hard military service in the obscurity of Egypt
and provincial France, Malus had already become closely associated with
the style of physics that prevailed at Arcueil and appears even to have
made contact with Laplace himself. In his earliest paper on light, which
he read in 1807, he did not hesitate to assume the existence of short-range
attractive forces,[53] and Laplace, in his own preliminary study of the
theory of double refraction, which he read to the First Class of the
Institute in January 1808, referred favorably to Malus' experimental work
on the subject.[54] Presumably with the competition in view, Laplace, in
this paper of 1808, also took the opportunity of pointedly laying down
certain principles that he took to be established beyond doubt. Huygens'
wave theory, he maintained, was inadequate for the explanation of double
refraction,[55] so that the task ahead was clear: it was simply to devise a
new explanation in terms of those short-range attractive and repulsive
molecular forces that Newton had invoked in explaining ordinary refraction.[56] Since Laplace himself had already made inroads into the prob-

[52] See, for example, R.J. Haüy, *Traité de minéralogie,* 5 vols. (Paris, 1801), *1,*
229–235, and *2,* 38–51 and 196–229. For a brief account and further references see
Burke, *op. cit.* (note 22), pp. 139–140. Haüy gave cautious support to Newton's
theory of double refraction.

[53] See his "Traité d'optique," *Mémoires présentés à l'Institut National des Sciences,
Lettres et Arts par divers savans et lus dans ses assemblées. Sciences mathématiques
et physiques, 2* (1811), especially 265–266. Read 20 April 1807. Laplace was one of
the referees appointed by the First Class for this paper; see *Procès-verbaux, 3,* 516
and 606–607 (20 April and 19 October 1807).

[54] Laplace, "Mémoire sur les mouvemens de la lumière dans les milieux diaphanes,"
Mémoires . . . de l'Institut, 10 (1809), 300–342, especially pp. 302 and 309.

[55] *Ibid.,* pp. 301 and 303.

[56] *Ibid.,* p. 304.

lem, as he showed in his paper, and since he left no doubt as to his own belief in the physical reality of the molecular forces, it would have been an act of remarkable folly to attack the question set in December 1807 in any other way than Laplace's, and, as we should expect, Malus duly conformed, producing the vindication of the Laplacian position that was expected of him.[57]

As patron Laplace had been well served; specific problems had been answered for him, and the position of his program had been strengthened. And Malus had more than just his prize-winning paper to contribute to the Laplacian program. For example, after his discovery of the polarization of light in the autumn of 1808 he had given his now considerable authority to the view that the new phenomenon could not be explained by Huygens' wave theory, and instead had given an explanation in terms of the various forces which he supposed, in the best Laplacian fashion, to act on the particles of light.[58] In fact, Malus did so much for Laplacian physics in general and for the furtherance of the Laplacian program in particular that his premature death from consumption in February 1812, after some five years of intensive research, must have come as a grievous blow to Laplace.

However, there was no question of abandoning the program, and there were still outstanding issues to be settled. Foremost among these were the problems concerned with the behavior of elastic surfaces. Laplace had already stated that the theory of such surfaces might be established in terms of short-range intermolecular forces of repulsion,[59] and in 1809 a competition on the subject had been set by the First Class of the Institute.[60] Although the competition was stated to have been suggested by Napoleon, who had been greatly impressed by a demonstration of the experiments of Chladni, we may be sure that Laplace, whose zeal for his program was then at its height, had had a hand in the matter. But, presumably to his disappointment, by the closing date on 1 October 1811 no entry of sufficient merit had been received, and the competition had to be set twice more, with closing dates in October 1813 and October 1815, before the

[57]His paper was published in *Mémoires présentés . . . par divers savans*, 2 (1811), 303–508, with the title "Théorie de la double réfraction." For Malus' attempt to explain double refraction in terms of short-range forces of the Laplacian type see especially pp. 489–496.

[58]Malus, "Sur une propriété des forces répulsives qui agissent sur la lumière," *Mémoires de physique et de chimie de la Société d'Arcueil*, 2 (1809), 254–267, especially pp. 260–267.

[59]See the first passage quoted in note 35.

[60]*Mémoires . . . de l'Institut*, 9 (1808), 240–241 of the "Histoire de la Classe" for 1808.

prize was eventually awarded to Sophie Germain in January 1816.[61] This
award represented something of a defeat for Laplacian interests, since
Germain's approach was fundamentally different from that of Poisson,
whose paper on elastic surfaces, read to the First Class of the Institute on
1 August 1814, used short-range forces in the orthodox Laplacian
manner.[62] And, insofar as it was a defeat, it has significance as a mark of
Laplace's declining influence and reputation after the downfall of
Napoleon. But this is a point I shall return to in Section 3.

Another outstanding problem that was settled in the last years of
Napoleon's rule, though this time with rather more success from the
Laplacian point of view, concerned the theory of heat. Again the issue was
raised in a prize competition of the Institute, set in January 1811. As was
made clear by the committee that proposed the subject, it was hoped
above all that the competition, which asked for a detailed experimental
study of the specific heats of gases, would lead to a decision on an important point in the caloric theory.[63] Since the point was one that had to
be resolved before even the simplest mathematical treatment of the theory
could be undertaken, it was of obvious interest to Laplace, and naturally
he sat on the committee that set the competition.[64] The aim was to
decide whether it was possible for some caloric to exist in a body in a
latent, or combined, state (i.e., without being detected by a thermometer)
or whether all of the caloric was present in its "sensible" form and therefore as a contribution to the body's temperature. The issue, which had
been very much a live one since the 1780's, though never more so than in
the first decade of the nineteenth century, had divided calorists into two
groups; the supporters of the former view looked chiefly to Lavoisier,
Laplace, and Joseph Black as their authorities, while those who advocated
the latter, such as Adair Crawford and John Dalton, followed the Scottish
pupil of Black, William Irvine. Predictably enough, victory went, in
January 1813, to two young men, François Delaroche and Jacques Étienne

[61]On the history of this competition see I. Todhunter, *A History of the Theory of
Elasticity and the Strength of Materials from Galilei to Lord Kelvin*, 2 vols. (Cambridge, 1886–1893), *1*, 147–149.

[62]Poisson, "Mémoire sur les surfaces élastiques," *Mémoires . . . de l'Institut*, Pt. 2,
13 (1812), 167–225, especially pp. 171–172 and 192–225.

[63]*Mémoires . . . de l'Institut*, Pt. 2, *11* (1810), xcv of the "Histoire de la Classe"
for 1810.

[64]See *Procès-verbaux*, *4*, 399 (3 December 1810). He did not sit on the adjudi-
cating committee but was well represented by Berthollet and Gay-Lussac; see
Procès-verbaux, *5*, 105 (12 October 1812).

Bérard, who had performed their experiments at Arcueil and who, by upholding the distinction between latent and sensible caloric, vindicated the position long favored by Laplace.[65] One cannot help feeling that the only other competitors, Nicolas Clément and Charles Bernard Desormes, who deviated from the Laplacian view, simply had no chance and were fortunate to receive even the "honorable mention" that was accorded them.

Despite a certain slackening of corporate research activity from 1812, Laplacian physics was still, to all appearances, in a strong position in France toward the end of the Napoleonic period. Certainly the fact that regular meetings of the Society of Arcueil stopped in 1813[66] did not augur well for the future. But Laplacian influence was still just as great in French scientific education as it had long been in research and, because of strong administrative centralization, it was just as easily exercised. An examination of syllabuses, textbooks, and sets of lecture notes of the period shows clearly that pure Laplacian physics was being taught as standard doctrine both in science courses in the lycées[67] and, what is even more important, in the courses that mattered most for the future of French physical science, those at the École Polytechnique. Here at the École Polytechnique, as we know from Hachette's handbook for students of 1809,[68] from the annually published outline syllabuses,[69] and even more clearly from some notes taken (by none other than Auguste Comte)

[65] The paper by Delaroche and Bérard appeared in *Annales de chimie et de physique, 85* (1813), 72-110 and 113-182, with the title "Mémoire sur la détermination de la chaleur spécifique des différens gaz." For a detailed discussion of the competition and the issues at stake in the caloric theory see Fox, *op. cit.* (note 8), especially pp. 25-32 and 104-150.

[66] Crosland, *op. cit.* (note 4), pp. 2-3.

[67] See, for example, R.J. Haüy, *Traité élémentaire de physique,* 1st ed. (Paris, an XII [1803]); 2nd ed. (Paris, 1806); 3rd ed. (Paris, 1821); all three editions are in two volumes. This excellent textbook was written at Napoleon's direction for the use of teachers in the *lycées.*

[68] J.N.P. Hachette, *Programme d'un cours de physique; ou précis de leçons sur les principaux phénomènes de la nature, et sur quelques applications des mathématiques à la physique* (Paris, 1809), especially pp. 1-8 and 49-220.

[69] See, for example, *Programmes de l'enseignement de l'École Royale Polytechnique arrêtés par le Conseil de Perfectionnement, dans la session de 1815-1816* (Paris, 1816), pp. 34-40, where it is clear that even in 1816 heat, electricity, magnetism, and optics were still to be taught in terms of the appropriate imponderable fluids. On the changes in the published syllabus that began to appear in the volume for 1817-1818 see Section 3 and notes 141 and 142.

at Petit's lectures of 1814–1815,[70] the existence of the imponderable fluids—caloric, light, electricity, and magnetism—went virtually unquestioned, and other aspects of Laplacian physics, notably the treatment of capillary action, were given great prominence.

So this was the situation in French physical science up to the end of the reign of Napoleon. Laplace and Berthollet stood for and fostered a unified program for their disciplines, based on a coherent set of traditional doctrines that had originated in Newton's comments on matter and molecular forces in the *Opticks* and had then come down to them through the eighteenth century. And by their zeal and power within the scientific community of France they had established a situation in which they were able to give at least certain of the main branches of physics and chemistry a remarkable degree of uniformity.

Now it must be stressed that the uniformity imposed by the program was by no means complete. In the first place, there were those, outside the Arcueil circle and usually outside the Parisian "establishment" of science, who opposed Laplace. I have referred already to the "outsiders" Clément and Desormes, and in the next section I shall argue that Fourier's mathematical treatment of the distribution of heat in solids appeared as a major challenge as early as 1807. Moreover, by the time of the Bourbon Restoration, Fresnel was already working, in almost total isolation, toward his critique of the corpuscular theory of light. And there were many more whose research was not opposed to Laplacian principles but independent of them. For example, the work of Gay-Lussac and Thenard (both members of the Arcueil circle) on the alkali metals and electrochemistry, Gay-Lussac's experiments on the combining volumes of gases (though they had subversive implications), and J. P. Dessaignes' study of phosphorescence (which won the Institute's prize competition for physics in 1809)[71] simply did not bear on the Laplacian program or, in any direct

---

[70] The notes, in six notebooks bearing the heading "École polytechnique—Cours de physique de M^r. Petit—Comte," cover all the main branches of physics with the exception of light. Comte was in his first year at the École Polytechnique in 1814–1815, and light was always studied in the second-year physics course. The notes are now kept at the Maison d'Auguste Comte, 10 rue Monsieur le Prince, Paris 6e. I am grateful to the resident archivist, D. Cantemir, for allowing me to examine the notes and for providing me with a copy of them.

[71] For accounts of this work see, in addition to the standard histories, Crosland, *op. cit.* (note 4), pp. 354–365, and, by the same author, "The Origins of Gay-Lussac's Law of Combining Volumes of Gases," *Annals of Science,* 17 (1961), 1–26. Dessaignes' prize-winning paper appeared as "Mémoire sur les phosphorescences," in *Journal de physique,* 68 (1809), 444–467, and 69 (1809), 5–35.

way, on the theories on which the program was based. The same may be said of the brilliant crystallography of this period, although the fact that Haüy, like Romé de l'Isle before him, chose to concentrate on the description and measurement of crystals, rather than on the theory of crystallization,[72] which could have included a study of the forces binding together crystals and their constituent molecules,[73] may be seen as a missed opportunity for Laplace and his program.

But, despite these deviations and independent traditions of work and despite Laplace's own occasional vacillation,[74] the uniformity in the physical science of Napoleonic France is striking. Even if in chemistry there had been little progress, the program in physics had been pursued for a decade with vigor and, to all appearances, success. And certainly in France at the beginning of 1815 there seemed no reason why the dominant orthodoxy that had emanated from Arcueil since the early years of the century should, at least in the foreseeable future, be abandoned.

3. THE REJECTION OF LAPLACIAN PHYSICS

Yet within about ten years, by the mid-1820's, the intricate structure of Laplacian physical science had collapsed, leaving just a few increasingly isolated diehards to pursue the chimera that the program and its attendant beliefs were then generally recognized to be. In these ten years of revolt against Laplacian orthodoxy, the tradition that had gone almost unchallenged in the physical sciences in the Napoleonic period was abandoned. To the men who led the revolt it undoubtedly seemed that a new and more glorious era was dawning. Indeed, it may lead us to see Napoleonic science in a somewhat less favorable light if we accept, as I believe we must, that these men shared much of that feeling of exhilaration and liberation of the intellect which Guizot, Edgar Quinet, Lamartine, and so many

[72]On the style of Haüy's work see Metzger, *op. cit.* (note 22), pp. 170–206, and Burke, *op. cit.* (note 22), pp. 78–106.

[73]For example, Guyton de Morveau had discussed the cause of crystallization at some length in the "Essai physico-chimique sur la dissolution et la crystallisation," in his *Digressions académiques, ou essais sur quelques sujets de physique, de chymie & d'histoire naturelle* (Dijon, 1762), pp. 271–359, and in his *Élémens de chymie théorique et pratique,* 3 vols. (Dijon, 1777–1778), *1,* 49–78. With acknowledgments to the work of Clairaut and Buffon, he treated the problem in terms of the same intermolecular forces by which he accounted for other chemical phenomena.

[74]See below, pp. 111 and 127, and notes 93 and 143.

110   THE RISE AND FALL OF LAPLACIAN PHYSICS

others associated with the Bourbon Restoration.[75] The men of science may not have gone so far as to see the Empire as an intellectual "desert," as Quinet did,[76] but, contrary to general belief, I do feel that most of them participated fully in the new optimism which so many Frenchmen experienced in those early Restoration years. Sadly, this enthusiasm and spirit of optimism, which was to be so fruitful in literature and the arts, in science came to nothing. For, as we can now see, what emerged from the ruins of Laplacian orthodoxy was not the new, revivified physical science that the early years of the Restoration had seemed to promise, but only a burst of creativity whose duration was no less brief, and whose decline was even more drastic, than that of Napoleonic physical science.

The men chiefly responsible for the revolt of the decade 1815-1825 were Joseph Fourier, Pierre Dulong, François Arago, Augustin Fresnel, and Alexis Thérèse Petit. Of these only Fourier, born in 1768, was over thirty in 1815, so that, with this one exception, they had learnt their science and performed their earliest work in the period when Laplacian principles had enjoyed their greatest success in France. Dulong, Arago, Fresnel, and Petit had all been thoroughly indoctrinated with these principles as students at the École Polytechnique;[77] and Dulong and Arago had even been members of the Arcueil circle (though, significantly perhaps, only since about 1810), and both had benefited in their careers from Arcueil patronage.[78] Petit, by contrast, was not a member of the circle, but his brilliant doctoral thesis of 1811 on capillary action[79] and his first lectures as professor of physics at the École Polytechnique in the winter of 1814-1815[80] were as Laplacian as they could possibly have been.

[75] F.P.G. Guizot, *Mémoires pour servir à l'histoire de mon temps*, 8 vols. (Paris, 1858-1867), *1*, 27-58; E. Quinet, *Histoire de mes idées. Autobiographie*, 7th ed. (Paris, 1895) [vol. 15 of the *Oeuvres complètes d'Edgar Quinet*], especially pp. 177-187 and 239-247; A.M.L. de P. de Lamartine, *Des destinées de la poésie* (Paris, 1834), in *Oeuvres complètes de Lamartine publiées et inédites*, 41 vols. (Paris, 1860-1866), *1*, 30-32. In certain ways, notably in matters of religion, the Restoration did not bring greater freedom, but generally the characterization of the period by Guizot, Quinet, and Lamartine seems just.

[76] Quinet, *op. cit.* (note 75), p. 241.

[77] In, respectively, 1801-1802, 1803-1805, 1804-1806, and 1807-1809. Because of ill health Dulong failed to complete the course.

[78] See Crosland, *op. cit.* (note 4), pp. 315-318.

[79] Petit, "Théorie mathématique de l'action capillaire," *Journal de l'École Polytechnique*, cahier 16, *9* (1813), 1-40. On his continued support of the Laplacian theory after 1815 see note 141.

[80] See note 70.

It is probably no coincidence that the challenge to the prevailing orthodoxy was first raised by the two members of the anti-Laplacian group, Fourier and Fresnel, who spent the greater part of the Napoleonic period in provincial obscurity far from Paris and hence far from the center of Laplacian control.[81] Fourier was nearly twenty years older than any other member of the group, and even when teaching at the École Polytechnique soon after its foundation in 1794 he had affiliated more closely with Monge than with Laplace and Legendre. With time taken from heavy prefectural duties at Grenoble (imposed on him by Napoleon in 1802 soon after his return from distinguished service in Egypt), he prepared a massive treatise on the distribution of heat in solid bodies, which he read to the First Class of the Institute in December 1807.[82] In this the entire Laplacian machinery of derivation of the basic equations by Newtonian principles was ignored by Fourier, who concentrated instead on his own methods for their derivation and solution. In 1811 he took his work further when he submitted a revised version of the 1807 paper for a prize competition set by the First Class of the Institute, and in January 1812 he was awarded the prize. But, despite evidence that he was receiving some favor in the eyes of Laplace himself,[83] the challenge was premature. His 1807 paper was published only in the form of an abstract, drawn up by a less than enthusiastic Poisson;[84] and even his great prize-winning paper was criticized by the judges (Lagrange, Laplace, Legendre, Malus, and Haüy) and, apparently as a result of this criticism (for which Lagrange was chiefly responsible), did not appear in print until 1824-1826,[85] when

[81] For the relevant biographical details of Fourier and Fresnel see the article by J.R. Ravetz and I. Grattan-Guinness (on Fourier) and that by R.H. Silliman (on Fresnel) in C.C. Gillispie, ed., *Dictionary of Scientific Biography* (New York, 1972), 5, 93-99 and 165-171; also I. Grattan-Guinness (in collaboration with J.R. Ravetz), *Joseph Fourier 1768-1830. A Survey of his Life and Work, based on a Critical Edition of his Monograph on the Propagation of Heat, presented to the Institut de France in 1807* (Cambridge, Mass., and London, 1972), pp. 14-25 and 441-459.

[82] *Procès-verbaux*, 3, 632 (21 December 1807). On the 1807 paper see J.R. Ravetz, "Preliminary Notes on the Study of J.B.J. Fourier," *Archives internationales d'histoire des sciences, 13* (1960), 247-251; I. Grattan-Guinness, "Joseph Fourier and the Revolution in Mathematical Physics," *Journal of the Institute of Mathematics and Its Applications, 5* (1969), 230-253; and, for a detailed study and critical edition, Grattan-Guinness, *op. cit.* (note 81).

[83] Discussed in Grattan-Guinness, *op. cit.* (note 81), pp. 444-452.

[84] [Fourier], "Mémoire sur la propagation de la chaleur dans les corps solides," *Nouveau bulletin des sciences par la Société Philomathique de Paris, 1* (1807-1809), 112-116 (in the issue for March 1808).

[85] Fourier, "Théorie du mouvement de la chaleur dans les corps solides," *Mémoires de l'Académie Royale des Sciences de l'Institut de France, 4* (1819-1820 [published

the whole spirit of French physical science was very different and when Fourier himself was a permanent secretary of the Academy of Sciences. So it was only when the sympathetic Arago became one of the joint editors of the reorganized *Annales de chimie et de physique* in 1816 that the public had the opportunity of learning any details of Fourier's achievement (by means of a lengthy summary that appeared in the *Annales*[86]).

It was about this time that Fresnel, too, began to make his mark on the Parisian scientific scene. In October 1815, having recently gained some months of leisure for research (a leisure that he owed incidentally to his expulsion from office for his royalist sympathies during the Hundred Days), he deposited his first paper on the diffraction of light at the Institute.[87] In this he gave powerful support to the wave theory of light and in doing so exposed serious shortcomings in the rival corpuscular theory. Immediately he won over Arago and Petit, hitherto good Laplacians, and by December 1815 these two new converts had even performed some experiments on refraction in gases which they interpreted in such a way as to support Fresnel.[88] Ampère, whose commitment to the anti-Laplacian cause became really apparent only in the early 1820's, had

---

1824]), 185–555, and 5 (1821–1822 [published 1826]), 153–246. On the criticism and delay in publication see G. Darboux's introduction to the *Oeuvres de Fourier*, 2 vols. (Paris, 1888–1890), *1*, vi–viii; also Arago's *éloge* of Fourier in *Mémoires de l'Académie des Sciences, 14* (1838), cxii–cxiii.

[86]"Théorie de la chaleur. Par M. Fourier. (Extrait)," *Annales de chimie et de physique, 3* (1816), 350–375. The volume from which the summary was made, described in a footnote as a quarto volume of 650 pp., never appeared. It would have been normal for Arago himself to prepare the summary, but it displays such familiarity with Fourier's work that Fourier himself was probably the author. This is the conclusion reached in Grattan-Guinness, *op. cit.* (note 81), p. 460n. If the summary was written by someone else, Sophie Germain may have been responsible, as Ravetz has suggested to me.

[87]Fresnel, "Mémoire sur la diffraction de la lumière, où l'on examine particulièrement le phénomène des franges colorées que présentent les ombres des corps éclairés par un point lumineux," *Annales de chimie et de physique, 1* (1816), 239–281. Presented to the First Class of the Institute on 23 October 1815 (*Procès-verbaux, 5*, 562).

[88]Arago and Petit, "Sur les puissances réfractives et dispersives de certains liquides et des vapeurs qu'ils forment," *Annales de chimie et de physique, 1* (1816), 1–9. Read to the First Class of the Institute on 11 December 1815. Two letters that illustrate Arago's enthusiasm for Fresnel's ideas during his period of conversion (both of them from Léonor Mérimée to Fresnel) are in *Oeuvres complètes d'Augustin Fresnel,* 3 vols. (Paris, 1866–1870), *2*, 831–833. According to Léonor Mérimée's letter of 20 December 1814 (*ibid., 2*, 830–831), Arago first learnt of Fresnel's work in December 1814. On the support that the work of Arago and Petit gave to Fresnel see Fox, *op. cit.* (note 8), pp. 202 and 233–234.

been won over by May 1816,[89] and even Berthollet's former *protégé* Gay-Lussac, now beginning to take on the mantle of his master as France's leading chemist, was sympathetic.[90] The interest in this challenge to the Laplacian position was enormous, and it was reflected most obviously in January 1817 in the decision of a committee of the Academy of Sciences, consisting of Laplace, Biot, Berthollet, Gay-Lussac, and the aged physicist J. A. C. Charles, to offer the prize in physics for a study of diffraction.[91] It seems clear that in this way the still powerful Laplacian party hoped to settle the issue finally in its own favor by bringing this important phenomenon of physical optics into line with polarization and double refraction, which had been explained so successfully in terms of the corpuscular theory. To Biot victory for a corpuscularian was an especially alluring prospect since, with the young C. S. M. Pouillet, he had recently been engaged in devising the corpuscular theory of diffraction which he described, with much other evidence likely to support the materiality of light, in his *Traité de physique* of 1816.[92] But the ruse—if such it was—backfired, for, despite the fact that among the five judges were Laplace himself and the two arch-Laplacians Biot and Poisson, Fresnel won the prize in March 1819 with a brilliant paper.[93]

[89] See Ampère's letter to Ballanche Fils, 19 May 1816, in *Correspondance du grand Ampère*, ed. L. de Launay, 3 vols. (Paris, 1936–1943), *2*, 511. Ampère was in no sense a typical or central figure in the anti-Laplacian group, but in his conflict with Biot in the early 1820's (see below, pp. 117) he attacked some of the fundamental beliefs of Laplacian physics. The central forces that were so important in his electrodynamic theory were decidedly, and significantly, not Laplacian in character, and a clear mistrust of Laplacian fluids can be seen in his *Théorie mathématique des phénomènes électro-dynamiques* (see below, pp. 117 and 128 and note 153). Moreover, in the 1820's Ampère experienced the direct opposition of Laplace concerning the possible identity of magnetism and electricity. In supposing the two to be identical Ampère was breaking with the view of Coulomb, which Laplace supported; see Ampère's letter to Davy, probably of 1825, in *Correspondance du grand Ampère*, *2*, 680.

[90] See Arago, "Éloge historique de T. Young," *Mémoires de l'Académie des Sciences, 13* (1835), cii–civ.

[91] *Procès-verbaux, 6*, 138 (13 January 1817).

[92] Biot, *Traité de physique expérimentale et mathématique*, 4 vols. (Paris, 1816), *4*, 743–775. In this volume of the *Traité* Biot wrote at great length on polarization (pp. 254–600), which he felt to be adequately explained by his (corpuscularian) theory of mobile polarization. As Frankel has pointed out to me, the *Traité* has great importance both as a restatement of Laplacian doctrines, especially in optics, and as evidence of continuing work on the program after the death of Malus.

[93] Fresnel, "Mémoire sur la diffraction de la lumière," *Mémoires de l'Académie des Sciences, 5* (1821–1822 [published 1826]), 339–445. On the identity of the judges see *Procès-verbaux, 6*, 345 (27 July 1818). The other judges were Fresnel's good

However, on the question of the nature of light the Laplacians did not give up easily. In 1837 William Whewell raised the possibility that Laplace, Biot, and Poisson were chiefly responsible for a seven-year delay in the publication of Fresnel's prize-winning paper, and it seems not inconceivable that they even resorted to such underhand methods as mislaying some of his other papers.[94] But they were fighting a losing battle, as even Biot finally recognized when, in the early 1820's after some protracted and acrimonious debate with Arago, he retired from the Parisian scientific community (in particular, from the Academy of Sciences) for several years,[95] and so, at least in the eyes of his contemporaries, conceded victory to his adversary.[96]

The successful attack on the corpuscular theory of light helped to

---

friend Arago and Gay-Lussac who, despite his closeness to Berthollet and Laplace, had already shown some sympathy toward the wave theory, as we have seen. It is interesting to speculate on the course of the discussions that gave Fresnel his victory. Arago and presumably Gay-Lussac would have supported Fresnel, while Poisson, always the most orthodox of Laplacians, and Biot, whose acrimonious public debate with Arago was now imminent, would surely have opposed him. Perhaps, therefore, it was Laplace himself who swayed the decision. It becomes less difficult to imagine Laplace supporting anti-Laplacians when we note how in 1822 he was to vote for Fourier rather than Biot in the election for a new permanent secretary of the Academy of Sciences; see note 143. And already he had shown some favor to the work of Fourier when he might have been expected to support Poisson; see Grattan-Guinness, *loc. cit.* (note 83).

[94] For Whewell's somewhat speculative account see his *History of the Inductive Sciences*, 3 vols. (London, 1837), *2*, 408–411; also his recollection in his paper "Comte and Positivism," *Macmillan's Magazine*, *13* (1866), 355–356, where it is stated that Arago had told Whewell that the Laplacian domination of French physical science had been so effective about 1815 that he had actually been afraid to voice his early support for Fresnel. The lengthy delays in the publication of the prize-winning papers by Fourier and Fresnel should, of course, be compared with the delay of little more than a year in the publication of the winning paper of Malus (see note 57), but it should also be noted that it was not only members of the anti-Laplacian group who suffered in this way. Cauchy, for example, had to wait more than ten years before his prize-winning paper of 1815 on water waves was published, with additions, in *Mémoires présentés par divers savans à l'Académie Royale des Sciences*, *1* (1827), 3–312.

[95] The *Procès-verbaux* for the period show that Biot's appearances at the meetings of the Academy of Sciences were infrequent from the autumn of 1822 until the early 1830's. Between the end of January 1823 and the end of 1824 he was present on only three occasions. About this time, however, he was not inactive and was engaged, for example, in preparing the third edition of his *Précis élémentaire de physique* (note 108).

[96] Arago was left, as Guglielmo Libri put it in the *Revue des deux mondes*, ser. 4, *21* (1840), 799, "maître du champ de bataille." See the passage quoted on pp. 123–124.

create an atmosphere in which it was natural that other Laplacian beliefs should be subjected to a new scrutiny. Once action at a distance was discredited in one branch of the Laplacian program, it became far easier to attack it in other branches; and the program in its strict and complete sense naturally collapsed completely. Moreover, the threat to Laplace and his school was heightened by another challenge that achieved success almost simultaneously. This was the challenge of Sophie Germain, whose victory in the Academy's prize competition in 1816 was, as I have already mentioned,[97] both a blow to Laplacian interests, as represented above all by Poisson, and a sign of diminishing Laplacian control. Like the criticism of the corpuscular theory of light, Germain's treatment of elastic surfaces stimulated a prolonged controversy, which lasted far into the 1820's and engaged Poisson in a bitter debate with a group of critics inspired by Fourier.[98]

In such conditions of mounting criticism it seems plausible to interpret the attack on the caloric theory which accompanied Petit and Dulong's announcement of their famous law of atomic heats in 1819[99] as a natural product of a questioning mood that had come to prevail in French science since 1815. In fact, such an interpretation seems necessary. For, despite the confidence with which Petit and Dulong stated their criticism of caloric,[100] the justification for their attack in terms of experimental facts was far from conclusive. They brought forward virtually no new evidence, and the attack was one that could have been made equally well ten years earlier.[101] However, as we know, it was not made at that time—and this in itself is strong evidence that the general intellectual atmosphere of the Napoleonic period was very different from that which existed within five years of Napoleon's downfall.

Another illustration of the changing atmosphere may be found in chemistry, where, at precisely the same time as the first attacks on Laplacian physics were being launched, there was a similar turning away from related

[97] See above, p. 106.

[98] See below, pp. 118-119.

[99] Petit and Dulong, "Recherches sur quelques points importans de la théorie de la chaleur," *Annales de chimie et de physique*, 10 (1819), 395-413, especially pp. 396-398 and 406-413.

[100] A confidence that is especially apparent in Dulong's letter to Berzelius, 15 January 1820, in *Jac. Berzelius Bref*, ed. H.G. Söderbaum, 6 vols. in 14 parts (Uppsala, 1912-1932), Pt. 1, 2, 13-14.

[101] For an account of the criticisms by Petit and Dulong see R. Fox, "The Background to the Discovery of Dulong and Petit's Law," *The British Journal for the History of Science*, 4 (1968-1969), 1-22, especially pp. 9-16.

principles that had gone virtually unchallenged in the Napoleonic period. In this challenge the break with the past was manifested not so much by an explicit, open attack on Berthollet's chemistry as by the gradual acceptance of Dalton's atomic theory, which directed attention away from molecular forces to combining weights. Again it is 1815, when France renewed close contact with Britain, which seems to be the turning point, for until that date the atomic theory, opposed by Berthollet, had made little headway in France.[102] Incidentally, I need hardly say that I consider it significant that the same man who appeared as the leading French critic of the caloric theory from 1819—Dulong—was also known, from 1816, as the most enthusiastic supporter of the atomic theory in chemistry.[103]

So the attack on the caloric theory in 1819 and the new support for the atomic theory both seem to reflect the critical spirit that was abroad in French science during the early years of the Restoration, insofar as neither depended essentially on any startling new discoveries or observations. Yet discoveries and experimental evidence did contribute to the weakening of the Laplacian position. It was a great blow to the corpuscular theory when, in 1819, experiment confirmed the prediction based on Fresnel's wave theory that there should be illumination at the center of the diffraction pattern of a small opaque disc,[104] and the discovery of Dulong and Petit's law did much to strengthen the atomists' case.

Perhaps the best illustration of the importance of a discovery in weakening Laplacian physics is in electromagnetism. Following Oersted's observation of the magnetic effect of a wire carrying an electric current in 1820, French physicists zealously engaged in the investigation of the new phenomenon, and Biot and Ampère were quickly among the most prominent of them. There were, of course, problems for the Laplacians. For example, electromagnetism introduced a rotational force which had no obvious connection with the central forces of Laplacian physics; and Coulomb, whose views on the electrical and magnetic fluids had become part of the Laplacian orthodoxy, had denied the possibility of an interaction between

[102]The early history of the theory in France is well described in M.P. Crosland, "The First Reception of Dalton's Atomic Theory in France," in Cardwell, op. cit. (note 32), pp. 274-287.

[103]See Fox, op. cit. (note 101), pp. 16-18.

[104]On this crucial experiment, which was suggested by Poisson in his capacity as one of the judges for the competition on diffraction but which did little to shake his confidence in the corpuscular theory, see Ronchi, op. cit. (note 49).

electricity and magnetism.[105] But Biot, then at his most belligerent, was undaunted, and inevitably, it seems, there was conflict.

Ampère's theory of electromagnetic interaction contained much that Biot found objectionable.[106] In particular, he protested at Ampère's attempt to reduce not only electromagnetic phenomena but even the forces between magnets to interactions between current-carrying conductors; magnetic forces, in his view, had been explained perfectly well in terms of Coulomb's two fluids of magnetism. And the fact that Ampère retained fluids of electricity was no consolation, for Ampère's fluids were thoroughly un-Laplacian and apparently had more in common with Fresnel's ether than with the fluids of Coulomb.[107] Between 1821 and 1824 Biot put forward his alternative explanation, while pursuing a policy of faint praise, misrepresentation, and open criticism toward the work of Ampère.[108] By 1824, when he published the third edition of his *Précis élémentaire de physique expérimentale,* he had developed fully a theory in which the forces of electromagnetism were explained in terms of magnetic interactions between tiny magnets which he supposed to be arranged in a circular fashion around the current-carrying wire.[109] For Biot, convinced of the correctness of Coulomb's explanation, the cause of magnetic interactions was, of course, not in doubt, so that his electromagnetic theory appeared to him a highly satisfactory one. But the model did not withstand the scrutiny of Ampère, who quickly demonstrated its weaknesses, while defending himself against the charges that his own theory was un-Newtonian.[110] So discredited, Biot's theory was soon forgotten, and the attempt to treat the exciting new phenomena in accordance with Laplacian principles had failed.

[105]See L.P. Williams, "Ampère's Electrodynamic Molecular Model," *Contemporary Physics, 4* (1962), 113-114.

[106]For my treatment of Biot's response to the work of Ampère I am greatly indebted to E. Frankel, *Jean Baptiste Biot: The Career of a Physicist in 19th-Century France* (Univ. of Princeton Ph.D. thesis, 1972), of which the author has kindly allowed me to see the relevant chapters.

[107]See Williams, *op. cit.* (note 105), pp. 118-122.

[108]See especially the cursory treatment of Ampère's work in Biot's paper "Sur l'aimantation imprimée aux métaux par l'électricité en mouvement," *Journal des savants* (1821), pp. 221-235, and the overt criticism in his *Précis élémentaire de physique expérimentale,* 3rd ed., 2 vols. (Paris, 1824), *2,* 771-772.

[109]Biot, *Précis* (note 108), *2,* 766-771.

[110]See the many references to Biot in Ampère, *Théorie mathématique des phénomènes électro-dynamiques, uniquement déduite de l'expérience* (Paris, 1826), especially pp. 180-188.

By the mid-1820's, then, the position of the Laplacian orthodoxy had been gravely weakened. The attacks had been directed, for the most part, against established beliefs in various branches of physical science, but there is evidence of diminishing Laplacian authority in mechanics also. The relevant debate dates principally from the 1820's, although several of the issues had been raised some years earlier, notably in the prize competition won by Sophie Germain in 1816. It concerned the methods to be used in rational mechanics, particularly in the study of elastic media.[111] Underlying the debate, in which Fourier's *protégés* Navier and Germain were opposed most frequently by Poisson, there was a fundamental opposition between the "physical mechanics" (*mécanique physique*) advocated by Poisson and the style of the Fourier school, which Poisson described as "analytical mechanics" (*mécanique analytique*) and (incorrectly) associated with the name of Lagrange.[112] Although he had been using his style of mechanics as early as August 1814,[113] it was in a paper of April 1828 that Poisson stated its principles most clearly, when he supported it in the following terms:

> Let me add that it would be desirable for geometers to re-examine the leading problems of mechanics from this point of view, which is at once physical and consonant with nature. In order to discover the general laws of equilibrium and motion, it was necessary to treat them in a completely abstract manner; and, as far as treatments of this general and abstract kind are concerned, Lagrange went as far as anyone could imagine when he replaced the physical connections between bodies by equations relating the coordinates of the various positions they occupied. It is this that constitutes analytical mechanics. But besides this wonderful conception we can now establish physical mechanics, the sole principle of which is to reduce everything to the molecular actions which

[111]For a factual account see Todhunter, *op. cit.* (note 61), *1*, 133–160 and 277–285.

[112]For discussions of this opposition see P. Duhem, "L'évolution de la mécanique," *Revue générale des sciences* (1903), pp. 127–132; L. Brunschvicg, *L'expérience humaine et le causalité physique* (Paris, 1922), pp. 327–337; J.W. Herivel, "Aspects of French Theoretical Physics in the Nineteenth Century," *The British Journal for the History of Science, 3* (1966–1967), 121–125. In various ways Fourier and his associates, Navier and Germain, were related back to the Basel-St. Petersburg school of rational mechanics of the earlier eighteenth century, a school that embraced the Bernoullis and Euler. The line of descent to the nineteenth century touches Lagrange only incidentally and altogether bypasses Laplace.

[113]In Poisson, *op. cit.* (note 62).

convey from one point to another the effects of the given forces and are the agents maintaining the equilibrium between these forces. If we proceeded in this way, it would no longer be necessary to draw up special hypotheses when one wanted to apply the general rules of mechanics to particular questions. Thus, in the problem of the equilibrium of flexible strings ·the tension that is introduced to achieve a solution would be the direct result of the actions of the molecules on one another when they are displaced slightly from their natural positions. In the case of an elastic membrane, the bending moment of elasticity would be a consequence of these same actions, taken throughout the whole thickness of the sheet, and the expression for it would be established without any hypothesis. And, finally, the pressures exerted by fluids both internally and on the walls of the vessels containing them would also be the resultant of the actions of the molecules on the surface under pressure, or rather on an extremely thin layer of fluid in contact with each surface.[114]

In seeking to explain phenomena in terms of the attractive and repulsive forces operating on the molecular scale, Poisson was of course declaring his allegiance to those principles which had dominated the *Système du monde* and the *Mécanique céleste*, but by the late 1820's, with Laplace now dead, he was increasingly isolated. He had no school to support him, and through the 1830's the advocates of Fourier's style of mechanics, led by Lamé and encouraged by Comte, carried all before them.[115]

Given this evidence, it is probably not too much to say that by the early 1820's there had emerged in France certain anti-Laplacian principles, not only in physics but also in chemistry and mechanics, to which all critics of Laplacian science could subscribe. Of these, skepticism toward the traditional imponderable fluids, sympathy for Dalton's atomic theory, the new rational mechanics of Fourier and his followers, and Ampère's electrodynamics were the most obvious. That those who sought to break with Laplace and his school had so many shared beliefs and operated on such a broad intellectual front is striking. For instance, Petit, as well as evidently sharing Dulong's skepticism toward caloric and his enthusiasm for the atomic theory, was among the earliest supporters of Fresnel's wave theory,

[114]Poisson, "Mémoire sur l'équilibre et le mouvement des corps élastiques," *Mémoires de l'Académie des Sciences, 8* (1829), 361–362. Read 14 April 1828.
[115]Brunschvicg, *op. cit.* (note 112), p. 331.

as were Dulong and Ampère.[116] Fresnel, for his part, was a critic of the traditional caloric theory and influenced Ampère in his work on electromagnetism,[117] and Arago not only championed Fresnel in his difficult early years after 1815 but also took a keen and highly favorable interest in the work of Petit and Dulong on heat.[118] Likewise, Fourier, whose role may be seen as that of a benign, influential, but rather detached patron of the new generation, expressed his support for the wave theory[119] and, by his extreme caution on the question of the nature of heat, notably in his *Théorie analytique de la chaleur* (1822), implied unmistakable criticism of caloric.[120] Such unanimity is, in fact, hardly surprising, for the members of the anti-Laplacian group were in close, almost daily contact in the scientific circles of Paris,[121] where they all lived and worked. Between some of them the relationship was especially close. Petit and Dulong were most intimate friends (until Petit's untimely death in 1820), and the same may be said of Ampère and Fresnel and of Arago and Fresnel. Arago, moreover, became Petit's brother-in-law when Petit married in November 1814.

Throughout the Restoration period the reaction of the Laplacian party to the growing criticism was complex. If we are to judge by Biot's response to Fresnel's wave theory, for example, the party felt the attacks keenly. But in their publications, at least, they gave the impression that little had changed. A comparison of the third edition of the *Précis élémentaire* (1824) with the *Traité de physique* (1816) shows that Biot was willing to make few concessions with regard to the imponderable fluids.[122] And

116See Petit and Dulong, *op. cit.* (note 99), p. 396; Dulong's letters of 15 January 1820 and 10 November 1825 to Berzelius, in *Berzelius Bref* (note 100), Pt. 1, 2, 13 and 64; and his comment in *Annales de chimie et de physique, 31* (1826), 180-181. Also see above, pp. 112-113.

117See his "Complément au mémoire sur la diffraction," dated 10 November 1815, in *Oeuvres complètes d'Augustin Fresnel* (note 88), *1*, 59-60; also his letters of 5 July 1814 and 11 July 1814 to Léonor Fresnel, *ibid.*, *2*, 820-822 and 827-829. The connection between Fresnel and Ampère is discussed in Williams, *op. cit.* (note 105), pp. 118-120.

118See Fox, *op. cit.* (note 101), p. 2.

119In a report, written jointly with Ampère and Arago, on a paper by Fresnel on double refraction. See *Annales de chimie et de physique, 20* (1822), 337-344.

120The book began (on p. i) with the words: "First causes (*les causes primordiales*) are unknown to us; but they are subject to simple, unvarying laws which can be discovered by observation and the study of which is the object of natural philosophy."

121Fresnel is, in part, an exception since until 1818 his visits to Paris were restricted to periods of leave from the Corps des Ponts et Chaussées.

122Certainly, in the *Précis* (note 108), *1*, 466, and *2*, 2, Biot confessed ignorance

Laplace and the ever-loyal Poisson continued to work out their programs in accordance with the principles laid down two decades earlier.[123] Even in the fifth edition of the *Système du monde*, published in 1824, Laplace gave no indication that he had modified his view in any way,[124] and between 1821 and 1823 he confidently proceeded to devise and publish what was easily the most detailed of all versions of the caloric theory accounting for the physical and thermal properties of gases in terms of those short-range forces that he still supposed to govern all phenomena on the molecular scale.[125] Indeed, it was in a paper on the subject published early in 1822 that he first gave the classic statement of the Laplacian program, one version of which is reproduced at the head of this paper. In his work on caloric in the 1820's Poisson showed a similar disregard for the criticisms of Petit and Dulong (and of others, including that important ally of the anti-Laplacian cause Berzelius).[126] And even as late as

of the true nature of heat, electricity, and magnetism, but he denied that the corpuscular theory had been discredited and maintained that the evidence still favored it (*2*, 130-132 and 452-463). In his *Traité de physique* (note 92), *1*, 66-68, he admitted that the existence of a fluid of heat was not certain, but the existence of fluids of electricity and magnetism was "very probable" (*1*, 7-8), and the materiality of light was "beyond doubt" (*3*, 148-149). This confident statement concerning light conflicts with his assertion, in the dedication of the *Traité* to Berthollet (*1*, xx-xxiii), that it was impossible to know its nature "with certainty." The caution that Biot displays, even in the *Traité*, is striking and it has to be compared with his criticism of fluids, referred to in note 43. Having regard to his optical work and his vigorous defense of the corpuscular theory, however, I feel (with Frankel, *op. cit.* [note 106]) that his caution was formal and that it does not convey the true measure of his conviction which is apparent in his major interpretative papers.

[123]On Laplace's changing attitude to Fourier, however, see below, p. 127, and notes 93 and 143.

[124]See, for instance, the "Avertissement" to his *Exposition du système du monde*, 5th ed., 2 vols. (8vo) (Paris, 1824), *1*, v, where Laplace wrote that he intended to make molecular forces the subject of a special supplement. Work on this project was never completed, as we see from the "Avertissement" to the quarto version of the sixth edition, published posthumously in 1835, and in this edition a chapter on molecular attraction which had appeared on pp. 315-357 of the (quarto) fourth edition (Paris, 1813) was simply reinstated (as Chapter XVIII of Book IV, on pp. 323-364). The chapter did not appear in the octavo versions of the sixth edition published in Paris and Brussels in 1827.

[125]This theory appeared first in a series of papers in the volumes of the *Connaissance des tems* for 1824 and 1825 (published respectively in 1821 and 1822) and was given its definitive form in April 1823 in Book XII of the *Mécanique céleste* (note 1), *5*, 87-144.

[126]See, for example, his "Mémoire sur les équations génerales de l'équilibre et du mouvement des corps solides élastiques et des fluides," *Journal de l'École Polytechnique*, cahier 20, *13* (1831), 1-174, especially pp. 4-8. Read to the Academy of

1835 we find him publishing a lengthy work, the *Théorie mathématique de la chaleur*, in which the existence of caloric and its traditional properties were taken as no less axiomatic than they would have been, say, thirty years before.[127] But by 1835 Poisson's book, although its author does not seem to have recognized the fact, was a relic of a bygone age, an anachronism in terms both of its physics and of its laborious and inelegant mathematics. And, to judge by the almost complete silence in which it was received, it was seen as such by his contemporaries.[128] By the 1830's Poisson was a lone, almost pathetic figure, clinging vainly to an ideal of a "physical mechanics," based on Laplacian principles, which was unrealizable. When he died in 1840, the mathematician Guglielmo Libri wrote of his funeral, in a notice of singular warmth and affection:[129] "Never, since the death of Cuvier, had anyone seen such general sorrow nor a cortège accompanied by so many demonstrations of grief of every kind."[130] But in reality there were few men of note to mourn him. He had no official *éloge*, and even the biographical memoir by the permanent secretary of the Academy of Sciences, Arago, who had been for so long the scourge of the Laplacian school, was never read in full.[131]

Sciences on 12 October 1829. On Berzelius as an ally of the cause see Fox, *op. cit.* (note 8), pp. 241–243 and 246–248.

[127]Poisson, *Théorie mathématique*, especially p. 7. The problem of heat diffusion in solids, which was treated in the *Théorie mathématique*, had been of great interest to Poisson for over twenty years. As is pointed out in Grattan-Guinness, *op. cit.* (note 81), pp. 466–470, his labored contributions on the subject are in sharp contrast with the elegant and incisive treatments of Fourier, and in a paper published in July 1823 Poisson did go so far as to acknowledge Fourier's priority with regard to most of his own results, though presumably with some reluctance; see his "Mémoire sur la distribution de la chaleur dans les corps solides," *Journal de l'École Polytechnique*, cahier 19, *12* (1823), 1–2. However, by way of justification of his own work, he stressed that his methods for deriving the results were different from Fourier's and that he had used Laplace's assumption that heat transfer within a solid was a short-range phenomenon (pp. 2–6).

[128]The rare comments which have been found, such as that by J.D. Forbes in his *Review of the Progress of Mathematical and Physical Science* (Edinburgh, 1858), p. 154, and the anonymous review in the Swiss *Bibliothèque universelle, 59* (1835), 144–166, are generally critical.

[129][Libri], "Lettres à un Américain sur l'état des sciences en France—III. M. Poisson," *Revue des deux mondes*, ser. 4, *23* (1840), 410–437.

[130]*Ibid.*, p. 429.

[131]And even when extracts from the memoir were read at the public meeting of the Academy, on 16 December 1850, Arago was absent; see the *Comptes rendus hebdomadaires des séances de l'Académie des Sciences, 31* (1850), 840. The whole memoir was printed in the *Oeuvres complètes de François Arago*, 17 vols. (Paris and Leipzig, 1854–1862), *2*, 593–689.

Naturally enough purely intellectualist factors were not alone in bringing about the move from Laplacian science. This was not simply a case of new principles being measured against old ones and being found superior, although there was something of this in the situation, especially with regard to the debate over the nature of light. Other relevant factors include the weakening of the authority of the Arcueil circle after regular meetings had ceased in 1813.[132] And possibly even the personal unpopularity incurred by Laplace in the early years of the Bourbon Restoration played its part. On this unpopularity, in which political considerations seem to have loomed large, Libri wrote (with reference to the debate between Biot and Arago on the nature of light):

> M. Biot and M. Arago were among the first to participate. Unfortunately, instead of serving to strengthen the ties that bound them, the fact that they were engaged in the same field of study became the source of lively exchanges which culminated in a dramatic break between them; and the Academy was frequently moved by the strife between these two rivals who, in their heated debates, sometimes allowed themselves to be unduly carried away, especially when discussing questions of priority, which are always so delicate. Other *savants* joined in these discussions, and since Laplace, a man who wanted problems to be treated geometrically rather than in any other way, had appeared to take sides against Arago, enemies were raised to oppose him on every side; Legendre was put up deliberately as an adversary; the hand of friendship was offered to anyone who attacked the results contained in the *Mécanique céleste;* and all the liberal press was aroused and directed against those of whom we had once been so proud, men who, it was said, were now just old idols that had to be destroyed. Because the geometer Laplace had become the Marquis de la Place and on the pretext that some other academicians belonged to the Société des Bonnes Lettres,[133] these men were pro-

[132]On the decline of Arcueil, to the death of Berthollet in November 1822, see Crosland, *op. cit.* (note 4), pp. 395-428.

[133]The Société des Bonnes Lettres was a mainly literary society founded in the early years of the Restoration by Louis Fontanes and Chateaubriand, both champions of the Bourbons. Taking it as their aim "to revive the taste for good doctrines and good literature," most of the members supported traditional religion and monarchy, with the result that the society quickly became a byword for antiliberalism and hence the object of a good deal of popular suspicion. Under the influence of Chateaubriand and Charles Nodier, it helped to strengthen the early association between royalist sentiments and the new romanticism in literature. Among other leading members were the antiquaries Désiré Raoul Rochette and Quatremère de Quincy,

claimed ignoramuses, in the name of the Charter, in all the newspapers. It was then, as I have already said, that members of the public began to be admitted to the Academy,[134] and there they became the supporters of those who did not wish to excel solely by science. Laplace was put to silence, M. Biot stayed away from the Institute for several years, and M. Arago remained master of the battle field.[135]

Laplace's name, as Libri suggests here, seems to have become a byword for illiberalism in certain quarters (notably in the circle of the liberal writer P. L. Courier) in the early years of the Restoration, and it remained so until long after his death. Indeed, in the freer atmosphere of the Orléans monarchy, which did little to encourage restraint, criticisms of his "pliability" (souplesse) in political and personal matters and of his failure to defend the freedom of the press became common.[136] Perhaps the criticisms in his own lifetime were not so severe as to hasten his death, as the author of one standard biographical sketch maintained in 1834,[137] but there is sufficient evidence to dispel any image of Laplace living his last

the orientalists Antoine Léonard de Chezy and Jean Pierre Abel de Rémusat, and the writer Eugène Destains. The activities of the society, which included poetry readings by Victor Hugo, are best studied in the thirty-three volumes of its official publication, the Annales de la littérature et des arts, which appeared between 1818 and 1829. For a brief account see C. Dejob, L'Instruction publique en France et en Italie au dix-neuvième siècle (Paris, n.d.), pp. viii, 210-225, and 441-444. Dejob points out that the society was established as a royalist answer to the Parisian Athénée, founded in 1781 by Pilâtre de Rozier, where ideas more in keeping with the traditions of the eighteenth-century Enlightenment were discussed. Libri's comparison of Laplace with Legendre, who suffered, for political reasons, at the Restoration, was an obvious one to make.

[134]The ease with which journalists and the general public could gain access to the meetings of the Academy of Sciences remained a source of grievance long after this date; see, for example, Biot's comments in his Mélanges scientifiques et littéraires (note 43), 2, 257-264 (first published in the Journal des savants for February 1837).

[135][Libri], "Lettres à un Américain ... ," Revue des deux mondes, ser. 4, 21 (1840), 798-799.

[136]See, for example, the entry on Laplace in A. Rabbe, V. de Boisjolin, and Sainte-Preuve, Biographie universelle et portative des contemporains, 5 vols. (Paris, 1834), 3, 151-153, especially p. 151. Valentin Parisot was equally critical in his article on Laplace in the Bibliothèque universelle, ancienne et moderne, 83 vols. (Paris, 1811-1853), 70 (supplément), 237-260, especially pp. 239-244. A most unflattering description of Laplace's shifting political views appeared in the article on him by E. Merlieux in the Nouvelle biographie générale, ed. F. Hoefer, 46 vols. (Paris, 1855-1866), 29, cols. 533-534. Poisson's behavior was subject to similar criticism in the article on him in Rabbe, et al., op. cit., 5, 591.

[137]Rabbe, et al., op. cit. (note 136), 3, 151.

years as a universally respected elder statesman of French science. By 1827 his reputation was severely tarnished.

And as Laplacian influence waned, so inevitably the leading members of the new anti-Laplacian generation were able, if only by virtue of age and seniority, to gain control of the still centralized scientific community of Paris. It was important, for example, that when the need for a wholesale reorganization of the *Annales de chimie* was felt, following the death of the secretary to the journal, Collet-Descotils, in December 1815, it was Arago (rather than, say, Biot) who became one of the two new editors.[138] Since this happened just after Arago's conversion to the wave theory of light and at a time when his relations with Biot were about to worsen rapidly, the appointment was crucial in the transfer of power from the Laplacian group. The publication of the work of Fourier and Fresnel, which followed with remarkable (and significant) rapidity,[139] soon gave a clear intimation of the changing allegiance of the *Annales* and, because of the established authority of the journal, did much to strengthen the anti-Laplacian position.

Almost as important as this new domination of the most prominent of the French research journals was the way in which critics of Laplace were able to exert influence at the École Polytechnique after 1815. Petit, for instance, had been made professor of physics there in 1814—as a good Laplacian and for reasons quite unconnected with the subsequent debate[140]—and he remained in the post until his death in 1820. Petit was followed in his turn by Dulong, who remained as professor until 1830, when he became Director of Studies for the École Polytechnique as a whole. The chief results of this sixteen-year tenure of the chair of physics by Petit and Dulong were, first, a marked rise both in the quality and the amount of physics taught and, second, some predictable changes of doctrine. In the published syllabus for 1817–1818, for example, the state-

[138]Crosland, *op. cit.* (note 4), pp. 404–406. Arago had special responsibility for physics, while Gay-Lussac, the other editor, was responsible for contributions on chemistry.

[139]Fresnel's first paper on diffraction appeared in the issue for March 1816 (see note 87), and the December issue contained the lengthy summary of Fourier's work cited in note 86. Moreover, the very first issue of the new series, in January 1816, began with an account of the recent experiments by Arago and Petit that strongly supported Fresnel; see note 88.

[140]The cause of Petit's promotion to the chair after some five years as a teaching assistant (*répétiteur*), first in analysis and from 1810 in physics, was the unsatisfactory standard of the teaching of J.H. Hassenfratz, who had been professor of physics since the foundation of the École in 1794; see Fox, *op. cit.* (note 8), pp. 231–232.

ment that light would be "treated as an emission from luminous bodies," which had appeared in earlier syllabuses, was omitted.[141] And in the syllabus for 1821–1822 all references to "caloric" (*calorique*) were eliminated and replaced by references to "heat" (*chaleur*).[142]

Changes that told against Laplacian interests also took place in the Academy of Sciences. Arago, then an orthodox Laplacian, had been elected back in 1809, and Ampère became a member in November 1814; but it was only after the final overthrow of the Empire that they were joined by the men who were to become their chief allies. Fourier, for example, was not elected to the place that he had so long deserved until 1816, and Dulong and Fresnel followed only in 1823, Navier in 1824. However, the really decisive election at the Academy dates from November 1822, when the anti-Laplacian cause gained its most glorious victory through the defeat of Biot by Fourier for the post of permanent secretary for the mathematical sciences.[143] The vote, thirty-eight to ten, was not overwhelming, but from that point Laplacian science was doomed, and the election of Arago to replace Fourier as permanent secretary in 1830 only sealed its fate, ushering in a period that sympathizers of Laplace seem to have resented bitterly.[144]

With the crumbling of the power of the Laplacian group, each of its members adopted his own strategy for survival. As has already been noted,

---

[141]*Programmes de l'enseignement de l'Ecole Royale Polytechnique, arrêtés par le Conseil de Perfectionnement, pour l'année scolaire 1817–1818* (Paris, n.d.), p. 35; cf. *Programmes . . . arrêtés . . . dans la session de 1815–1816* (Paris, 1816), p. 39. Naturally Petit did not effect a wholesale rejection of Laplacian doctrines, and in *Annales de chimie et de physique*, 5 (1817), 404–406, he even defended Laplace against criticism of his theory of capillary action.

[142]*Programmes . . . arrêtés . . . pour l'année scolaire 1821–1822* (Paris, n.d.), pp. 31–32; cf. *Programmes . . . arrêtés . . . pour l'année scolaire 1820–1821* (Paris, n.d.), pp. 31–32. References to the fluid of electricity were far slower to disappear. In fact, it was only in *Programmes pour l'admission et pour l'enseignement à l'École Polytechnique, arrêtés par la commission nommée en exécution de la loi du 5 juin 1850, et approuvés par le Ministre de la Guerre* (Paris, n.d.), p. 93, that earlier references to "electric fluid" (*fluide électrique*) were replaced by references to "electricity" (*électricité*); cf. *Programmes pour l'admission et pour l'enseignement . . . arrêtés . . . pour l'année scolaire 1849–1850* (Paris, 1850), p. 29, where "electric fluid" is used.

[143]Arago was also a candidate in this election but openly gave his support to Fourier. It is interesting to note that, despite a show of strict impartiality, Laplace appears to have voted for Fourier rather than for Biot; see *Oeuvres complètes de François Arago* (note 131), 1, 100–101; see also note 93.

[144]Libri, for example, launched a violent personal attack on Arago on pp. 796–812 of the first of his (unsigned) "Lettres à un Américain" of 1840 (cited in note 135). His main charge was that, as a result of Arago's consistent abuse of his position, the activities of the Academy of Sciences had become increasingly trivial.

Biot chose to retire from the scientific scene of Paris in the early 1820's, but in the 1830's he seems to have mellowed and he returned to resume a valuable career in which his patronage of Pasteur in his early researches on crystals was perhaps his most important contribution. Poisson, by contrast, remained stubbornly loyal to the doctrine of Laplace until his death in 1840. In fact, he seems to have pursued the program with even greater zeal than the master himself, who, at least on certain issues, showed some signs of trying to adjust to the winning side. For example, after Fourier had produced an estimate for the age of the earth from geothermal considerations in 1819, Laplace wrote of his work in a decidedly complimentary manner.[145] And we must not forget that it was apparently Arago rather than Biot who had Laplace's vote in the momentous election at the Academy of Sciences in 1822.[146] Laplace's support for Fourier and his behavior in 1822 could, of course, only serve to alienate Poisson, and in view of the shifting allegiances of the 1820's it is remarkable only that so many personal friendships survived.[147]

So by the mid-1820's the style of science that had appeared so right and unassailable in the Napoleonic period had been abandoned by the leading figures in a new generation; and the peculiar organizational structure centered on Arcueil, which had provided essential support for the old science, had collapsed, leaving power in new hands.[148]

4. THE NEW AGE

It remains now to examine what was built on the ruins of Laplacian physics. The problem is a difficult one and simple statements are not possible. For, despite the solidarity of those who turned against Laplace and his disciples in the years after 1815, there emerged no single well-defined new style of science that was capable of filling the gap left by the

[145] See especially Laplace, "Sur la diminution de la durée du jour par le refroidissement de la terre," *Annales de chimie et de physique, 13* (1820), 416–417.

[146] See note 143.

[147] For example, when Dalton visited Arcueil in July 1822 there was every sign of friendship in a gathering for dinner that included Berthollet, Laplace, Biot, Fourier, and Arago; see the accounts in H.E. Roscoe, *John Dalton and the Rise of Modern Chemistry* (London, 1895), pp. 178–181. This, of course, was some four months before the election for the new permanent secretary.

[148] Ravetz has pointed out to me that a similar transfer of power took place in the early 1820's in the Société Philomathique of Paris, with Poisson and Biot giving way to non-Laplacians, notably H.M.D. de Blainville (soon to become a follower of Comte) and Fresnel; see Fox, *op. cit.* (note 8), pp. 272–273.

128   THE RISE AND FALL OF LAPLACIAN PHYSICS

old and of yielding a clear program for the future. And this should not surprise us, since a certain diversity of approach was a natural enough product of a period of reaction against the Laplacian orthodoxy, and diversity in any case reflects a more normal situation in science than the one that had prevailed under Napoleon.

It has been suggested (with an eye on Poisson's distinction between "physical mechanics" and "analytical mechanics") that what has here been described as Laplacian physics was followed by a turning toward a positivist approach, and the emergence of the positivist strain of the 1820's and 1830's onward has been cited as the beginning of the end of French theoretical physics.[149] Now Fourier's program for the science of heat, modeled on the traditional rational mechanics in which the causes are taken as given, could indeed be interpreted as positivist in the sense the term later acquired in the philosophy of Comte (although the program was conceived by 1807 and so was not truly Comtean in any sense[150]). Moreover, a mathematical study of heat transfer based on the principles laid down most prominently by Fourier in his *Théorie analytique de la chaleur* of 1822[151] did continue through the 1830's, with Lamé and Duhamel as its most distinguished exponents.[152] More evidence for a positivist trend in French physics is to be found in Ampère's refusal to discuss causes in his *Théorie mathématique des phénomènes électro-dynamiques, uniquement déduite de l'expérience* of 1826[153] and in his decision to present his theory in terms only of observed phenomena; i.e., in terms of forces such as those that were known from his own experiments to exist between two current-carrying conductors. And in any discussion of positivist science it is obviously impossible to omit Comte himself, who expressed the prevailing skepticism toward the Laplacian imponderable fluids in an extreme form when he wrote on the subject in the mid-1830's.[154]

[149]Herivel, *op. cit.* (note 112), especially pp. 121–132.

[150]On Fourier's work in 1807 see above, p. 111.

[151]The style of Fourier's treatment is conveyed in the opening words of the "Discours préliminaire" quoted in note 120.

[152]See G. Bachelard, *Étude sur l'évolution d'un problème de physique. La propagation thermique dans les solides* (Paris, 1928), pp. 89–132.

[153]See especially pp. 4–8 of the book. It is hardly necessary to point out that Ampère did not adopt this positivist stand in all his work. On the two faces of Ampère see L.P. Williams, *Michael Faraday. A Biography* (London, 1965), pp. 143–144.

[154]I.A.M.F.X. Comte, *Cours de philosophie positive,* 6 vols. (Paris, 1830–1842), 2 (1835), 438–445.

But positivism, with its variants, was not the only, or even the dominant, philosophy to gain favor, at least in the early years of the attacks on Laplacian science. It is important to observe that of the men who were most closely involved in the revolt of 1815–1820, Dulong, Arago, Fresnel, and Petit were emphatically not positivists, although they appear to have been no less concerned at the errors perpetrated by the Laplacians than were Fourier and Ampère. Certainly they advocated caution, and they rejected many Laplacian doctrines; but they all wanted to substitute new theories for the ones they were criticizing, and they championed their theories—the wave theory of light, the atomic theory, the vibrational theory of heat—with enthusiasm and utter conviction. Moreover, they provided the basis for research traditions that were anything but positivist. Perhaps the best example of such a tradition, which grew from the work of Fresnel, was the search for a model for an all-pervading fluid ether that possessed at once the high elasticity of a solid and also the capacity to allow solid objects, such as planets, to pass through it unhindered.[155] This problem, which quite defied solution until the work of Stokes in the 1840's, together with a number of related problems engaged some of the great men of nineteenth-century physics from the late 1820's until the rise of electromagnetic theory in the 1880's. With mathematical physicists of the stature of MacCullagh, Green, William Thomson, and Maxwell involved, the tradition was by no means exclusively French, but major French contributions were made by Navier and, more particularly, Cauchy.

So positivism was certainly not the one philosophy that rose to take the place of the Laplacian principles prevailing in the Napoleonic period; and still less was it a *cause* of the rejection of these principles. Positivism, in fact, did not emerge formally as a recognizable strain in French physical science until after the short period of creativity which itself followed the discrediting of Laplace and his school. It was at most a symptom, and not a cause, of the state of physics in France after the mid-1820's. And in any case, as I have argued, it did not have the philosophical field to itself, even by the mid-century.

Despite the confusing diversity that characterized French physics for several decades following the abandonment of Laplacian orthodoxy, one thing is clear: the enthusiasm, zeal, and confidence of the decade 1815–1825 were quickly lost. And it is in this loss of intellectual impetus, how-

[155]On this work see E.T. Whittaker, *A History of the Theories of Aether and Electricity,* 2nd ed. (London, 1951), pp. 128–169.

ever caused, that I believe we must see one of the great turning points in the history of physical science in nineteenth-century France. In the last eighteen years of his life, for example, Dulong never again openly voiced his support for the modern vibrational theory of heat, despite the extreme confidence with which he had expressed it (and scorned caloric) in 1820.[156] There was no going back to caloric, of course, but, in the absence of the energy conservation principle, there was no going forward to a new theory of heat either. There was, in fact, widespread agnosticism, as is clear from the textbooks of the day.[157] By the mid-1830's the same was true of the atomic theory, which Dulong had championed so uncompromisingly as an anti-Laplacian doctrine some twenty years before. By this time and on this particular issue Dulong was not alone in his caution with regard to the physical reality of atoms, as we see from the guarded comments being made in the face of dauntingly complex problems concerning the determination of atomic weights by Jean Baptiste Dumas.[158] And there is a similar story of exasperation and unfulfilled hopes in the history of the wave theory of light, where the complex problems arose in the 1820's, 1830's, and 1840's in the search for a satisfactory model for the fluid ether.[159]

So the second quarter of the century in France seems to have been characterized by a failure to consolidate the gains of 1815-1825, and for a variety of reasons the men who once seemed so certain that they were retrieving the physical sciences from error and initiating a new golden age ceased to give French physical science the leadership it needed. Dulong, for one, was beset by frustrations and ill health and quickly lost heart for

[156]For evidence of Dulong's support for the vibrational theory see his letter of 15 January 1820 to Berzelius, cited in note 100 and quoted in Fox, *op. cit.* (note 101), p. 13, and *op. cit.* (note 8), p. 244.

[157]See Fox, *op. cit.* (note 8), pp. 275-277.

[158]J.B.A. Dumas, *Leçons sur la philosophie chimique* (Paris, 1837), pp. 231-290 (6th and 7th lessons). For accounts of the growing difficulties in the atomic theory and Dumas' mounting despair see G. Buchdahl, "Sources of Scepticism in Atomic Theory," *The British Journal for the Philosophy of Science, 10* (1959), 120-134, and Fox, *op. cit.* (note 8), pp. 282-295. As J.H. Brooke points out in his Cambridge University Ph.D. thesis "The Role of Analogical Argument in the Development of Organic Chemistry" (1969), pp. 61-68, the difficulties did not lead Dumas to positivism, even in the mid-1830's; indeed, Dumas can safely be described as a consistent realist with regard to chemical theory. But positivism did become prominent in French chemistry in the 1840's with the work of Charles Gerhardt; see Brooke, *op. cit.*, pp. 114-149.

[159]See note 155 for reference.

the struggle,[160] while Arago, like so many others, soon found the world of politics, in his case both academic and national, more alluring than the laboratory bench.[161] And, saddest blow of all, Petit, Fresnel, and Fourier were all dead by 1830; and Ampère died in 1836, just when Britain was beginning to assume European supremacy in the study of electricity and magnetism. Of these only Fourier had succeeded in establishing anything resembling a school of disciples to carry on the tradition of his work into the 1830's and beyond,[162] so that by the 1840's the thread of continuity with the early 1820's was tenuous indeed. It is significant in this respect that the most esteemed French physicist of the later period, Victor Regnault, had his intellectual roots not in the exciting years that followed the rejection of the Laplacian orthodoxy but in the subsequent period of diminishing impetus. In fact, the massive, dreary compilation of data which earned Regnault his high reputation and for which he is now best known[163] was begun in 1840 in answer to a plea from Dumas, who had urged a full experimental investigation of specific heats in an attempt to remove the notorious anomalies in current values for atomic weights.[164] Not surprisingly, the attitude to scientific investigation that Regnault's work reveals is strikingly similar to Dumas' at this time, being cautious but not truly positivist.

Regnault illustrates as clearly as any one man can the state of French physics at the middle of the century. Far from being an outsider like

[160]On Dulong's work between 1820 and 1838, when he died, see Fox, *op. cit.* (note 8), pp. 248-270.

[161]Arago was engrossed in his political activities, at the Academy of Sciences and as a liberal deputy, from 1830 until his death in 1853. Politics similarly enticed Dumas from science in later life, and Gay-Lussac and Thenard both gave much time to politics and public affairs. Among the less prominent men of science with strong political interests were Sadi Carnot, Charles Dupin, Galois, Desormes, Raspail, and Poncelet. The lure of public life in nineteenth-century France is discussed more fully in R. Fox, "Scientific Enterprise and the Patronage of Research in France 1800-1870," *Minerva, 11* (1973)

[162]See above, pp. 119 and 128.

[163]H.V. Regnault, *Relation des expériences entreprises par ordre de Monsieur le Ministre des Travaux Publics, et sur la proposition de la Commission Centrale des Machines à Vapeur, pour déterminer les principales lois et les données numériques qui entrent dans le calcul,* 3 vols. (Paris, 1847-1870). As well as being published separately in this form, the *Relation des expériences* also occupied almost the whole of three large volumes of the *Mémoires* of the Academy of Sciences (vol. 21, 1847; vol. 26, 1862; vol. 37, parts 1 and 2, 1868-1870). The pagination is identical in the two versions.

[164]On the work of Regnault and his debt to Dumas, see Fox, *op. cit.* (note 8), pp. 283-302 and 315-317.

Sadi Carnot, Galois, Laurent, or Gerhardt, he was in every sense a man of the scientific establishment, as the honor that was accorded him both inside France and, to a lesser extent, in the rest of Europe shows clearly enough;[165] and it is for this reason that a study of his career is unusually revealing. In nearly all respects he was ideally placed to participate in one of the most exciting developments in nineteenth-century physics—the discovery of the principle of the conservation of energy in the 1840's. Unlike many of his contemporaries in France, he had every material facility, in terms of laboratory equipment and assistance, that he could possibly have desired (because of generous government sponsorship), and by 1840, owing to his membership in the Academy of Sciences and his chair in chemistry at the École Polytechnique, he had the eminence and prestige to make his views felt. Moreover, he had the close familiarity with the operation of steam engines that seems to have been one of the most important elements in the intellectual makeup of the discoverers of the energy conservation principle.[166] Yet he failed; and he failed not only to make the discovery of the principle himself but also to appreciate its true significance when it had been made by others. Even by the mid-1850's Regnault recognized all too clearly the harm he had done himself by his preoccupation with experimenting, when he saw, sadly but too late, that the main course of physics had passed him by.[167] He was, I believe, a tragic figure, and he knew it only too well.

5. CONCLUSION

I have argued in this paper that between the end of the First Empire and the middle of the nineteenth century there occurred a highly significant change of style in French physical science. The change began, im-

---

[165]See Fox, *op. cit.* (note 8), pp. 299-300; also my article on Regnault in a forthcoming volume of the *Dictionary of Scientific Biography* (note 81). Regnault's was one of the few laboratories in France to attract students from abroad about the middle of the century. In addition to the students listed in Regnault, *op. cit.* (note 163), *2*, ix, the young William Thomson worked there in 1845; see S.P. Thompson, *The Life of William Thomson, Baron Kelvin of Largs,* 2 vols. (London, 1910), *1*, 122-133.

[166]See T.S. Kuhn, "Energy Conservation as an Example of Simultaneous Discovery," in M. Clagett, ed., *Critical Problems in the History of Science* (Madison, 1959), pp. 329-336.

[167]His disappointment can be observed in his comments in Regnault, *op. cit.* (note 163), *2*, iii and iv. Here, as also in vol. 1, p. 12, it is apparent that Regnault had seriously underestimated the magnitude of his task. His tone was certainly not that of someone content with his achievements.

mediately after 1815, with a sharp reaction against the leading tradition of the Napoleonic era. This reaction seems to have stimulated rather less than a decade of great creativity in physics, in which period scientists, along with many other French intellectuals of the day, apparently breathed more freely than they had done before 1815. But, in a way that remains to be analyzed in detail, enthusiasm and excitement were quickly dissipated and, despite the initial high promise, consolidation of the achievements of the first ten years of the Restoration period was not achieved.

This interpretation of the course of French physics in the first half of the nineteenth century leads naturally to a somewhat equivocal view of the achievements of the Napoleonic period. Of course, it is impossible to discredit completely an approach to physics that was capable of stimulating equally experimental work (such as Delaroche and Bérard's determination of the specific heats of gases, which remained standard until the 1820's), the highly sophisticated theoretical studies of Biot, Poisson, Malus, and Laplace himself, and at least one major discovery (polarization). Nor can it be doubted that these achievements owed much to the effectiveness of the highly centralized organizational structure that was developed in the period and that allowed Laplace, Berthollet, and their *protégés* to work together as one of the most closely knit schools in the whole history of science. But by the same token we must also observe certain grave weaknesses both in the content and in the strategy of Laplacian physics, as this was pursued under Napoleon. These weaknesses are to be seen in the excessively firm adherence to doctrines which, because their basic principles were open to such serious objection, were almost bound to be attacked and discredited once the influence of the Arcueil group waned. Certainly to scientists working during the First Empire, as to Pasteur a half century later, the years of Napoleonic rule were glorious ones for science.[168] But the historian, I believe, must take a different view. For he would merit no charge of writing Whig history if he were to assess many of the theoretical studies of the great mathematical physicists of the school of Laplace as little more than spectacular *tours de force* based on models that had largely outlived their usefulness. In fact, I would maintain that, far from being uniformly glorious, the period of Laplacian domination was

[168]The point cannot be missed in, for example, the annual reports on the work of the First Class of the Institute, published in the *Mémoires* of the Class. And it is explicit in J.B.J. Delambre, *Rapport historique sur les progrès des sciences mathématiques depuis 1789, et sur leur état actuel* (Paris, 1810), especially pp. 1-3 and 40-42, and in Cuvier's companion volume for the *sciences naturelles,* also published in 1810, especially pp. 389-394.

one in which French physics (and to a lesser extent chemistry) suffered from the imposition of theories, notably the theories of imponderable fluids, which were not only incapable of internal development but also quite inappropriate for the stimulation, or even the effective study, of the new experimental results that were transforming the physical sciences in the early years of the nineteenth century. It is worth noting, for example, that although, over a forty-year period that embraced the First Empire, Coulomb and Poisson between them were able to formulate the classic theories of static electricity and dipole magnetism, thereby completing a characteristically eighteenth-century research tradition, the exciting new field of electrochemistry had its origins not in France but in Britain, Germany, and Scandinavia. Significantly, too, it was in England that the freer, if less stimulating, intellectual climate allowed serious criticism of the imponderables of heat and light to get under way by the first years of the nineteenth century (in the writings of Rumford, Davy, and Young).

Hence I would suggest that the physical scientists of Napoleonic France had their notable successes when grappling with the outstanding problems of the eighteenth century. In the newer fields of research, by contrast, it is the paucity of their contributions that is remarkable.

In later life Arago recalled how the orthodoxy at Arcueil had been so rigid that at first he had not even dared to voice his support for the wave theory of light.[169] And in his biographical sketch of Gay-Lussac, read in 1851, he said of the Arcueil circle:

> For young men beginning in science it was a distinctly flattering situation to have, as the first judges and advisers in their work, men of European renown, such as Laplace, Berthollet, Humboldt, etc. But could one be sure that some preconceived ideas, which the best minds adopt more readily in what I may term an intimate gathering than before a large audience, were not such as to stifle the spontaneity of genius and to limit research to generally agreed problems? Also, was it not inevitable that the wish to display a fertile mind in the presence of the most famous *savants* of the day would sometimes lead men of lively intellect to commit themselves to rash theories?[170]

These comments, like much that Arago wrote, have a strong element of contentiousness and they cannot be accepted uncritically as reliable evidence. They ignore some of the great strengths of Arcueil. There were

[169]See note 94.
[170]*Oeuvres complètes de François Arago* (note 131), *3*, 33–34.

great leaders (albeit leaders steeped in the traditions of an age that was rapidly coming to a close), and under them there was a distinguished school of young men whose common objective gave their work at least coherence and momentum. But, in pointing to the way in which exclusiveness could lead all too easily to the perpetuation of "preconceived" doctrines that inhibited new research, Arago surely identified the Achilles' heel of Napoleonic science.

Of course, the weaknesses were not evident in the years of the Empire. The long-acknowledged distinction of Laplace and Berthollet, their influence, and their considerable personal fortunes (for which they had Napoleon, above all others, to thank[171]) allowed their control over teaching, research, and the careers of the rising generation to go unchecked and gave them every opportunity to pursue their own ambitions in science and to create an indebtedness and natural allegiance on the part of the young men who formed their school. But the situation in which they held such power, both directly (by the provision of research facilities at Arcueil) and indirectly (by their ability to manipulate the Institute and teaching institutions for their own advantage), was an unstable one, as was proved when, from 1815, the authority of the Arcueil school was seriously challenged for the first time.

In its suddenness and in its effectiveness the challenge of the years immediately after 1815 was remarkable, and it can only be explained by reference to a quite fortuitous combination of circumstances. I have referred already to Laplace's personal unpopularity following the Restoration; and the diminished income of the masters of Arcueil also had an effect, in that it made private patronage of research on a large scale impossible.[172] But perhaps the most important changes that accompanied

[171]On the incomes of Laplace and Berthollet, which exceeded 50,000 francs p.a. during the Empire, see Crosland, *op. cit.* (note 4), pp. 69-74. This figure should be compared with the 6000 francs paid to the permanent secretaries of the First Class of the Institute and to professors at the École Polytechnique. According to a letter cited in J.B. Morrell, "Science and Scottish University Reform: Edinburgh in 1826," *The British Journal for the History of Science, 6* (1972-1973), 51 (note 54), John Leslie was impressed to find Laplace and Berthollet with incomes of between £5000 and £6000 each when he visited Paris in 1814. At the current rate of exchange, this suggests an income in each case of over 100,000 francs p.a. Salaries paid to them as Senators (positions accorded them by Napoleon in 1799) were their chief source of income.

[172]By February 1816 Berthollet's income was reduced to 24,000 francs; see Crosland, *op. cit.* (note 4), p. 400. For a comment on the lack of research schools after the Restoration see note 115 of Fox, "Scientific Enterprise and the Patronage of Research" (note 161).

the Restoration were those that resulted simply from the passage of time. From 1815, when he was sixty-six, until his death in 1822, Berthollet increasingly felt the burden of old age and ill health;[173] and time had the even more important effect of bringing maturity, status, and the possibility of independent thought and action to men who once would have seen the favor and patronage of Arcueil as the surest way to success in research and in their careers as teachers.

However, even these changed conditions would not have sufficed to cause the sudden rejection of Laplacian orthodoxy. For this overthrow to come about, the old order had to be tested, and it was unfortunate for Laplace and his remaining disciples that so many major issues arose over such a short period of time. It was quite by chance, of course, that Fresnel emerged from isolation just when influential encouragement and a sympathetic audience were awaiting him in Paris. And the discovery of Dulong and Petit's law and Oersted's discovery of the magnetic effect of an electric current were likewise chance events which could easily have occurred before 1815 but which then would not have had the corrosive effect that they did have between 1815 and 1825.

So it was that one of the most distinguished schools in the history of physical science collapsed. As a research school it was impressive both for the boldness of the program that gave it coherence and purpose and for the enterprise with which its leaders sought to give it institutional strength. As an illustration of the power of "totalitarianism" in science, it is perhaps without equal, and for the historian seeking to explain, for example, why the French persisted so long and so keenly in their adherence to the theories of imponderable fluids, it displays the dark side of totalitarianism. For in the perpetuation of ideas that were kept immune from rigorous criticism at a time when reappraisal could have been beneficial there lay the dangers of the orthodoxy that Laplace and his followers tried to impose on French science.

[173]Crosland, *op. cit.* (note 4), pp. 398–401. See also Berthollet's letters to Berzelius, dated 20 December 1819 and 4 September 1820, in *Berzelius Bref* (note 100), Pt. 1, *1*, 70 and 73.

X

The Fire Piston and Its Origins in Europe

I. *Introduction*

In previous accounts of the ingenious fire-making device usually known as the fire piston,[1] anthropologists and historians of technology alike have paid special attention to those specimens of the instrument which have been found in use in southeast Asia over the last hundred years. Understandably enough, they have been attracted above all by the intriguing possibility that the device was invented in that region quite independently of European influences and possibly long before the 1860's, when it appears to have been first observed there by European travelers. Consequently they have tended to regard the fire piston's appearance in Europe in the early years of the nineteenth century as the result either of a separate, though rather less interesting, process of invention in the West[2] or, alternatively, of direct importation from those parts of Asia where the device was already commonly used.[3] Although such accounts have been valuable, if only in preserving the fire piston from neglect, they have conveyed a view of the instrument's history which is not only incomplete (by virtue of their scant treatment of the European version of the instrument) but which is also supported by inadequate evidence on certain important points.

Dr. Fox, lecturer in the history of science at the University of Lancaster, is the author of a forthcoming book on the caloric theory of gases.

[1] Other names which have been applied to the fire piston include the tachypyrion, the aerophore, the pyrophorus, and the fire syringe.

[2] This is the view to which Henry Balfour inclines in his classic paper "The fire-piston," in *Anthropological Essays Presented to Edward Burnett Tylor in Honour of his 75th Birthday, Oct. 2, 1907*, hereafter cited as *Anthropological Essays* (Oxford, 1907), pp. 17–49, of which see especially pp. 39–46. Balfour's conclusion is tentatively adopted also in C. Singer *et al., A History of Technology* (Oxford, 1954), I, p. 228, and in A. Leroi-Gourhan, *Évolution et techniques. L'homme et la matière* (Paris, 1943), pp. 68–69.

[3] The view adopted in J. Needham, *Science and Civilisation in China* (Cambridge, 1965), IV, Part 2, p. 140 n., and with rather less conviction in W. Hough, *Fire as an Agent in Human Culture*, United States National Museum Bulletin 139 (Washington, 1926), p. 110.

Reprinted from *Technology and Culture*, Volume 10, No. 3, July 1969. © 1969 by The Society for the History of Technology. Used by permission.

356

Of course, the almost complete absence of documentation concerning even nineteenth-century technology in southeast Asia makes it virtually impossible to dismiss the traditional accounts entirely, and this I shall not attempt to do. But although my conclusions concerning the origins of the fire piston in Asia are somewhat conjectural for this reason, I shall discuss, rather more sympathetically than most earlier writers on the subject, the case for believing that the instrument's appearance in Asia was the result of European influences. My chief concern in this paper, however, is with the early history of the fire piston in Europe and, more specifically, with the task of establishing beyond doubt that the device was not imported but invented there independently.

II. *The Scientific Background*

In its commonest form (see Plate I, items 2 and 3) the European fire piston consists of a metal cylinder, closed at one end and usually no more than 6 inches in length and ½ inch in diameter, in which air can be compressed by means of a closely fitting piston. In an efficient instrument a single rapid stroke of the piston compresses and hence also heats the air sufficiently for a piece of dry tinder placed in a cavity in the end of the piston to be ignited, whereupon the piston is quickly withdrawn and the glowing tinder is applied as required. To modern eyes the process provides a striking demonstration of the conversion of mechanical work into heat, the work in this case being performed in the compression of the air. And in fact it was as a demonstration experiment in physics that the fire piston survived long after it had fallen out of common use as a fire-making device.[4]

Yet the invention of the fire piston in Europe took place some fifty years before the establishment of the principle of the conservation of energy, at a time when the possibility that work might be converted into heat was perceived only dimly and incoherently, if at all, by a few men such as Count Rumford and Humphry Davy. Indeed, it came at a time when the view that heat was a weightless, highly elastic fluid known as caloric had won an almost total victory over its main rival, according to which the phenomena of heat were explained in terms of the vibrations or, as in the case of gases, the rectilinear motion of the particles of ordinary ponderable matter. It might be expected that this situation

---

[4] See, for example, A. Ganot, *Traité élémentaire de physique expérimentale et appliquée* (Paris, 1851), pp. 321–22; and J. Tyndall, *Heat Considered as a Mode of Motion* (London, 1863), p. 29. Large glass fire pistons made specifically for demonstration purposes exist in the Science Museum, London (which has two examples), and in the museum of the Conservatoire National des Arts et Métiers, Paris (which has one).

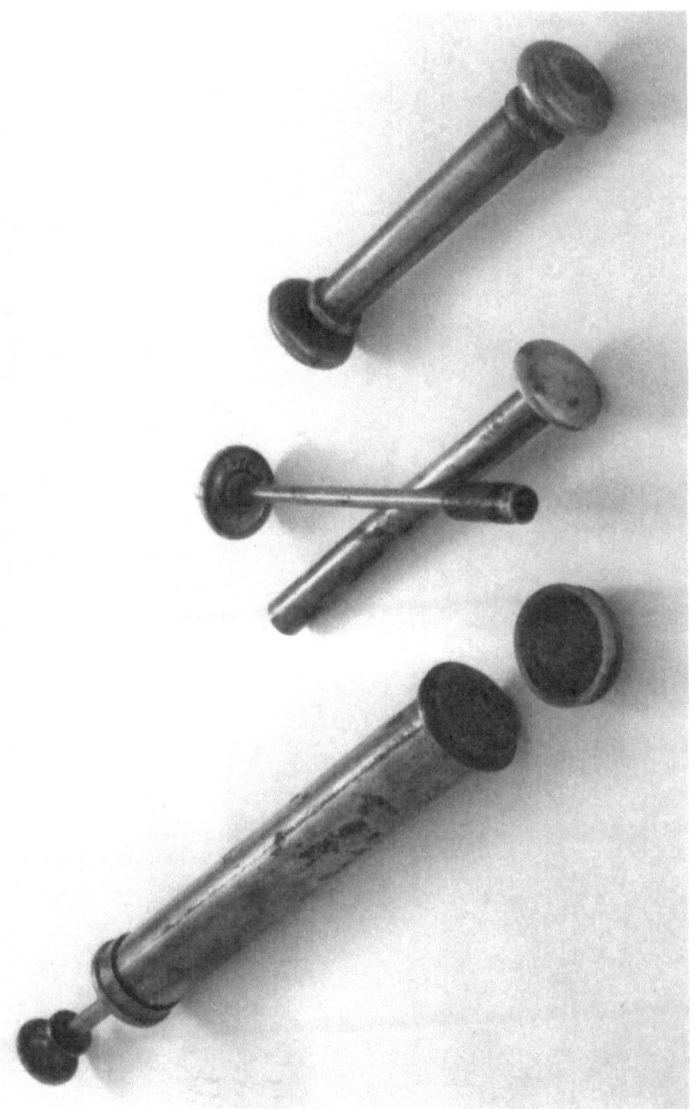

Plate I Fire pistons of the early nineteenth century. The instruments in the centre and on the right are of conventional design, with the tinder held in the end of the piston. In the instrument on the left, the air passes through a number of small holes in the end of the cylinder before encountering the tinder, which is placed in a removable cap. Lent by Bryant and May. Reproduced by courtesy of the Trustees of the Science Museum.

x

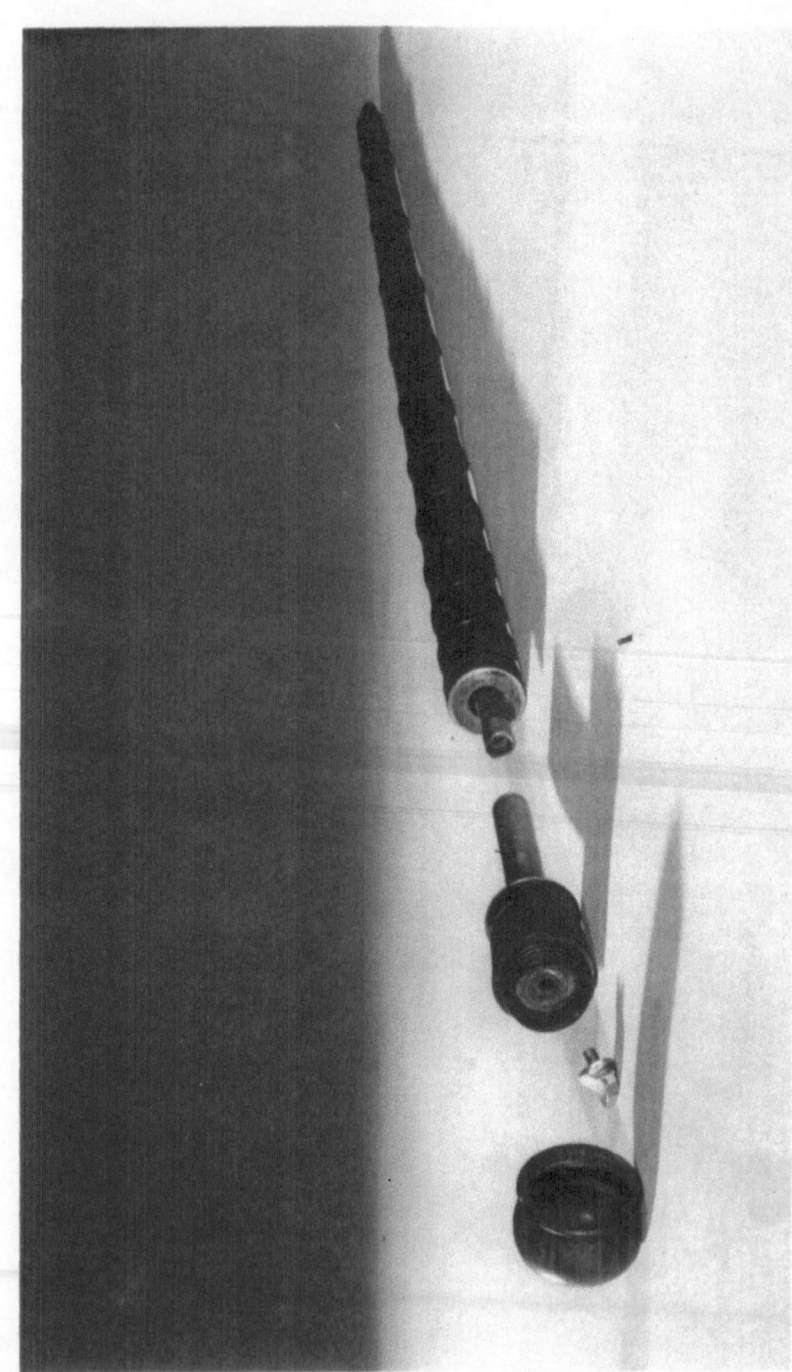

Plate II   Fire piston contained in a walking stick. For a full description, see Appendix, item 5. Lent by Bryant and May. Reproduced by courtesy of the Trustees of the Science Museum.

would have inhibited the study of the heating and cooling effects of which a knowledge was fundamental to the understanding, if not to the discovery, of the fire-piston principle, but, as Thomas S. Kuhn has shown,[5] this was by no means the case. The effects had been known and studied, though they had often been misinterpreted, since the middle of the eighteenth century, and in the early years of the nineteenth century they were attracting an intense and unprecedented interest. As I shall argue, this interest seems to form an important part of the general intellectual background against which the fire piston was invented in Europe, so that the fact that it was most evident in France and the fact that the fire piston first appeared there and not elsewhere in Europe are almost certainly related.

It was in 1802 that the thermal effects to which I have referred began to attract serious attention in France. In April of that year Jean-Baptiste Biot, then a young protégé of the great French mathematician Pierre Simon Laplace, read before the First Class of the Institute in Paris a paper[6] in which he put forward a highly plausible (and essentially correct) explanation of the notorious discrepancy between the experimental value for the velocity of sound in air and the figure obtained by applying Newton's theoretical expression.[7] Laplace, it seems, had asked Biot to investigate how the theoretical figure would be affected by the heat and cold which he suggested might be produced in the successive regions of compression and rarefaction constituting the sound wave. Hence it is Laplace and not Biot who must take the greater credit for the resulting demonstration that theory and experiment could be brought into agreement by the assumption that slight temperature changes did in fact occur or, in other words, that Boyle's law, the truth of which had been axiomatic in nearly all earlier work on the subject, could not be applied to a gas through which a sound wave was passing. Laplace's masterly insight was very properly acclaimed, above all, for its bearing on the velocity of sound problem; yet his work exerted a great influence in other respects also, and it

[5] T. S. Kuhn, "The caloric theory of adiabatic compression," *Isis*, XLIX (1958), 132–40.

[6] J. B. Biot, "Sur la théorie du son," *Journal de Physique*, LV (1802), 173–82. Thus the paper was not published until the month of Fructidor of the republican year X (August-September 1802). On the date of the reading of the paper see *Académie des Sciences. Procès-verbaux des séances de l'Académie tenues depuis la fondation de l'Institut jusqu'au mois d'aôut 1835*, II, 487 (11 Germinal, an XI). The latter work, published in ten volumes at Hendaye between 1910 and 1922, is cited hereafter as *Procès-verbaux*.

[7] As given in the *Philosophiae naturalis principia mathematica* (London, 1687), pp. 369–72. In subsequent editions the proof remained unchanged.

is this latter, rather less obvious influence which seems especially relevant to the events described in this paper.

In the first place, Laplace's discovery provided striking confirmation of the fact that changes in temperature did occur when gases underwent rapid expansion or compression. In view of the numerous investigations which had been undertaken by the British, Swiss, and Germans since the temperature changes had first been observed by the Scottish chemist William Cullen in 1755,[8] it may be thought that such evidence would have been unnecessary. But there were still those who doubted whether the fluctuations in the thermometer reading really were the result of a rise or fall in temperature, as opposed, for example, to some mechanical effect due to variations in gas pressure.[9] Moreover, it is a curious fact that the effect, whether interpreted correctly or not, appears to have been almost completely unknown in French scientific circles until the very end of the eighteenth century, in fact until 1798, when Marc Auguste Pictet, professor of natural philosophy at the University of Geneva but a regular visitor to Paris, communicated to the French *Journal de Physique* an account of the cold observed on suddenly releasing air from a compression pump. It is a measure of the ignorance on the matter which prevailed in France that the editor of the journal, J. C. Delamétherie, who was presumably the author of the unsigned note describing Pictet's communication,[10] treated the observation as a case of cooling by evaporation, having apparently missed the point entirely. The correct interpretations of this and other similar observations which did appear, very occasionally, in the French scientific literature during the three years preceding Laplace's great discovery[11] probably did little to remedy this situation. It can hardly be doubted, therefore, that the paper of 1802, apart from solving a notorious problem, was also drawing attention to an effect of which the great majority of readers were ignorant.

There is one other respect in which Biot's paper seems to have been

[8] On which see Kuhn, *Isis*, XLIX (1958), 133–35.

[9] Thus John Dalton, speaking in June 1800, felt it necessary to refute this suggestion explicitly by reference to experiments which he had conducted for the purpose. See *Memoirs of the Literary and Philosophical Society of Manchester*, V, Part 2 (1802), 523–24. *Cf.* also the objections which were raised against Joseph Mollet's interpretation of his early experiments (see below, Section III).

[10] [J. C. Delamétherie], "Note sur un froid considérable produit par la sortie prompte de l'air atmosphérique fortement comprimé," *Journal de Physique*, XLVII (1798), 186. The note was in the issue for Fructidor, an VI (August–September 1798).

[11] See, for example, A. N. Baillet, "Lettre . . . sur la glace produite par l'expansion de l'air comprimé," *Journal de Physique*, XLVIII (1799), 166–67.

relevant to the emergence of the fire piston. This is that it gave the first intimation of the true magnitude of the temperature changes taking place in expansion and compression, a magnitude which had previously been masked by heat exchange between the gas and its surroundings. According to Biot's calculation,[12] if the experimental and theoretical figures for the velocity of sound were to agree, it followed that the compression of a mass of air to, say, one-half of its initial volume, under conditions such that no heat exchange occurred, would produce a rise in temperature of the order of 100° C. Since most experiments performed before this date had suggested that the temperature change in such a case would not exceed a few degrees at the most,[13] Biot's conclusion was a startling one and it was not likely to be overlooked. As it happened, any slight possibility that it might be ignored soon disappeared completely when, toward the end of 1802, experimental confirmation that the temperature changes had indeed been grossly underestimated by eighteenth-century writers became available in an important and widely read paper by John Dalton of Manchester.[14]

III. *The Discovery of the Principle*

It was just at the time when the temperature changes accompanying the expansion and compression of gases were attracting a great deal of attention in France and when the true magnitude of the changes was beginning to be recognized that the fire piston emerged. Clearly the fact that its invention was, as we shall see, the direct consequence of a chance observation rather than a deliberate attempt to apply newly acquired scientific knowledge to the long-established problem of fire making, must diminish to a certain extent the importance of this scientific background. But the background cannot be ignored completely. Without it the ignition (by the rapid compression of air) of a small piece of linen lodged in the exit tube of the condensing pump of

[12] Biot, *Journal de Physique*, LV (1802), 182.

[13] See, for example, M. A. Pictet, *Essais de physique* (Geneva, 1790), p. 20.

[14] J. Dalton, "Experiments and observations on the heat and cold produced by the mechanical condensation and rarefaction of air," *Memoirs of the Literary and Philosophical Society of Manchester*, V, Part 2 (1802), 515–26 (read June 27, 1800). The first extracts to appear in French journals were those in *Journal des Mines*, XIII, 257–60, and in *Annales de Chimie*, XLV, 103–7, both published in the issues for Nivôse, an XI (December 1802–January 1803). On the date of publication in England see T. Thomson, *A System of Chemistry* (4th ed.; Edinburgh, 1810), I, 488. In his paper Dalton described experiments by which he claimed to have shown that a rise in temperature of 50° F resulted when air was allowed to rush into a space evacuated to one quarter of atmospheric pressure and that a fall in temperature of a similar magnitude occurred when air at a pressure twice that of the atmosphere was allowed to escape rapidly from a closed vessel.

an air gun, which was noticed by a workman at the armory in Saint-Étienne (Loire) in 1802, might well have been overlooked or at least misinterpreted. Moreover, it was almost certainly the background of growing interest and understanding that encouraged the subsequent development of the fire piston after this initial chance observation had been made.

It was Joseph Mollet, at this time professor of physics at the École Centrale in Lyons,[15] who brought the workman's observation to the attention of the scientific world by a letter which he sent to the Institute in Paris and which was read before the First Class on December 29, 1802.[16] In the letter Mollet described not only the ignition of the piece of linen in the compression pump but also another observation made by the same workman at Saint-Étienne, namely, the appearance of a flash of light whenever an air gun was discharged in the dark. Mollet's communication was probably received with some skepticism, and his case was certainly not helped by the fact that the two referees appointed by the Institute, J. A. C. Charles and L. Lefèvre-Gineau, were unable to reproduce either of the observations.[17] Only Pictet, who submitted a brief report on Mollet's letter to Alexander Tilloch's *Philosophical Magazine* in London,[18] appears to have thought the matter

[15] For brief biographical sketches of Mollet (1758 1829) see J. B. Dumas, *Histoire de l'Académie Royale des Sciences, Belles-Lettres et Arts de Lyon,* hereafter cited as *Histoire de l'Académie* (Lyons, 1839), II, 149–50, and *Biographie universelle, ancienne et moderne* (Michaud), (85 vols., Paris, 1811–1862), LXIV, 173. Dumas's *Histoire* also contains lists of Mollet's writings, both published and unpublished, in Vol. II, pp. 23–24 and 605.

[16] *Procès-verbaux,* II, 606 (8 Nivôse, an XI). On the contents of the letter see J. Mollet, *Mémoire sur deux faits nouveaux, l'inflammation des matières combustibles, et l'apparition d'une vive lumière, obtenues par la seule compression de l'air,* hereafter cited as *Mémoire sur deux faits* (Lyons, 1811), especially pp. 3 and 6. The workman is identified as Citizen Chauvain in *Procès-verbaux,* III, 96 (17 Floréal, an XII).

[17] Mollet, *Mémoire sur deux faits,* p. 3. That the mere discharge of compressed air from an air gun did not in fact cause the appearance of light was demonstrated by Thenard in 1823 (see L. J. Thenard, "Sur la lumière produite par la décharge du fusil à vent," *Annales de Chimie et de Physique,* XXII [1823], 436–39).

[18] M. A. Pictet, *Philosophical Magazine,* XIV (1803), 363–64. Pictet was in Paris at the time and, as an associate member of the First Class, he may well have heard Mollet's letter read (see "Journal d'un Genevois à Paris," *Mémoires et documents publiés par la Société d'Histoire et d'Archéologie de Genève,* 2d series, V [1893–1901], 109). Unfortunately Pictet's diary covering this period contains no entry for December 29, 1802. A copy of the diary made by Edmond Pictet, which is in the possession of the *Société d'Histoire et d'Archéologie de Genève,* was consulted on my behalf by Monsieur Ph. Monnier, keeper of manuscripts at the *Bibliothèque publique et universitaire,* Geneva.

worthy of public comment, but even this summary provoked nothing more than a statement from William Nicholson to the effect that the flash produced on discharging the air gun had been observed about eighteen months previously by a Mr. Fletcher at a weekly meeting "for philosophical experiments and conversations" held at Nicholson's house.[19]

Apparently undeterred by the general neglect of his communication, Mollet continued his experiments in collaboration with Ennemond Eynard, a doctor and member of the science section of the *Académie des Sciences, Belles-Lettres et Arts* in Lyons,[20] and with two other residents of Lyons—Haex[21] and Gensoul.[22] Although he never managed to reproduce the flash associated with the discharge of the air gun to his complete satisfaction. Mollet now discovered that a similar flash occurred when air was compressed in a cylinder by means of a closely fitting piston.[23] He described the new observation and the already well-established ignition of an inflammable material, such as tinder, by the same method, first in a communication to the Institute in Paris in November 1803, then to the Lyons Academy in January 1804, and finally

---

[19] W. Nicholson, "Flash from an air-gun," in Nicholson's *Journal of Natural Philosophy, Chemistry, and the Arts*, 2d series, IV (1803), 280.

[20] For a brief biographical sketch of Eynard (1749–1837) see *Le deuxième centenaire de l'Académie Nationale des Sciences, Belles-Lettres et Arts de Lyon, 1700–1900* (Lyons, 1900), p. 113. Lists of Eynard's published and unpublished writings appear in Dumas, *Histoire de l'Académie*, II, 29–31 and 607.

[21] Little is known of Haex, although he is described in the *Bulletin de Lyon*, No. 55 (10 Germinal, an XII), 218, as an "artiste," his name there being spelled "Haez." Monsieur H. Hours, archivist of the city of Lyons and general secretary of the *Classes des Lettres* of the Lyons Academy, has suggested to me that this Haex may be Thibaud Haess, a German-born turner who died in Lyons in 1812 at the age of 67.

[22] It was these three colleagues who were responsible for the earliest experiments and also for bringing the workman's observations to the notice of Mollet. See Dumas, *Histoire de l'Académie*, II, 233 n., and Mollet, *Mémoire sur deux faits*, pp. 7–8. In a private communication Monsieur Hours of Lyons, referred to in note 21, has identified Gensoul as Joseph Ferdinand Gensoul, who was born at Conaux (Gard) and who died in Lyons in 1833 at the age of 67. The work of Gensoul, who was an engineer especially noted for his work in copper, is described briefly in F. F. A. Potton, *Notice historique sur la vie et les travaux du docteur Joseph Gensoul* (Lyons, 1861), pp. 11–12. I am indebted to Monsieur Hours for this reference.

[23] Mollet, *Mémoire sur deux faits*, pp. 25–27. That the appearance of light was not the result of the compression of air was demonstrated by Thenard in 1830 (see L. J. Thenard, "Observations sur la lumière qui jaillit de l'air et de l'oxigène par compression," *Annales de Chimie et de Physique*, XLIV [1830], 181–88).

at the Academy's public meeting on March 27 of the same year.[24] In Paris, the Institute's referees, Charles and A. F. Fourcroy, reported favorably on this occasion, and the reading of their report in May 1804 was even accompanied by a successful performance of Mollet's experiments.[25] In Lyons, on the other hand, there was still some skepticism, but Mollet seems to have been successful in answering suggestions that the cause of the ignition of the tinder was something other than a rise in temperature of the air undergoing compression.[26] That heating resulted from friction either between the piston and the cylinder or between the tinder and the air rushing over its surface was among the alternative explanations which were offered about this time. Others were that the compression brought about oxidation of the surface of the metal cylinder and hence also the emission of heat, and that the ignition of the tinder was to be associated with the presence of the lubricating oil used to facilitate the movement of the piston. That such objections, however few, should have been raised at all may seem somewhat surprising, but they do serve to emphasize the ease with which Mollet's experiments could be misinterpreted and hence also the importance of the scientific background described above. It is significant, perhaps, that such objections as were made came apparently from the scientific community of Lyons rather than from that of Paris, where the impact of Laplace's success with the velocity-of-sound problem was almost certainly more keenly felt.

IV. *Commercial Manufacture*

The principle of the fire piston, as we have seen, was discovered accidentally in the course of work on the condensing pump of an air gun, an instrument which itself was not suitable for the purpose of fire making. Yet it was not long before apparatus for the ignition of tinder by compression began to be devised specially. In January 1804, for example, in the course of Mollet's paper to the Lyons Academy, Haex demonstrated a simple pump consisting of a piston and a cylinder,

[24] On these three communications see *Procès-verbaux*, III, 28 (6 Frimaire, an XII); Dumas, *Histoire de l'Académie*, II, 232–35; and *Bulletin de Lyon*, No. 55 (10 Germinal, an XII), 218. Successful practical demonstrations by Haex accompanied the reading of Mollet's papers in both January and March 1804. The text of the paper read on March 27, 1804 appeared in 1811 as the *Mémoire sur deux faits* cited fully in note 16 above. There is no evidence that changes had been made since the paper had been delivered, although brief introductory and concluding notes were added.

[25] *Procès-verbaux*, III, 95–96 (17 Floréal, an XII).

[26] See Dumas, *Histoire de l'Académie*, II, 234–36; and Mollet, *Mémoire sur deux faits*, pp. 8–14 and 30.

the closed end of which could be unscrewed after the compression stroke to expose the glowing tinder.[27] Whether at this stage there was any thought that such an instrument might be used specifically for fire making is not known, although the possibility was certainly mentioned in March 1804 by Mollet.[28] However, it seems that it was not until 1806 that the fire piston became available commercially.

In February 1806 the *Journal de Physique* contained a brief announcement that the celebrated instrument maker Dumotiez of Paris[29] was producing what was termed the "briquet pneumatique" in various sizes and designs.[30] That Dumotiez was the manufacturer is hardly surprising when we note that it was in his workshop, in February 1804, that tinder was first successfully ignited in Paris by the compression of air.[31] Since the experiment of February 1804 seems to have been conducted strictly in accordance with Mollet's instructions, it is reasonable to suppose that the apparatus used was similar to that which Haex had demonstrated before the Lyons Academy in the previous month, and this would certainly account for the fact that, in Dumotiez's early models at least, the tinder was held in the closed end of the cylinder, which therefore had to be rapidly unscrewed before a light was obtained.[32]

Although no example of a fire piston by Dumotiez seems to have survived, his early instruments were probably similar to one patented in London by Richard Lorentz of Brook Green, near Hammersmith, in February 1807,[33] and the "foreigners residing abroad" who communicated the invention to Lorentz may well have been connected with Dumotiez. Lorentz's version was unusual not merely for its size (the working part was contained inside a walking stick, and the length

[27] Dumas, *Histoire de l'Académie*, II, 233.

[28] Mollet, *Mémoire sur deux faits*, p. 13.

[29] "Briquet pneumatique par Dumotiez," *Journal de Physique*, LXII (1806), 189. See also Mollet, *Mémoire sur deux faits*, p. 30.

[30] According to the announcement Dumotiez's first pistons were typically 6 inches in length and approximately ¼ inch in diameter, but in 1811 Mollet wrote that one model sold by Dumotiez was in the form of a walking stick (see Mollet, *Mémoire sur deux faits*, p. 30. Also *cf.* Lorentz's first piston, described below, and the English instrument shown in Plate II). The identity of the "M. Dumotiez" referred to in the announcement is not entirely clear, since the firm of Dumotiez was in fact controlled by two brothers, Louis Joseph and Pierre François Dumotiez, who seem to have worked in close collaboration. On the Dumotiez brothers see M. Daumas, *Les instruments scientifiques aux XVIIe et XVIIIe siècles* (Paris, 1953), pp. 378–79.

[31] Mollet, *Mémoire sur deux faits*, p. 3.

[32] *Ibid.*, p. 30.

[33] The patent specification, the reference number of which is 3007 (1807), has been consulted by courtesy of the Patent Office, London.

of the piston stroke was about 12 inches) but also for the fact that the compressed air passed through a narrow aperture before meeting the tinder. Only two surviving examples of fire pistons constructed on this principle are known, and even in these cases there are certain important departures from the patent specification.[34] We may be certain therefore, that Lorentz's designs, like Dumotiez's, never came into common use and that Mollet's preference for a second type, which he expressed in 1811,[35] was widely shared even by that date.

In this second type, which became easily the most common, the tinder was contained in a small cavity in the end of the piston, so that after the compression stroke it was necessary simply to withdraw the piston before the glowing tinder could be applied. The credit for this design, no less than for that adopted by Dumotiez and Lorentz, lies with the men of Lyons, although who should be named as its inventor is not entirely clear. Certainly the first man to patent and manufacture fire pistons of the second type was a metalfounder of the Rue St. François in Lyons by the name of Dubois, who worked from a particularly efficient prototype constructed by Eynard and demonstrated by him at a public meeting of the Lyons Academy on August 26, 1806.[36] Whether Eynard had lodged the tinder in the piston rather than in the end of the cylinder or whether this was an innovation by Dubois is unfortunately not known, but, as has already been mentioned, we may be sure that the superiority of instruments of the type constructed by Dubois quickly became apparent. Thus the fact that in 1811 Dumotiez's fire pistons could be cited by an independent observer as the best known[37] may indicate only that Dumotiez, even by that date, had already adopted Dubois's design in preference to his own.

The extent to which fire pistons were used in Europe is extremely difficult to ascertain, but there is every reason to believe that they never displaced the conventional tinder box as the most popular firemaking device and that they remained something of a scientific curiosity. The fact that so few instruments of European origin are to be found in British museums today (see Appendix) is striking evidence for this conclusion, the more so when we consider that the European fire piston was virtually indestructible, being constructed entirely of metal with the exception of some padding around the piston. On the other hand, it must be pointed out that in scientific books and papers,

[34] The instruments referred to are shown in Plate I (item 1) and Plate II. See also Appendix, items 1 and 5.

[35] Mollet, *Mémoire sur deux faits*, p. 31.

[36] Dumas, *Histoire de l'Académie*, II, 236.

[37] J. P. Dessaignes, *Journal de Physique*, LXXIII (1811), 50.

if not in other literature, frequent references to the fire piston continued to be made until about 1830. Such authors as Humphry Davy, Louis Jacques Thenard, Sadi Carnot, and Jöns Jacob Berzelius[38] all mention the instrument in terms which imply considerable familiarity. However, the fact that the fire piston was well known to the scientists of the day is almost certainly an indication not of its widespread use for domestic purposes but rather òf its suitability as a means of illustrating an effect which would have formed part of any early nineteenth-century course on heat.

The introduction in Europe of cheap and efficient friction matches in the late 1820's dealt a severe blow to the fire piston, which by comparison with the new matches was both clumsy and unreliable.[39] Quite suddenly references to the fire piston became far less frequent, and there is some evidence for believing that even by the early 1830's the instrument was beginning to be forgotten, at least in England.[40] Somewhat later, probably about the middle of the century, it appears to have been in common use in certain rural parts of England, notably among shepherds in the Lake District and to a lesser extent in the Yorkshire Dales,[41] and about 1900 manufacture of the instrument was briefly resumed in France.[42] But such evidence of its use after 1830 is rare,

---

[38] H. Davy, *Elements of Chemical Philosophy* (London, 1812), p. 90; L. J. Thenard, *Traité de chimie élémentaire* (Paris, 1813), I, 82; N. L. S. Carnot, *Réflexions sur la puissance motrice du feu* (Paris, 1824), p. 30 n.; J. J. Berzelius, *Lehrbuch der Chemie*, trans. F. Wöhler (Dresden, 1825), I, 60.

[39] The main defects of the fire piston as a domestic fire-making device were summarized very neatly in 1834 by the author of the unsigned article on "Fire" cited in the Appendix, under item 1. On p. 286 of the article it is stated: "Some modification of this instrument may be found useful, but in its present state it is inferior to the common tinder-box:—it requires considerable strength,—is equally slow in getting a light,—requires a match to be lighted after the tinder has taken fire, and is easily put out of order." To this list we may add the fact that the tinder has to be very dry before the fire piston will operate effectively.

[40] This seems to be implied by a letter from E. J. Mitchell of Bradford, Yorkshire, published in *Mechanics' Magazine*, XVII (1832), 328–29. In the letter Mitchell described what he termed the "instantaneous light-giving syringe" (*i.e.* the fire piston) as an invention which "though not new, is, perhaps, not generally known." Mitchell believed (probably correctly) that the instrument was better known on the Continent.

[41] J. Greenop, "A contrivance for producing fire, formerly used in the English Lake District," *Transactions of the Cumberland & Westmorland Antiquarian & Archaeological Society*, n.s., VII (1907), 206–8. I am grateful to John Anstee, assistant curator of the Museum of Lakeland Life and Industry, Abbot Hall, Kendal, for drawing my attention to this paper.

[42] See Appendix, item 7. According to Balfour (*Anthropological Essays*, p. 47) this fire piston was purchased in Paris, apparently about 1900.

so that the history of the fire piston as a useful fire-making device effectively ends in Europe only a quarter of a century after its invention.

V. *An Outstanding Problem*

The familiarity of the fire piston to Europeans is clearly a relevant factor in one major problem to which reference has already been made, namely, how a similar device, usually somewhat smaller than its European counterpart and made of wood or horn, came to be in common use in parts of southeast Asia later in the nineteenth century and certainly before 1865.[43] If the fire piston really was well known before that date in Europe, albeit chiefly in scientific circles, it seems difficult to discount quite so readily, as most anthropologists in the past have done, the possibility that the instrument was introduced from the West. It is surely significant that the British, by virtue either of trade or conquest, exerted a very considerable influence in southeast Asia in general during the first three decades of the nineteenth century, at the time when we might reasonably expect the fire piston to have been introduced there by Europeans. And, to take the point still further, we cannot overlook the fact that the fire pistons observed by visitors to the area later in the nineteenth century were found only in Java, which was occupied by the British from 1811 until 1816, and in parts of Burma, Laos, the Malay peninsula, Sumatra, North Borneo, and the Philippines, which either had been under direct British influence themselves or which, alternatively, had been in close contact with regions that had.[44]

This, of course, in no way implies that the possibility of an independent process of invention in Asia can be ruled out. Certainly, in the light of the well defined and clearly documented pattern of discovery and early development which has been outlined above for the European fire piston, we can now hardly accept the extreme view of this problem, according to which the fire piston was not only invented independently in Asia but was also introduced from there into Europe.[45] Yet the more moderate claims, such as those of Henry Balfour,[46] which allow that the fire piston could have emerged quite independently in

[43] On the dates of the earliest accounts of the fire piston in Asia see Balfour, *Anthropological Essays*, pp. 27–28.

[44] On the exact provenance of the Asian fire pistons see Balfour, *Anthropological Essays*, pp. 23–29 and 46–49. The possibility that the French were responsible for introducing the fire piston into southeast Asia has not been considered here. By comparison with that of the British, French influence in the region was slight at the time when the instrument was most probably introduced.

[45] For references see note 3.

[46] For reference see note 2.

the West as well as in Asia, are far more difficult, if not impossible, to eliminate and they still deserve very serious attention.

In considering the circumstances which might have led to the discovery of the fire-piston principle in southeast Asia, we shall not expect to find anything resembling the tradition of purely scientific inquiry which seems to have been at least one factor contributing to the emergence of the device in Europe. We may be certain, therefore, that the Asian fire piston, if it was in fact indigenous in Asia, would have been invented as the result of a chance observation. How this observation might have been made has been the subject of much speculation, but almost inevitably it remains an unsolved question. Balfour, for example, argued convincingly against toy popguns, water syringes, and the pestle and mortar commonly used in Asia for crushing betel nuts as possible antecedents of the fire piston, but, as an alternative, he was able to do no more than suggest tentatively that the chance ignition of some combustible material might have occurred while a muzzle-loading cannon was being cleaned with a closely fitting cleaning rod.[47] As Balfour recognized only too well, the lack of adequate documentation insures, and always will insure, that none of these possibilities can be rejected outright, however improbable they may appear. Nevertheless, in any assessment of the likelihood of an independent process of invention in Asia it is worth emphasizing that in none of the devices mentioned above was a well-fitting piston moving smoothly in a cylinder an essential element of the design, as it was in the fire piston.

Of the products of Asian technology to which this last objection does not apply the most likely candidate as a possible antecedent of the fire piston is the piston bellows,[48] an ancient device in which a good fit between the walls of a cylinder of uniform cross-section and a piston moving in it was almost as important as in the fire piston itself.[49] The case with regard to this instrument is made particularly strong by the fact that it had long been in common use in those parts of southeast Asia where the fire piston was found by nineteenth-century travelers from the West. The bellows consisted of a cylinder, held vertically and usually made of a hollowed-out tree trunk, in which a piston was moved by hand. The size appears to have been very variable, but a typical cylinder would be rather less than 5 feet in length and

[47] Balfour, *Anthropological Essays*, pp. 44–46.

[48] These two instruments are particularly closely associated in, for example, Needham, *Science and Civilisation in China*, IV, Part 2, 140–41.

[49] For an account of the piston bellows see T. Ewbank, *A Description and Historical Account of Hydraulic and Other Machines for Raising Water, Ancient and Modern* (New York, 1842), pp. 244–58.

between 4 inches and 6 inches in internal diameter. At its lower end, which was usually held in contact with the ground, it was closed and a small hole in the wall of the cylinder, near the bottom, allowed the air compressed by the descending piston to pass to the furnace by way of a short pipe. A leather washer or other packing, which was commonly fitted to the piston, insured that the fit was a reasonably good one, yet the extent to which pressure could be built up inside the cylinder would obviously depend on the ease with which air could escape through the exit hole. If the device was to be efficient, this hole and the attached pipe would certainly not be narrow; it seems unlikely, therefore, that any temperature increment large enough to be noticed, to say nothing of one sufficient to ignite tinder, would have been obtained in the course of normal usage. And even if effective compression had occurred, possibly as a result of some temporary blockage in the exit pipe, the force necessary to reduce the air to one-fifth its original volume (the degree of compression found necessary in operating a purpose-built fire piston, even with perfectly dry tinder[50]) would certainly have been far greater than any single man could apply, being many times greater in a piston bellows of typical dimensions than in any known fire piston.[51] Moreover, any slight heating which might have occurred would surely have been associated only with the proximity of the furnace and so ignored.

But even if the piston bellows did not lead to an independent discovery of the fire piston in Asia, the instrument may not be wholly irrelevant to this account. It is at least conceivable that the Europeans who brought fire pistons, or more probably a knowledge of them, to the East saw in the established methods for the construction of piston bellows a ready-made technology suitable for the manufacture of a form of fire piston using materials easily obtainable in the region. The fact that piston bellows were in common use in those areas where fire pistons were found in the nineteenth century would be wholly consistent with this view; it would also go some way toward answering two points raised by Balfour as objections to the possibility that the Asian fire piston was introduced originally from Europe. These points are, first, that no instrument of European origin has ever been found in Asia, and second, that if the fire piston was introduced by Europeans in the early years of the nineteenth century, then knowledge of it must have spread with astonishing rapidity for it to be in use in areas as far apart as Burma, Java, and the Philippines by the 1860's. The weight of both of

[50] J. L. Gay-Lussac, "Sur le froid produit par la dilatation des gaz," *Annales de Chimie et de Physique*, IX (1818), 308.

[51] The force being proportional to the cross-section area of the cylinder.

these objections would be greatly diminished if we could assume that the techniques required for the production of the fire piston were derived directly from the long-established and widely known procedures used in the manufacture of the piston bellows. If this was the case, would importation from Europe ever have been necessary, and would it be surprising that the ability to manufacture fire pistons spread so quickly through Asia?

VI. *Conclusion*

The case for seeing Europe rather than Asia as the unique home of the fire piston is, I must admit, scarcely advanced by this paper. My attempt to answer some of the weightier objections to such a view, my examination of the history of the fire piston in Europe, and the suggestion of a possible role for the piston bellows do not constitute conclusive evidence, and they are not offered as such. When we recall that even in France the principle of the fire piston belongs wholly to the class of discoveries made, as Bacon put it, "when men were not seeking them but were busy about other things,"[52] it would be rash indeed to state that the device could not have emerged in Asia as the result of a pattern of development quite different from that followed in Europe. Certainly, the patterns of development which have been suggested so far all have a strong element of implausibility, but very few of them can be ruled out entirely, and, in any case, there are undoubtedly others which could be suggested in their place.

Concerning the origins of the fire piston in Europe, however, there is far less room for doubt. The evidence establishes conclusively, I think, that credit for the invention of the European instrument must now be given to Joseph Mollet and his ingenious friends and colleagues in Lyons.

APPENDIX

Domestic fire pistons of European origin have been located in the following museums:

Science Museum, London
   1. Fire piston in brass (length 5½ inches), modeled on one described in an unsigned article on "Fire," *The Penny Magazine*, III (1834), 286, and stated there to have been manufactured in France. In this instrument, which was probably made in London, the air is forced through a number of small holes in the end of

---

[52] *The Works of Francis Bacon*, ed. J. Spedding, R. L. Ellis, and D. D. Heath (London, 1860), IV, 98. The phrase is in the first book of the *Novum Organum*, aphorism 108.

the cylinder before encountering the tinder. See Plate I, item 1. (Bryant & May collection, No. 1228.)

2. Fire piston of conventional design with cylinder in brass and piston in steel with brass head (length 4⅛ inches). British. See Plate I, item 2. (Bryant & May collection, No. 1231.)

3. Conventional fire piston with cylinder in brass and piston in steel with brass head (length 3 inches). British. See Plate I, item 3. (Bryant & May collection, No. 1230.)

4. Conventional fire piston in brass (length 5½ inches). British. (Bryant & May collection, No. 1229.)

5. Fire piston contained in a walking stick of dark mahogany or rosewood (length 2 feet, 5½ inches, but reduced from about 2 feet, 10 inches through use as a poker!). British. In Plate II the instrument is shown dismantled. It consists of (left to right): a knob which unscrews to disclose a compartment for storing tinder, a brass screw for holding the stored tinder in position, the handle of the stick from which there projects a short hollow brass cylinder (length 3 inches), and finally the main body of the walking stick bearing a short projection which just fits inside the brass cylinder. This projection has a perforated screw-on cap under which the tinder is placed. Ignition is effected by placing the cylinder over the projection and then forcing it down rapidly (the point of the stick being struck sharply on the ground). (Bryant & May collection, No. 1232.)

The Science Museum instruments are all described in detail in M. Christy, *The Bryant and May Museum of Fire-Making Appliances. Catalogue of the Exhibits* (London, 1926), p. 91, and in the supplement (London, 1928), p. 266. A photograph of item 4 (above) appears on p. 233 of this catalogue.

Pitt Rivers Museum, Oxford

6. Conventional fire piston in brass (length 5½ inches). Probably British.

7. Fire piston in white metal of a type which was introduced in France *ca.* 1900 (length 3 inches). See note 42.

Museum of the History of Science, Oxford

8. Conventional fire piston with cylinder in brass and piston in steel with brass head (length 6½ inches). Probably British.

(The lengths given are those of the cylinders only.)

XI

THE CHALLENGE OF A NEW TECHNOLOGY : THEORISTS AND THE HIGH-PRESSURE STEAM ENGINE BEFORE 1824

RÉSUMÉ

En France, l'étude de la théorie de la machine à vapeur a pris un nouvel élan en 1815, au moment où furent introduites les nouvelles machines *compound* d'Arthur Woolf, qui venaient de remporter un grand succès en Angleterre. Les travaux des théoriciens français de 1815 à 1824 sont examinés en détail dans cet article. L'auteur tente de reconstituer la période d'intenses recherches qui précéda la publication des *Réflexions* de Carnot et de définir les problèmes auxquels celui-ci dut faire face. Ses prédécesseurs s'étaient surtout préoccupés de la quantité de charbon consommée par les machines qui, comme celles de Woolf, fonctionnaient à des pressions supérieures à celle de l'atmosphère. Ils cherchaient notamment à comprendre le rendement remarquable des machines de Woolf. Etait-ce que la vapeur avait une propriété thermique particulière ou bien était-ce la manière dont elle était utilisée, en application du principe de Watt, dans presque toutes les machines à haute et à moyenne pression ? Etait-il possible d'améliorer le rendement, par exemple en élevant encore la pression ?

Ce genre d'étude nous permet d'apprécier l'originalité de Carnot, tout en reconnaissant combien il devait aux recherches faites par ses prédécesseurs, surtout par son ami Nicolas Clément, dans les dix dernières années. C'est en connaissant mieux les idées reçues des théoriciens de la machine à vapeur au début des années 1820 que l'on pourra comprendre l'oubli où tomba l'œuvre de Carnot.

For some forty years before the publication of the *Réflexions sur la puissance motrice du feu* the experimental and theoretical study of the thermal properties of matter in the gaseous state had been one of the most exciting and fruitful traditions in physical science ([1]). It was a tradition that yielded important laws, like Gay-Lussac's law of expansion for gases (1802), and results, like those of Delaroche and Bérard on the specific heats of gases (1813), which were to exert a marked and sometimes misleading influence on physical theory until the definitive work of Victor Regnault in the middle of the nineteenth century. It also yielded theoretical advances of lasting importance ; for example, Laplace's explanation of the notorious discrepancy between the experimental value for the velocity of sound and the value predicted in accordance with the theory of Newton (1802) was an early consequence of an interest in adiabatic phenomena which remained strong and profitable, especially among French physicists, until well into the 1820s.

The stimulus for most of this work appears to have been theoretical, and the results that attracted most attention were those that touched on theory. It is entirely characteristic, therefore, that the choice of the subject for the French Institute's prize competition of 1811-13, which led to the experiments of Delaroche and Bérard, was made in response to a theoretical problem : it was hoped that a study of the specific heats of gases would resolve the vexed question whether all the heat (or caloric) in a body existed in a 'free' state (i.e. in a state in which its presence would be detected by a thermometer) or whether some of it was 'latent' or 'combined' ([2]). However, practical incentives were not totally absent, especially in work on the properties of steam. Dalton's study of the variation of the pressure of steam with temperature (1802) was almost certainly a consequence of the rapidly increasing use of steam power in industrial Manchester, and there can be no doubt that the motives for the experiments on steam by Watt in the 1780s, Bétancourt (1790), John Southern (1803), and John Sharpe (1806) were similarly utilitarian.

Not surprisingly, practical considerations tended to be more prominent in Britain than in France until about 1815. But with the resumption of peace and the sudden growth of French interest in steam engines, it was the French who took the lead in investigating the applications of an understanding of the thermal properties of gaseous matter. As is well known, the advances that had been made in British steam power technology between the 1790s and 1815 came as a revelation to French engineers once contact between Britain and France was renewed ; for the economist Jean-Baptiste Say, the steam engine was the symbol of the new, industrialized Britain which contrasted so sharply with the very different country he had known before the wars. As he wrote in 1815 :

---

(1) For an account of this tradition, see Robert FOX, *The caloric theory of gases from Lavoisier to Regnault* (Oxford, Clarendon Press, 1971), especially chapters 1-5. The best study of the aspects of the tradition that are treated in this paper is in D. S. L. CARDWELL, *From Watt to Clausius. The rise of thermodynamics in the early industrial age* (London, Heinemann, 1971), p. 153-81, a work to which I am greatly indebted.

(2) See the announcement of the competition, dated 7 January 1811, in *Mémoires de la classe des sciences mathématiques et physiques de l'Institut Impérial de France*, xi (1811), p. xcv ; also FOX, *op. cit.* (1), p. 134.

Partout les machines à vapeur se sont prodigieusement multipliées. Il n'y en avait que deux ou trois à Londres il y a trente ans ; il y en a des milliers à présent. Elles sont par centaines dans les grandes villes manufacturières ; on en voit même dans les campagnes, et les travaux industriels ne peuvent plus se soutenir avec avantage qu'au moyen de leur puissant secours ([3]).

But for those who were concerned with the practical problems of power technology it was the design, rather than the number, of the British engines which aroused interest. In particular, the French were amazed by the economy claimed for the two-cylinder compound engines of the Cornish engineer Arthur Woolf, which incorporated Watt's expansive principle and operated at about three atmospheres. Although the essential elements in their design had been patented as early as 1804 ([4]), Woolf's engines only began to operate effectively in the autum of 1814 ; yet within a few months they were no less a sensation in France than in Britain ([5]).

For the reputation of the Woolf engine and for those in France who made their fortunes by it — like Humphrey Edwards, a former associate of Woolf who began importing the engine from Britain in 1815 — it was a happy circumstance that the reports of its finest performance coincided almost exactly with the renewal of easy contact between Britain and France ([6]). In fact, Edwards' early activities as an importer and builder of Woolf engines were blessed with remarkable good fortune, and, with coal as always a scarce commodity in France, it is not surprising that his engines were still preferred by most French engineers in the 1820s, although by then there were several rivals ([7]) and, at least in Britain, growing scepticism about the capacity of the Woolf engine to sustain its performance over a long period.

---

(3) J. B. SAY, *De l'Angleterre et des Anglais* (Paris and London, 1815), p. 30-1.

(4) "Woolf's improvements in the construction of steam engines", British patent specification no. 2772 ; dated 7 July 1804.

(5) On the history of the Woolf engine, see T. R. HARRIS, *Arthur Woolf the Cornish engineer 1766-1837* (Truro, D. Bradford Barton Ltd., 1966), and the discussion in CARDWELL, *op. cit.* (1), p. 155-9. The most important primary source is the second volume of John FAREY, *A treatise on the steam engine, historical, practical, and descriptive* (Newton Abbot, David & Charles, 1971), which is very largely devoted to the Woolf engine. Although page proofs were ready about 1840, this second volume of Farey's book remained unpublished and little-known until 1971. The first volume had appeared in 1827. It is interesting to speculate on the reasons for the remarkable performance of the Woolf engine. Clearly a slight increase in thermodynamic efficiency was gained by operating at pressures above that of the atmosphere, but, as Professor Cardwell has pointed out (CARDWELL, *op. cit.*, p. 180), the most important improvements may well have been those concerned with constructional details (in the boiler, for example, or in thermal lagging).

(6) According to Lean's monthly *Engine reporter*, which gave independent reports on the performances of the steam engines of Cornwall, the famous Woolf engine at the Wheal Abraham copper mine achieved its greatest economy (a 'duty' of 56.9 ft lb for every bushel of coal burnt) in May 1816, but it had been performing impressively since October 1814. A 'duty' of between 20 and 30 ft lb per bushel was considered highly satisfactory for Watt engines of comparable horse power.

(7) The most exciting of these was the engine of the American-born Jacob Perkins, which operated at 35 atmospheres and first became known in Britain and France early in 1823. The claims made on its behalf were quickly found to be exaggerated and its inflated reputation did not survive the mounting scepticism of 1823-4. But there is an indication of the excitement that it caused in the *Réflexions sur la puissance motrice du feu* (1824 edition), p. 99-100n.

In the early years of the Restoration it was widely accepted in France that, by comparison with the best Watt engines of comparable power, Edwards' Woolf engines could effect a saving in fuel of about 50 per cent. It is true that according to tests on an engine erected by Edwards at le Gros Caillou in Paris, which Riche de Prony conducted in 1821-2 ([8]), the saving was much smaller than this (probably no more than 13 per cent), but there was a good deal of contrary evidence, not least from satisfied clients of Edwards, which suggested that Prony had underestimated the advantages of Woolf's design. If we are to judge by Edwards' continued success in the 1820s, it was this contrary evidence which carried the greater conviction among French engineers and industrialists. In fact, for nearly a decade before the publication of the *Réflexions* the superior economy of the Woolf engine went virtually unquestioned, so that the problem for those of Carnot's contemporaries who took an interest in the theory of steam power was not to prove that a saving was obtained by using the engine but to account for it.

Woolf himself had based his design on experiments of his own which, as he reported in his patent specification of 1804 ([9]), convinced him that if steam at a pressure of, say, three lb per sq inch above atmospheric pressure were allowed to expand until its pressure fell to that of the atmosphere, its volume would increase three-fold ; for an excess pressure of 5 lb per sq inch the increase in volume would be five-fold, and so on, provided always that the temperature remained constant during expansion. Hence for a very slight increase in the pressure of steam above atmospheric pressure, a large effect could be obtained in free expansion. But Woolf's explanation, based on results for which there was no independent support of any kind, had understandably convinced no one, and subsequent attempts to explain the economy achieved by using his and other high-pressure expansive engines took no account of it.

Until about 1814, when the merits of the Woolf engine were so suddenly and startlingly demonstrated, there was little incentive for theoretical discussion of the merits of high-pressure engines. The prevailing opinion, founded chiefly on the engines of Trevithick and Vivian, seems to have been that operating at pressures above one atmosphere might lead to compactness and portability, but it was rarely suggested that a saving in fuel was to be looked for. As the American engineer Oliver Evans put it in 1808 : 'it has been generally believed that fuel is required in direct proportion to the power obtained' ([10]) — and this regardless of the pressure.

---

(8) Reported in PRONY, "Rapport fait à MM. les Président et Conseillers de la Cour royale séante à Paris, sur la nouvelle et l'ancienne machine à vapeur établies, à Paris, à Gros-Caillou, à l'occasion du procès pendant au tribunal de la Cour royale de Paris, entre M. Edwards, vendeur, et M. Lecour, acquéreur de la nouvelle machine", *Annales des mines*, xii (1826), 3-100. Prony's tests were occasioned by a law suit between Humphrey Edwards and Louis-Didier Lecour, described as 'un entrepreneur du chauffage des pompes à feu du Gros Caillou et de Chaillot'. Lecour had purchased a Woolf engine from Edwards for erection at le Gros Caillou, chiefly for pumping, in 1817, but, on finding that the saving in fuel, by comparison with Périer's old double-acting engine, was far less than the 50 per cent claimed, he refused to pay the 45,000 francs that he still owed to Edwards. Tests conducted in 1818 and 1819 by P. S. Girard had supported Edwards' claims for his engine, and in 1819 the court ordered Lecour to pay the outstanding 45,000 francs and the costs of the case. Prony's tests, which were far more thorough than Girard's, followed an appeal by Lecour. Many of the papers concerning the case are in MSS. 2758 and 2766 at the Ecole Nationale des Ponts et Chaussées in Paris.

(9) *Op. cit.* (4), p. 2-4.

(10) See Evans' letter to Robert R. Livingston, 20 November 1808, in Greville and Dorothy BATHE, *Oliver Evans. A chronicle of early American engineering* (Philadelphia : Historical Society of

Evans, however, was one of the few who did not share the view he was summarizing ; indeed, it was economy that he stressed as the chief merit of his own high-pressure engine, which he had been advocating since 1804 ([11]). For my purpose, his discussion is particularly interesting, for he rested his claims not on the performance of his engines but on theory, albeit rather inadequate theory. He had been impressed, apparently through reading Dalton's work on vapour pressures, by the great increases in the pressure of steam which accompanied comparatively small increments in temperature. For an increase in temperature from 212°F (100°C) to 242°F (117°C), for example, the pressure of steam was doubled from one to two atmospheres, and for a further increase of only 60°F (33°C), to 302°F (150°C), it increased to eight atmospheres. The pressure, in short, rose at a far greater rate than the temperature and, as Evans assumed (wrongly and without justification), than the fuel supplied to the steam. Hence by using higher and higher pressures it seemed possible to effect large increases in the power obtained (which depended on the pressure) for only modest increases in the expenditure of fuel. From the point of view of economy, the prospect appeared an exciting one.

But, as Evans pointed out, the advantage of his engine lay not only in the ease with which high-pressure steam could be produced. Through operating at between six and ten atmospheres, Evans was also able to use the expansive principle that Watt had discovered as early as 1769, patented in 1782, but seldom used in his own engines, in which the pressure scarcely exceeded one atmosphere. By this principle, the supply of steam from the boiler was cut off during the stroke (usually when the cylinder was about one quarter full of steam), and the steam was then allowed to expand freely, with diminishing pressure, until the stroke was completed. The power of the stroke was obviously less than if the steam had remained at its initial high pressure throughout, but the saving obtained by only partially filling the cylinder far outweighed the loss in power. As John Robison showed in his important article 'Steam engine' in the third edition of the *Encyclopaedia Britannica* in 1797, by cutting off the supply of steam

---

(10) Suite

Pennsylvania, 1935), p. 151. As I point out later (see note 22), Peter Ewart supported the view cited by Evans very strongly in the same year ; on Ewart's position and those of Wollaston and Playfair about the same time, see CARDWELL, *op. cit.* (1), p. 162-164. Prony too had expressed a similar view in 1796, though he stressed the care that was needed in order to ensure that the full effect of the fuel was obtained. See his *Nouvelle architecture hydraulique* (2 vols., Paris, 1790-6), ii. 148 : 'L'effet des machines à feu est en général, et toutes choses égales d'ailleurs, proportionnel à la quantité de combustible consommée. Le temps entre nécessairement et implicitement dans cette évaluation ; car, pour tirer le plus grand parti possible d'une masse donnée de charbon, il faut faire en sorte que le chauffage ne soit ni trop rapide ni trop lent, sans quoi on perdrait sur la vaporisation et sur le produit de la machine'. The prevailing opinion in the early years of the nineteenth century seems to have been that there was a maximum effect that could be obtained in principle from a given quantity of fuel but that in the machines actually in use the maximum effect was not obtained, for purely practical reasons of the kind cited by Prony. Theoretical discussions aimed at redefining the ideal conditions under which steam should be used were rare by comparison with the period after 1814, when the performance of the Woolf engine stimulated an unprecedented interest in theory.

(11) For Evans' own account of his engine, see his book *The young steam engineer's guide* (Philadelphia, n.d. [1805]), especially p. 12-21 and 28-36 ; translated into French by I. Doolittle, as *Manuel de l'ingénieur mécanicien constructeur des machines à vapeur* (Paris, 1821). The letter cited in note 10 is also interesting.

154

one quarter of the way through the stroke, it was possible to obtain a saving in fuel of 75 per cent while reducing the effect of the stroke by only a half ([12]).

I have discussed Evans' argument at some length not because it was influential (indeed, it was virtually unknown in Europe until it was translated into French in 1821 ([13])) but because it identified the two possible causes of the economy of the high-pressure engine that were under discussion in the years when Carnot was developing his own theory. In fact, it was only in October 1816 that an analysis similar to that of Evans, which clearly treated both sources of economy, became available in Europe. The analysis appeared in an article in the *Annales de chimie et de physique* by A. R. Bouvier, a civil engineer and graduate of the Ecole Polytechnique ([14]). Bouvier's chief purpose was to demonstrate the merits of a rotational engine of unusual design, incorporating several cylinders in each of which high-pressure steam was allowed to expand freely after cut off. Using an argument that was essentially identical to that given by Robison in 1797 and discussed below (though with no acknowledgement either to Robison or to any other authority), Bouvier was able to demonstrate quite straightforwardly the economy that was obtained by expansive working. However, his discussion of the advantages that accrued from the thermal properties of high-pressure steam was far more suspect. In this part of the paper Bouvier ignored the effect of expansion altogether, arguing that two volumes of steam, V and V', with pressures P and P' such that $PV = P'V'$, were capable of yielding the same mechanical effect (the effect for a non-expansive engine being the product of the pressure and volume of the steam used). He then invoked the evidence of cooling by adiabatic expansion to argue that the steam with the higher pressure (and the smaller volume) contained less heat, and so could be produced more cheaply, than the steam at the lower pressure ; hence there seemed to be no limit to the increase that could be effected in the ratio $\left(\dfrac{\text{effect obtained}}{\text{fuel expended}}\right)$ simply by using ever smaller volumes of steam at very high pressures. In 1816, in the absence of any notion of the equivalence of heat and mechanical work, the argument based on adiabatic phenomena would have appeared convincing enough for the case of gases. For it was axiomatic in the caloric theory that if the temperature of a gas undergoing any expansion were to remain constant, heat had to be added to it ; if no heat were added (in the adiabatic case), the temperature would fall. But the weakness of Bouvier's position lay in the absence of any evidence that this argument could be applied to vapours. In the first place, it was very far from certain that the relationship PV = constant held good for a given mass of vapour at different temperatures ([15]), and, as Biot had already pointed out in his *Traité de physique* earlier

---

(12) [Robison], "Steam-engine", in *Encyclopaedia Britannica* (3rd edn., Edinburgh, 1797), xvii. 765-6.

(13) See note 11.

(14) A. R. BOUVIER, "Note sur les machines à vapeur, et description d'une de ces machines propre à produire immédiatement le mouvement de rotation", *Annales de chimie et de physique*, iii (1816), 177-92.

(15) See note 28.

in 1816 ([16]), there was as yet (at least in France) no reliable information on the dependence of the heat content of a saturated vapour on its temperature.

Understandably, Bouvier's article appears to have convinced few of its readers, though it might conceivably have done something to further the view that there were two distinct explanations of the economy of the high-pressure engine to be considered. Certainly this view, associated with a new interest in the theoretical advantages of expansive operation, did gain ground in France in the early years of the Bourbon restoration, although until about 1819 it was still the nature of steam rather than the expansive principle which tended to attract attention.

Both in Britain and in France the property of steam that was most often discussed was its heat content (or 'total heat') at different pressures — a discussion that usually arose in the context of the crucial question whether more heat was required to fill a cylinder with steam at a high pressure than at the pressure of the atmosphere. If more heat was required, then at least part of the benefit gained by using higher pressures would be offset. Although surprisingly few writers seem to have known of his work in the early years of the nineteenth century, Watt had gone some way towards answering this question long before. In 1765 and again in the 1780s he had performed experiments which suggested that the amount of heat needed in order to produce a given mass of steam from water at a fixed initial temperature, say $0°C$, was independent of the temperature and pressure of the steam ([17]). It is true that Watt's experiments had been concerned with steam at pressures below that of the atmosphere; he had shown, for example, that, from water at $0°C$, as much heat was required in order to produce steam at $21°C$ and a pressure of about 1/40 atmosphere as when the same water was converted to steam at $100°C$ and at atmospheric pressure. But the inference that the amount of heat required would be the same for *all* pressures (commonly known as Watt's law) was a simple and obvious one ([18]). Although a full account of Watt's

---

(16) Biot's plea for experiments to determine the dependence of the total heat of a saturated vapour on its temperature is discussed later in this article. On the date of the publication of the *Traité*, see *Académie des Sciences. Procès-verbaux des séances de l'Académie tenues depuis la fondation de l'Institut jusqu'au mois d'août 1835* (10 vols., Hendaye, 1910-22), vi. 47, where it is recorded that Biot presented a copy of his book to the Académie on 22 April 1816.

(17) Watt's experiments are described most fully in John ROBISON, *A system of mechanical philosophy*, ed. D. Brewster (Edinburgh, 1822), ii. 5-10, but they had already been referred to in J. A. DE LUC, "Examen d'un mémoire de M. Monge, sur la cause des principaux phénomènes de la météorologie", *Annales de chimie*, viii (1791), 79-81, and in Joseph BLACK, *Lectures on the elements of chemistry*, ed. John Robison (Edinburgh, 1803), i. 190. The relevant parts of Robison's *System of mechanical philosophy* were in fact printed by June 1818 and were distributed by Watt to a number of his friends; see W. A. SMEATON, "Some comments on James Watt's published account of his work on steam and steam engines", *Notes and records of the Royal Society of London*, xxvi (1971), 35-42. Among those who received copies of this separate publication were Prony and Berthollet in France; see *Bulletin de la Société d'Encouragement pour l'Industrie Nationale*, XVIIIe année (1819), 255.

(18) The law was always viewed with suspicion, even though John Sharpe in 1806 and Clément and Desormes in 1819 rediscovered it, apparently independently. About 1830 Clément wrote with obvious bitterness about the opposition that had been voiced when he announced the law to the Académie des Sciences in 1819; Laplace had evidently been prominent among the academicians who dismissed Clément's discovery as absurd : see *Travaux scientifiques et industriels de M. Clément* (Paris, n.d. [1830]), p. 8. The law had been shown to be no more than approximately true by about 1827; on

experiments was not published until 1822, there was no reason why they should not have been known much earlier, for they had been referred to, albeit briefly, in 1791 in an article by J. A. De Luc in the *Annales de chimie* and in 1803 in John Robison's important edition of Joseph Black's *Lectures on the elements of chemistry* ([19]). But knowledge of the law seems to have spread slowly. When the Manchester solicitor John Sharpe raised the problem of explaining the apparent economy of the high-pressure engine in a paper to the Manchester Literary and Philosophical Society in 1806 ([20]), he announced his own independent rediscovery of Watt's law as if it were wholly original.

With a caution that Evans had singularly lacked, Sharpe declined to draw any firm conclusions on the question of economy. He pointed out that the effect gained from each stroke of a non-expansive engine was doubled by using steam at two atmospheres instead of steam at one atmosphere only if the same volume of steam was used in the two cases. This statement would not have been questioned, but, in the absence of reliable information about the density of steam at different pressures, it did little to explain the economy obtained by operating above atmospheric pressure. As Sharpe admitted, even when Watt's law was accepted, there was no way of ascertaining how much steam (by *weight*), and hence how much fuel, was required in order to fill the cylinder. If, as he thought likely, the pressure of the steam was proportional to its density (so that the weight of steam required to fill the cylinder at two atmospheres was twice that required at one atmosphere), it followed from Watt's law that no economy was possible. This rather disappointing conclusion was the one favoured by Dalton, a good friend of Sharpe's, in a note, dated October 1810, which Sharpe added to the published version of his paper ([21]). So the prevailing opinion referred to by Evans in 1808 — and expressed with great cogency in the same year by the Scottish engineer Peter Ewart, who was a good friend of both Dalton and Sharpe in Manchester ([22]) — appeared to be vindicated.

It was only with the remarkable performances of the Woolf engine in 1814-15 that interest in the theoretical problems raised by Evans and Sharpe was revived ; henceforth there could be no doubt that the use of high pressures was economical, and an explanation was called for. Understandably, it was now the French, rather than the

---

(18) Suite

the work of Despretz and Dulong which led to this conclusion, see notes 67 and 68 of FOX, "The background to the discovery of Dulong and Petit's law", *The British journal for the history of science*, iv (1968-9), 1-22.

(19) See note 17. The earliest reference to Watt's experiments in France appeared in DE LUC, *op. cit.*, (17), p. 79-81. De Luc recounted that he had observed the experiments seven years earlier ; they had shown, as he put it, that 'plus les vapeurs d'eau bouillante sont rares plus elles contiennent de feu latent'. He gave no further details, and Watt's results remained unknown in France until 1818 (see note 17).

(20) SHARPE, "An account of some experiments, to ascertain whether the force of steam be in proportion to the generating heat", *Memoirs of the Literary and Philosophical Society of Manchester*, 2nd ser. ii (1813), 1-14. Read 7 February 1806.

(21) *Ibid.*, p. 13-14.

(22) EWART, "On the measure of moving force", *Memoirs of the Literary and Philosophical Society of Manchester*, 2nd ser. ii (1813), 105-258 (168-9). Read 18 November 1808.

British, who began to make the most interesting contributions. Not only were the French engineers amazed by the novelty and economy of the new engines, but they could also call upon a tradition of theoretical study of machines which had its origins far back in the eighteenth century and which recently had been fostered with particular brilliance in the academic atmosphere of the Ecole Polytechnique.

At first, like their contemporaries in Britain, most French engineers looked for the economy of the Woolf engine in some property peculiar to steam, rather than in the use of the expansive principle. This was the approach adopted in one of the earliest French discussions, by Jean-Baptiste Biot in his *Traité de physique* in 1816 ([23]). Biot argued that the amount of heat required to fill the cylinder of an engine with steam at a temperature t°C and a pressure P was proportional to $\rho'c'$, where ρ' was the density of the steam at the temperature t, and c' was the amount of heat required to produce 1 gm of the steam from water at a standard temperature, say 0°C. Departing from the assumption made by Sharpe and Dalton, and adopting one that seems to have been more widely accepted in France (not least by Carnot), Biot claimed that steam obeyed all the normal gas laws ; hence he could put

$$\rho' = \frac{P\rho}{0.76 \left(1 + \dfrac{t}{267}\right)}$$

where ρ was the density corrected (notionally) to a temperature of 0°C and a pressure of one atmosphere (0.76 m of mercury). Putting the power obtained from an engine proportional to P — an assumption that could only be made for a non-expansive engine ([24]) — it could then be argued that the most advantageous pressure was that at which $\dfrac{c'}{(267 + t)}$ was a minimum. It followed that if c' increased more slowly than $(267 + t)$ with increasing t, there would be no limit to the economy gained by using steam at high pressures. To the extent that he inclined to the view that c' did indeed increase more slowly than $(267 + t)$, Biot thus came out as a supporter of the high-pressure engine, though he was not explicit in his support and strongly advocated the crucial experiments to determine the variation of c' with t.

It is a good indication of the irregularity of scientific contacts between Britain and France at this time that Watt's law and Sharpe's discussion of it, which provided just the information he sought, were obviously unknown to Biot in 1816. In fact, the law remained unknown in France until it was rediscovered independently by Nicolas Clément and Charles Bernard Desormes in 1819 ([25]). That the French knew nothing of

---

(23) BIOT, *Traité de physique expérimentale et mathématique* (Paris, 1816), iv. 739-41.

(24) Biot appears not to have realized that Woolf's design incorporated an expansive phase ; but see note 34 on his later discussion of the expansive principle.

(25) Clément and Desormes announced their results in a paper read to the Académie des Sciences on 16 and 23 August 1819. Unfortunately their paper was never published, although an unsigned article summarizing it did appear as "Mémoire sur la théorie des machines à vapeur ; par MM. Desormes et Clément.(Extrait.)", in *Bulletin des sciences, par la Société Philomathique de Paris,* new ser. vi (1819), 115-18. Clément communicated at least some of the results privately to the Société d'Encouragement pour l'Industrie Nationale in April 1819 ; see *Bulletin de la Société d'Encouragement,*

the law until then is clearly implied by discussions at meetings of the Société Philomathique and of the Council of the Société d'Encouragement pour l'Industrie Nationale which took place in, respectively, June and December 1818. At both meetings the need for an experimental investigation of the heat content of steam at different pressures was made glaringly apparent. At the Société Philomathique, on 6 June 1818, the engineer and former professor at the Ecole Polytechnique, J.N.P. Hachette, read a paper on the theory of the Woolf engine which appears to have provoked lively, even acrimonious discussion ([26]). Hachette argued that if the latent heat of steam was assumed to be independent of temperature – an assumption for which he offered no justification, although Watt's assistant John Southern had arrived at precisely this result by experiment in 1803 ([27]) – its total heat would rise steadily as the pressure (and temperature) rose. For example, according to Hachette and Southern, the amounts of fuel required to produce equal weights of steam at one and two atmospheres (i.e. at temperatures of $100°$ C and $122°$C) from water at $0°$C would be in the ratio $\dfrac{100+550}{122+550}$, or $\dfrac{650}{672}$, where 550 cal per gm (to use modern units) was the latent heat of steam. (By Watt's law, of course, this ratio would have been exactly unity.) It followed from Hachette's argument, and from his assumption that the density of steam was proportional to its pressure ([28]), that the amount of steam required in order to fill a cylinder with steam above atmospheric pressure rose more rapidly than the pressure ; hence if the engine operated non-expansively (i.e. with the effect of each stroke being

---

(25) Suite

XVIIIe année (1819), 255. It was evidently not until early in 1822 that Clément heard of Sharpe's work. His informant was Thomas Thomson, whom he met in Glasgow; see Thomson's account of the meeting in *Annals of philosophy*, new ser. iii (1822), 303.

(26) The argument of this paper is summarized in "Méthode pratique pour comparer les effets des machines à vapeur ; par M. Hachette", *Bulletin de la Société d'Encouragement*, XVIIe année (1818), 169-74, and in HACHETTE, *Traité élémentaire des machines* (2nd edn., Paris, 1819), p. 212-16 (also in the fourth edition, of 1828, p. 270-4). Accounts of the meeting and of the discussion that followed Hachette's paper are in "Tableau de M. Clément-Desormes, relatif à la théorie générale de la puissance mécanique de la vapeur. (Extrait.)", *Nouveau bulletin des sciences, par la Société Philomathique de Paris*, new ser. ii (1826), 53.

(27) Southern's results were announced in a letter to Watt, dated 26 March 1814 and published in ROBISON, *op. cit.* (17), ii. 160-73. The relevant pages of Robison's work had been printed by June 1818, so that it is just possible, though unlikely, that Hachette knew of Southern's work. See note 17 on the distribution of these pages by Watt.

(28) It should be noted how important this particular assumption was for the arguments of Hachette and other writers on the subject. Opinions were sharply divided between those who felt, like Hachette, Sharpe, and Dalton, that the pressure of saturated steam at different temperatures was proportional to its density (i.e. $PV = $ constant) and those like Biot, Clément, and Desormes who thought that the general gas law, $\dfrac{PV}{T} = $ constant, should be applied. The urgent need for experimental evidence was not answered until 5 November 1821, when C. M. Despretz described his work on the subject to the Académie des Sciences ; see his "Mémoire sur la densité des vapeurs", *Annales de chimie et de physique*, xxi (1822), 143-55. Despretz gave strong support to those who favoured the relationship $\dfrac{PV}{T} = $ constant, and it is not surprising to find his conclusion being generally adopted thereafter, not least by Carnot and Poisson.

proportional to the pressure), the economy actually deteriorated as the pressure increased. In a conclusion which appears to have been the source of the objections to his paper, Hachette deduced from this that the economy arose simply from the use of the expansive principle, the value of which he proceeded to demonstrate theoretically.

Conceivably Hachette was no more satisfied with his treatment than was his audience at the Société Philomathique, and any doubts he may have had were fully justified. It was obvious to everyone who tackled the problem that the theory of the high-pressure engine demanded an accurate knowledge both of the dependence of the latent heat of steam on temperature and of the relationship between the density and pressure of vapours. It is not surprising, therefore, that in December 1818, when interest in Edwards' engines was at its height, Hachette's proposal that the Société d'Encouragement should undertake an experimental investigation of latent heats was enthusiastically adopted ([29]). A four-man committee was set up and 300 francs were allocated for the purchase of fuel.

Unexpectedly, an answer to Hachette's problem became available before the committee could present its report. By April 1819 Nicolas Clément had performed experiments at Oberkampf's factory at Jouy which showed that the amount of heat needed to raise a given mass of steam was independent of its pressure, and by August 1819 this result, as well as the earlier work on the subject by Watt and Southern, was common knowledge in France ([30]).

If the paper of August 1819 in which Clément and his father-in-law Desormes described their work to the Académie des Sciences had been published in full – and thereby been formally recognized as a major contribution – the debate about the theory of the high-pressure engine could well have taken a different course in the 1820s. I say this partly because of the unusual theoretical discussion in the paper (to which I shall return later) but also because the experimental data of Clément and Desormes were both timely and stamped with an authority that was lacking in earlier work. If their results were accepted (i.e. if c' was taken to be independent of t), it followed that there was some advantage in using high-pressure steam, even without expansion, simply because they assumed, like Biot in 1816, that steam obeyed the general gas law

$$\frac{PV}{(267 + t)} = \text{constant}.$$ However, as I shall point out, their calculations also showed convincingly that the economy obtained in this way was small by comparison with the economy resulting from expansive operation ; it was, moreover, far smaller than that which was actually obtained in practice in a Woolf engine – a point also made by Poisson, working from identical assumptions, in 1823 ([31]).

Unfortunately the brief summary of the paper that was published in August 1819, in the *Bulletin* of the Société Philomathique, failed to do justice to the work of its authors, and, although their experiments on latent heats quickly became well known,

(29) "Expériences à faire sur la quantité d'eau évaporée dans la machine à vapeur de M. Edwards", *Bulletin de la Société d'Encouragement*, XVIIe année (1819), 385-6.

(30) See notes 17 and 25.

(31) POISSON, "Sur la chaleur des gaz et des vapeurs", *Annales de chimie et de physique*, xxiii (1823), 337-52 (348).

the theoretical parts of their paper were generally ignored. So it is not surprising that the belief that the economy of the high-pressure engine lay in some property of the steam rather than in the way it was used (expansively) remained a persistent one for some years to come. In 1821, for example, Laplace used his very suspect version of the caloric theory of gases to argue that the pressure of a gas was proportional to the square of the amount of caloric in a given volume – a conclusion which, in his view, explained the economy of high-pressure engines without reference to the expansive phase [32]. And in *Annals of philosophy,* between 1822 and 1825, Thomas Thomson, Charles Sylvester, and John Prideaux engaged in a debate that is notable chiefly for its lack of originality [33]. Surprisingly, it was Thomson, Regius Professor of Chemistry at Glasgow and editor of the *Annals,* who made the most confused contribution. Without any reference to experimental evidence or to expansive operation, he argued that the economy of high-pressure engines was an obvious consequence of the conclusions of Sharpe and Clément and Desormes, while also assuming that the pressure of steam at different temperatures was proportional to its density ; it is remarkable that he did not observe the incompatibility of these two positions, as his friend Dalton had done as early as 1810 and as Sylvester did almost immediately after Thomson's contribution had appeared. With Sylvester taking the view that no economy was possible, it was left for Prideaux to rehabilitate the (non-expansive) high-pressure engine, using an argument identical to that of Biot in 1816, though with the refinement that he could now state categorically that c' was independent of the pressure of the steam produced.

Despite these rather unfruitful discussions, the leading problem in the 1820s was clear : as Biot recognized by 1824 [34], it was to calculate the advantage gained by using steam expansively. The problem did not appear too difficult. In his *Encyclopaedia Britannica* article in 1797 Robison had given an expression for the 'duty' yielded by steam in the injection and expansion phases which was still widely used in the middle of the nineteenth century [35]. As Robison argued – in a treatment that depended heavily on Watt – the effect obtained during the injection of volume V_1 of steam at a pressure P_1 into a cylinder was $P_1 V_1$, while that obtained as the steam expanded freely from the

---

(32) P. S. de LAPLACE, "Sur l'attraction des corps sphériques, et sur la répulsion des fluides élastiques", *Annales de chimie et de physique,* xviii (1821), 186.

(33) THOMSON, "On the influence of humidity in modifying the specific gravity of gases", *Annals of philosophy,* new ser. iii (1822), 302-8 ; SYLVESTER, "Observations on the presence of moisture in modifying the specific gravity of gases", *ibid.,* new ser. iv (1822), 29-31 ; PRIDEAUX, "On the advantages of high pressure steam", *ibid.,* new ser. x (1825), 432-4.

(34) BIOT, *Précis élémentaire de physique expérimentale* (3rd edn., Paris, 1824), ii. 698-701. Using the same argument as he had used in his *Traité* in 1816 (though with the additional assumption that c' was independent of t), Biot showed that a large increase in pressure effected only a small saving in fuel – a point which Poisson too had made in the previous year (see the paper cited in note 31). Any economy that was achieved in high-pressure engines was therefore to be attributed to the use of the expansive principle. It is interesting that even in 1824 Biot was by no means convinced that the economy was sufficient to justify the practical difficulties and dangers of using high pressures.

(35) ROBISON, *loc. cit.* (12). On the later use of Robison's treatment, incorporating the assumption that the expansion after cut off took place isothermally, see note 49 of FOX, "Watt's expansive principle in the work of Sadi Carnot and Nicolas Clément", *Notes and records of the Royal Society of London,* xxiv (1969), 233-53.

volume V_1 to volume V_2 was $P_1 V_1 \ln\left(\frac{V_2}{V_1}\right)$. The sum of these two expressions, giving the total effect, was $P_1 V_1 \left(1 + \ln\left(\frac{V_2}{V_1}\right)\right)$. It is important to note that in deriving this expression Robison assumed that the expansion after cut off took place isothermally ; it was an assumption that, of all the writers on the subject in the first half of the nineteenth century, only Carnot and his friend Clément — and the very small number of writers whose opinions were derived from them ([36]) — were to question.

By the early 1820s a number of more refined versions of Robison's expression had been proposed ; in 1818, for example, Hachette had tried to take account of the pressure of the vapour arising from the condensing water, which would offer unwanted resistance to the piston ([37]). But the essentials of Robison's discussion were enshrined in all the standard treatments at the time when Carnot was writing, and, to the extent that they yielded (very roughly) the predictions that were expected of them, they were seen to be satisfactory. Above all, theory amply vindicated the merits of expansive operation, as Hachette demonstrated in his paper of June 1818 to the Société Philomathique ([38]). With a volume V of steam at an initial pressure of four atmospheres, he argued, it was possible to cut off the supply one quarter of the way through the stroke and then to allow the steam to expand (isothermally) until its volume became 4V and its pressure fell to one atmosphere. By so doing, the effect was some three times that obtained by using the volume 4V of steam at atmospheric pressure in a non-expansive Watt engine, although the amounts of steam and fuel used in the two cases were, as Hachette assumed, the same ([39]).

---

(36) It was only during the conference that I learned, from Mr Robert McKeon, that, in his lectures at the Ecole des Ponts et Chaussées in 1819-20, Navier had assumed that in an ideal steam engine the expansion after cut off should take place adiabatically. The relevant passage is on p. 82-3 of the third part of the lithographed notes of Navier's lectures : *Programme du cours de mécanique appliquée à l'Ecole des Ponts et Chaussées. Année scolaire 1819-1820.* Several references in the text make it clear that the paper of 1819 by Clément and Desormes was Navier's source. It is interesting to note that Navier had reverted to the more common assumption of an isothermal expansion by the time he wrote the paper of 1821 cited in note 43. Clapeyron's paper of 1834 is, of course, the best-known source in which the incorporation of a phase of adiabatic expansion in the discussion of an ideal heat engine can be attributed to the influence of the *Réflexions*. The vigorous endorsement, by the military engineer Antoine Vène, of the need to assume that the temperature fell after cut off — an endorsement based at least partly on the *Réflexions* (which is cited, in addition to Clapeyron's paper) — appears not to have been noticed previously ; see Vène, *Précis historique et pratique sur les forces industrielles et notamment sur les machines à vapeur* (Paris, n.d. [1837]).

(37) See his paper (cited in note 26), in *Bulletin de la Société d'Encouragement*, XVIIe année (1818), 171-3.

(38) *Ibid.* Hachette's argument in this paper refers specifically to the Woolf engine. In this engine 'cut off' takes place when the smaller of the two cylinders is filled with steam (at about three atmospheres) ; the steam is then allowed to expand into the larger cylinder, where the expansive phase occurs.

(39) Hachette's assumption was, of course, totally unjustified by the evidence that was available to him in 1818. In 1818 he lacked information on the way in which both the heat content and the density of saturated steam varied with its temperature.

About 1820 Hachette's discussion was well known in France, for after publication in the *Bulletin* of the Société d'Encouragement in June 1818, the essential elements were reproduced in 1819 in the second edition of his important *Traité élémentaire des machines* ([40]) ; hence Carnot would certainly have known of Hachette's work. However, it is not referred to by Carnot, and this omission is in my view a significant one. It seems to reflect not only the weaknesses in Hachette's argument but also the disparity between two quite different approaches to the theory of steam engines which need to be identified. Hachette was concerned with the immediate problems of practising engineers. By showing that the advantage gained by expansive operation increased with higher pressures, he was answering one of the most urgent problems of the day, and he did so by concentrating exclusively on the way in which engines actually behaved rather than on how they might behave under idealized conditions. In the latter respect he differed markedly from Carnot, of course, but from other theorists as well. Among these theorists, I regard Clément and Desormes as being particularly important for our understanding of Carnot's work.

The problem tackled by Clément and Desormes was precisely the problem of the *Réflexions :* what was the *maximum* effect that could be obtained from a fixed amount of fuel, disregarding all practical constraints ? It is true that Clément and Desormes, like Carnot, were not unaware of the realities of steam engine operation : stressing the difficulty and danger of using steam at very high pressures, they stated that in practice they preferred the more modest pressures used in the Woolf engine ([41]), and the very fact that they tackled their problem at all reflects an excitement that originated in the world of the practising engineers. But, like Carnot, Clément and Desormes did not believe that their theory was invalidated by the large (twelve-fold) discrepancy between their theoretical predictions and the actual performance of the engines they were describing ; as they saw it, their approach could not be expected to yield agreement.

In treating the injection and expansion phases separately and in putting the effect produced during the admission of steam equal to the product of its pressure and the volume that it occupied at cut off, Clément and Desormes were wholly conventional ([42]). Their originality — and their importance for Carnot — lay in the assumption that the maximum effect could only be obtained if the expansive phase took place adiabatically and hence, as was well known, with a fall in temperature. Moreover, according to Clément and Desormes, in an ideal engine the expansion should continue

(40) See note 26.

(41) The preference is stated in notes taken on 24 January 1825 by J. M. Baudot at one of Clément's lectures at the Conservatoire des Arts et Métiers. The relevant passage is on p. 43-4 of the second of two books of notes bound in volume i of MS. 8° Fa 40 (2) in the library of the Conservatoire des Arts et Métiers. For a full account of these notes, see Jacques PAYEN, "Une source de la pensée de Sadi Carnot", *Archives internationales d'histoire des sciences*, xxi (1968), 15-37. The calculations that led Clément to doubt the merits of using steam at very high pressures had been made by 1819 ; they showed that the economy rose less rapidly as the pressure increased (see graph 1).

(42) My reconstruction of the argument used by Clément and Desormes is based not only on the summary cited in note 25, but also on the notes referred to in note 41 and on some taken by Louis Benjamin Francoeur at Clément's lectures in 1823-4 ; Francoeur's notes are now MS. 407 in the library of the Ecole Nationale Supérieure des Beaux-Arts, Paris. For the details of my reconstruction, see FOX, *op. cit.*, (1), p. 180-1, and *op. cit.*, (35), p. 237 and 242-4.

until the pressure of the steam and its temperature were equal to the pressure and temperature of the vapour in the condenser ; then, and only then, should condensation take place. The similarity between this criterion for obtaining maximum effect and that used by Carnot should be obvious. It appears to have been a completely novel criterion in 1819, and even by 1824 only Navier (who knew Clément and Desormes's paper) had proposed one that resembled it. But even Navier, writing in 1821, stated merely that the pressure of the expanding steam should be allowed to fall to the pressure in the condenser ([43]) ; there was no question that the temperature (which in his view remained constant) should fall as well.

But for the law of total heats, announced as a new discovery in the 1819 paper, it would have been impossible to predict theoretically the effect obtained in adiabatic expansion. Using the law, Clément and Desormes could argue that, so long as there was no heat exchange with the surroundings, the heat content of the steam would remain constant and the initial state of complete saturation would be maintained throughout the expansive phase. Hence the temperature and pressure at any point would be related in accordance with the well-known tables of saturated vapour pressure, so that if (as Clément and Desormes assumed) the gas laws could be applied to saturated vapours, it was possible to give a complete description of the state of the vapour. It was then an easy, if somewhat laborious, task to calculate the motive power obtained.

Clément and Desormes's conclusions about the dependence of motive power on the pressure and temperature of the steam generally endorsed prevailing opinion. Their calculations left them in no doubt that it was economical to use high pressures and that the economy increased with increasing pressure, though less rapidly as the pressure rose (see graph 1). They also confirmed that the mounting advantage arose almost entirely from the expansive phase. However, we should beware of overlooking what was original in their work. The very least that can be said on their behalf is that no one else had investigated the dependence of motive power on pressure in such detail, and before 1824 the adiabatic phase is found nowhere other than in the 1819 paper.

My belief that Carnot owed a very significant debt to Clément and Desormes rests largely on the presence of an adiabatic phase both in the 1819 paper and in the *Réflexions,* and on the fact that Carnot's method for calculating the effect obtained in adiabatic expansion was very similar to that used by Clément and Desormes ([44]). But it

---

(43) C. L. M. H. NAVIER, "Note sur l'action mécanique des combustibles", *Annales de chimie et de physique,* xvii (1821), especially p. 364-7.

(44) I have argued that Carnot owed a significant debt to Clément in FOX, *op. cit.* (35), and "The intellectual environment of Sadi Carnot : a new look", *Actes du XIIe Congrès International d'Histoire des Sciences ; Paris, 1968* (Paris, 1971), iv. 67-72. The importance of Clément for our understanding of Carnot was first noted by Professor E. Mendoza (who kindly drew my attention to the sets of notes cited in notes 41 and 42) ; see his articles, "Contributions to the study of Sadi Carnot and his work", *Archives internationales d'histoire des sciences,* xii (1959), 377-96, and "Sadi Carnot and the Cagnard engine", *Isis,* liv (1963), 262-3. The connexion between the two men has also been explored in an important article by Jacques Payen ; see PAYEN, *op. cit.,* (41). The similarity between Carnot's treatment of the adiabatic phase and that of Clément and Desormes is apparent in the *Réflexions* (see especially, in the 1824 edn., p. 66-8) but even more so in the manuscript paper by Carnot that is reproduced and discussed in W. A. GABBEY and J. W. HERIVEL, "Un manuscrit inédit de Sadi Carnot", *Revue d'histoire des sciences,* xix (1966), 151-66.

rests also on the evidence that Carnot and Clément were close friends in the 1820s and that, as a result of their friendship, Carnot had read the full manuscript of the 1819 paper before he wrote the *Réflexions* ([45]). In fact, Carnot and Clément had so much in common, both personally and in their work on steam engines, that it is tempting to look for other ways in which Carnot may have been influenced by his visits to the Conservatoire des Arts et Métiers in Paris, where Clément was professor of industrial chemistry from November 1819. I shall mention just two possible debts that Carnot may have owed to Clément.

First, as I have pointed out elsewhere ([46]), we know that in his lecture course for 1823-24 (though almost certainly not in 1819) Clément adopted the practice of subtracting the effect of the condensation stroke from the combined effect of the phases of injection and expansion in order to calculate the total effect. Possibly, therefore, Clément was also the source for the third (condensation) phase of the Carnot cycle, which is not found in any other source, though by 1823-24, with the *Réflexions* nearing publication, Clément could equally well have been drawing on Carnot's work. Secondly, it is worth noting that Clément's treatment, as given in 1819 and summarized in graph 2, leads to the conclusion that the motive power obtained in a steam engine is very nearly proportional to the difference between the temperature of the boiler and that of the condenser, i.e. to what Carnot called the 'fall' of caloric. This result is, of course, one that Carnot considered very seriously ([47]), though, since there is no evidence that Clément even pointed out the proportionality himself, I can only indicate what may be no more than coincidence.

In this paper I have tried to set Carnot's work in the context of a continuing discussion of the theory of high-pressure engines which had been in progress for nearly a decade before the *Réflexions* was published. It was a tradition in which the abstract style of Carnot's treatment of an idealized engine was by no means out of place ; when Clément, Navier, and Charles Dupin noted the huge disparity between their predictions and the actual performance of the engines they were discussing ([48]), they did so with

---

(45) CARNOT, *Réflexions* (1824 edn.), p. 98n.

(46) FOX, *op. cit.*, (35), p. 245. It is clear from the similarity between the results cited in the 1819 paper and those which appear in a table published by Clément in 1826 that the effect of the condensation stroke was not considered in 1819 ; nor was it considered in the lectures which Baudot attended between 1825 and 1828. The 1826 table, which gives data concerning the pressure of saturated steam, its temperature, volume, and the effect ('puissance mécanique') that it produces when used under the idealized conditions specified by Clément and Desormes, is reproduced and discussed in Jacques PAYEN, "Deux nouveaux documents sur Nicolas Clément", *Revue d'histoire des sciences*, xxiv (1971), 53-60.

(47) See especially CARNOT, *Réflexions* (1824 edn.), p. 73-9n.

(48) As I have already mentioned, in the 1819 paper Clément and Desormes observed that the economy calculated for their ideal engine was about twelve times greater than that observed in even the best expansive engines of their day. Navier found a similar disparity between the predicted and actual performances in his discussion of 1821 ; see NAVIER, *op. cit.*, (43), p. 369. For Dupin's comment (1826), see F. P. C. DUPIN, *Géométrie et mécanique des arts et métiers et des beaux-arts* (3 vols., Paris, 1826), iii. 378. Carnot noted that even when it attained the peak of its performance, in May 1816, the great Woolf engine at Wheal Abraham yielded only about $\frac{1}{20}$ of the economy of which it was theoretically capable ; see CARNOT, *Réflexions* (1824 edn.), p. 117.

no sense of failure, for like Carnot, but unlike Hachette, for example, they were seeking to define a theoretical maximum which they knew to be unattainable. More specifically, it was a tradition that had touched repeatedly on the very problems tackled in the *Réflexions*. In particular, it had raised the question whether there was a limit to the effect that could be obtained from a given amount of fuel (or heat) or whether this effect could be increased indefinitely — by the use of higher pressures, for example, or (a possibility that I have not discussed in this paper) of working substances other than steam. The discussions of the period 1816-24 had allowed at least some fundamental points to be settled. By 1824 there was no doubt that the chief advantage of the high-pressure engine lay in its ability to make effective use of the expansive principle, and theory suggested that the economy gained in this way would go on increasing as the pressure and temperature of the steam produced in the boiler were raised, although the practical problems and dangers of operating at pressures greater than a few atmospheres were such that these predictions could not readily be tested.

Concerning the merits of different working substances, there was probably far less agreement, not least because the two most important discussions in the period in question had led to diametrically opposed conclusions. In 1818 the brilliant young physicist Alexis-Thérèse Petit had argued that there was a four-fold economy to be gained by the use of air rather than steam in an expansive engine, while in 1821 Navier had shown that the advantage (by a factor of more than five) lay with steam [49]. Unfortunately, by the time Navier's paper appeared, Petit was dead ; no debate ensued (in striking contrast with the discussions about the way in which steam should be used), and in 1824 this particular issue was still an open one.

In conclusion, and despite all I have said in this paper, I believe that, in Carnot's case, it is possible to place too much emphasis on the identification of traditions and precursors. As a result of recent research, we can now trace possible sources for most of the individual elements in his theory, but the originality of the way in which he synthesized these elements remains his unique achievement.

---

(49) See PETIT, "Sur l'emploi du principe des forces vives dans le calcul de l'effet des machines", *Annales de chimie et de physique*, viii (1818), 287-305 (294), and NAVIER, *op. cit.*, (43), p. 368.

Graphs

Both graphs are based on a table published by Clément in 1826. The table gives details of the pressure of saturated steam, its temperature, volume, and the mechanical effect ("puissance mécanique") that it is capable of producing when used under the idealized conditions described by Clément and Desormes in their paper of 1819 and subsequently by Clément in his lectures at the Conservatoire des Arts et Métiers. In compiling this table Clément did not take account of the condensation phase, the effect of which would in any case be slight.

The table is reproduced and discussed in Jacques Payen, "Deux nouveaux documents sur Nicolas Clément", *Revue d'histoire des sciences*, xxiv (1971), 53-60. A slightly abbreviated version of the table is in F.P.C. Dupin, *Géométrie et mécanique des arts et métiers et des beaux-arts* (3 vols., Paris, 1826), iii. 376 and 378.

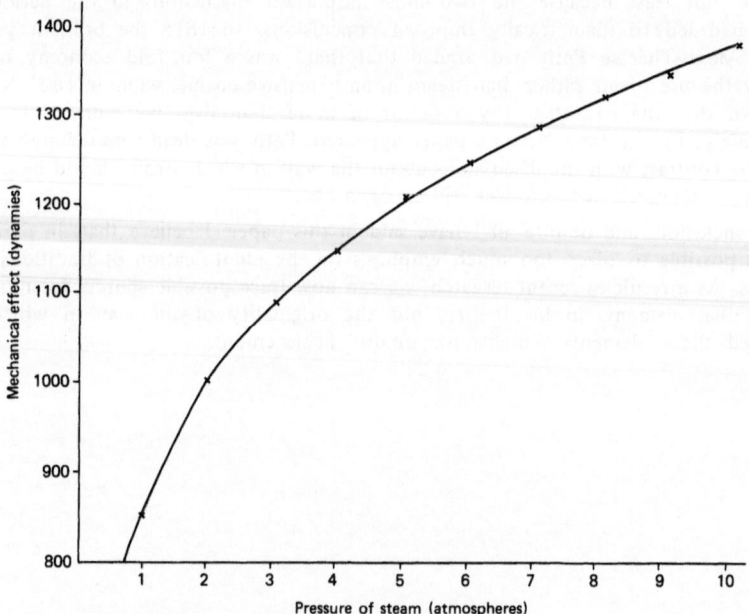

Graph 1. — The "puissance mécanique" is the total effect obtained in the injection and expansion phases by burning 1 kg of coal to produce the steam. It is measured in "dynamies", i.e. the number of cubic metres of water that can be raised through one metre. Expansion is continued until the initial pressure and temperature fall to $0.0141\left(\dfrac{1}{71}\right)$ atmosphere and 12°C, the pressure and temperature of the vapour in the condenser.

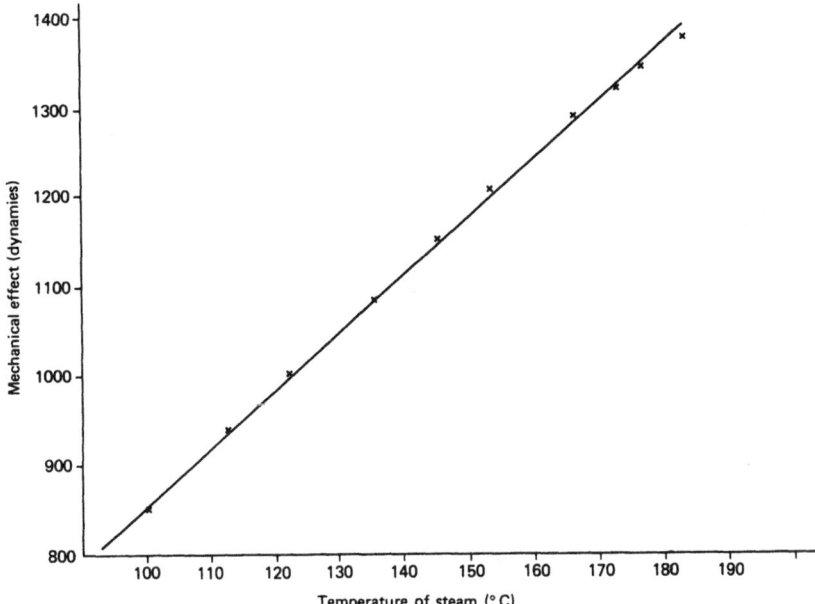

Graph 2. – The "puissance mécanique" is again in "dynamies", and the amount of steam used is that produced by burning 1 kg of coal. The straight line indicates that the effect is proportional to the difference between the temperature of the steam entering the cylinder and its temperature after expansion.

XII

WATT'S EXPANSIVE PRINCIPLE IN THE WORK OF SADI CARNOT AND NICOLAS CLÉMENT

I mentioned to you a method of still doubling the effect of the steam, and that tolerably easy, by using the power of steam rushing into a vacuum, at present lost . . . it is peculiarly applicable to wheel-engines, and may supply the want of a condenser where force of steam is only used; for, open one of the steam-valves, and admit steam until one-fourth of the distance between it and the next valve is filled with steam; shut the valve, the steam will continue to expand, and to press round the wheel with a diminishing power ending in one-fourth of its first exertion. The sum of this series you will find greater than one-half, though only one-fourth steam was used.

Watt to Dr William Small, 28 May 1769 (1)

My first new improvement in steam or fire engines consists in admitting steam into the cylinders or steam vessels of the engine only during some certain part or portion of the descent or ascent of the piston of the said cylinder, and using the elastic forces, wherewith the said steam expands itself in proceeding to occupy larger spaces, as the acting powers on the piston through the other parts or portions of the length of the stroke of the said piston; and in applying combinations of levers, or other contrivances, to cause the unequal powers wherewith the steam acts upon the piston, to produce uniform effects in working the pumps or other machinery required to be wrought by the said engine: whereby certain large proportions of the steam hitherto found necessary to do the same work are saved.

Watt's patent specification of 12 March 1782 (2)

INTRODUCTION

RECENT studies have had the effect of showing Sadi Carnot to be a far less isolated figure than he was thought to be only a few years ago. As recently as 1954, for example, Professor T. S. Kuhn could state unequivocally that Carnot, unlike most of the other pioneers of thermodynamics, 'stands alone' in his achievements (3); yet, significantly, this same scholar has since revised his opinion and has demonstrated most convincingly that Carnot owed a great debt to the scientific and, more particularly, the engineering traditions of his day (4). In 1965 the task of identifying possible sources for Carnot's ideas in the world of engineering was taken a stage further in an important article by Dr D. S. L. Cardwell, who pointed to the clear precedents which are to be found in the water-power technology of the eighteenth and early nineteenth centuries, notably in the theory of the column-of-water engine, a device with which Carnot was certainly familiar (5). It is chiefly as a result of these studies that we are now able to see Carnot not as a man who appeared, inexplicably, some two or three decades ahead of his time, but rather as one who found both his initial problems *and* many of his solutions in the writings and current practices of contemporary power engineers (6).

Even now, however, there are certain aspects of Carnot's debt to the engineering traditions which demand further examination, and in looking at the problem yet again in this paper I shall be seeking to fill one of these gaps rather than to dispute the view of Carnot's achievements already established so ably by Professor Kuhn and Dr Cardwell. In particular, I shall be concerned to show how James Watt's idea that benefit could be obtained by allowing the expansion stroke in a steam engine to continue after the cut off of the supply of steam (his 'expansive principle') influenced Carnot's work and how, in a modified form, it almost certainly led him to incorporate the phase of adiabatic expansion in his now celebrated cycle of operations for the ideal heat engine.

It should be made clear at this point that I am by no means the first to see the adiabatic phase in the expansion stroke, following the initial phase of isothermal expansion, as a development from Watt's expansive principle (7), but in this paper I hope to be able to put beyond doubt what has hitherto been little more than a suggestion. To do this I shall use new documentary evidence and argue in some detail that Carnot's innovation cannot be dissociated from the fact that high-pressure engines, in which steam was used expansively, had arrived in France for the first time only in the decade preceding the publication of the *Réflexions sur la puissance motrice du feu* in 1824 (8) (although Watt's principle had certainly been conceived by 1769 (9) —possibly even by 1767 (10)—patented in 1782 (11) and, if only in Britain, applied in Arthur Woolf's highly successful two-cylinder compound engine by 1804 (12)).

As it happens, this practical engineering background, though important, does not constitute the whole story, because, in so far as he maintained that the expansion after cut off should take place adiabatically, Carnot's treatment of the expansive principle represented not a simple extension of Watt's idea but what I would consider a major departure both from it and also from the standard treatments which were given by steam engineers at the time Carnot was writing. Hence it is crucial to my argument that nearly all of Carnot's contemporaries followed Watt in supposing that the expansion after cut off was isothermal, or very nearly so, and that only two men, so far as I am aware, departed from this quite reasonable assumption. These men were Nicolas Clément and Charles Bernard Desormes.

It is often very difficult to separate the contributions of Clément and Desormes (13), since as intimate friends, who, moreover, were related by marriage, they did much of their most important work together. But, as I have already argued in another paper (14), it was probably Clément who exerted the greater influence on Carnot's ideas and it is he rather than Desormes who seems to provide the essential link between Watt and Carnot in the theoretical study of the expansive use of steam.

CLÉMENT AND DESORMES

Clément and Desormes were men of many parts (15). They were both chemists of some note and in the early years of the nineteenth century they published several joint papers on a variety of chemical problems (16). They are also remembered for the fact that in 1812 they submitted a very worthy entry for the French Institute's prize competition in physics (17). They did not win the prize but their paper was widely read and praised, and one of the procedures which they adopted has since earned lasting fame as a standard method for the determination of γ, the ratio of the specific heat at constant pressure to the specific heat at constant volume. However, it is above all as industrialists that they deserve to be remembered (18). Their highly successful factory for the manufacture of alum, established in the reign of Napoleon, was only one of their numerous industrial interests, and Clément in particular seems to have made something of a fortune in the last twenty years of his life by acting as a consultant at a time of great expansion in French industry. It was clearly his talent for work of this type which, in November 1819, earned him the chair of applied chemistry (*chimie appliquée aux arts*) at the *Conservatoire des Arts et Métiers* in Paris, a position which he held until his death in 1841.

Neither the circumstances nor the date of Clément and Desormes's first meeting with Carnot are known, and it is not even certain, although it is highly probable, that Desormes knew Carnot personally. However, a reference in a footnote to the *Réflexions* (19) indicates that Carnot was certainly in touch with Clément by the early part of 1824, when the book was published (20). We know also that Carnot attended at least one of Clément's lectures at the *Conservatoire* in January 1825 (21), and the existence of a close friendship between the two men was later vouched for by Hippolyte Carnot (22), Sadi's younger brother and his best-known biographer. It must be pointed out that Hippolyte associated the friendship particularly with the *late* 1820s, but since his account was not given until 1878, we can hardly use this evidence to rule out the possibility that the friendship was already a close one even *before* the publication of the *Réflexions*.

Although the point cannot be proved definitively, there is good reason to believe that Carnot first met Clément at the *Conservatoire*, probably in the early 1820s (23). Elsewhere I have tried to show how both the aims of the courses at this institution (they were established in 1819 to provide education of a technological nature for working men (24)) and also its decidedly liberal tone would have appealed to Carnot (25), whose ardent republicanism and scorn of royalist governments had been undimmed by the Bourbon restoration of 1815 and who in any case was denied the more traditional outlets for his scientific interests, such as would have been provided by membership of the *Académie des Sciences* or a teaching post at the *École Polytechnique* (26). It is also important to point out that the attractiveness of the *Conservatoire* in Carnot's eyes would have been enhanced still further by the fact that one of the two men who were appointed, with Clément, to the newly founded chairs in 1819 was the same Charles Dupin (later Baron Charles Dupin) who had written vehemently in support of Carnot's father, Lazare, when the former 'organizer of victory' became an exile from France shortly after the downfall of Napoleon (27). Indeed, it may well have been Dupin who first introduced Carnot to Clément.

But whatever conclusions may be drawn on this matter, it is not essential to my argument to suppose that Carnot had established a *personal* contact with Clément before 1824, although for myself I should find it hard to believe that he had not done so. The important, and indisputable, point is that Carnot was already familiar with Clément's views on the theory of the heat engine before he wrote, or at least completed, the *Réflexions*.

Carnot had learnt of Clément's views, if not personally, then certainly through a joint paper by Clément and Desormes which had been read before the *Académie des Sciences* in August 1819 (28). Although this paper was never published, an extract from it appeared almost simultaneously in the monthly bulletin of the *Société Philomathique* in Paris (29) and a copy of the complete paper was made available to Carnot by Clément himself (30). The problem which was tackled in the paper was a familiar one, namely the theoretical determination of the maximum effect which could be obtained in a heat engine from a given mass of different working substances under various conditions of temperature and pressure (31). However, the solution, based on an ingenious thought experiment (32), contained much that was original. Clément and Desormes imagined the working substance under examination, whether it was steam, some other vapour or a gas, being introduced, in the form of a bubble, at the bottom of a tall vessel of uniform cross section, which was filled to the brim with water (see sketch in plate 1). As more of the working substance was introduced and the bubble grew bigger, water would overflow from the vessel and in this way work would be performed or, to use the terminology of the paper, mechanical power (*puissance mécanique*) would be expended (33). In calculating the amount of mechanical power made available in this process Clément and Desormes postulated that the pressure of the working substance in the bubble remained constant and equal to the pressure exerted by the column of water in the vessel. Hence, as any engineer of the day would have appreciated, the mechanical power expended during the introduction of the working substance could be given either as the product of the height of the vessel and the weight of water expelled, or as the product of the pressure of the working substance (p) and the volume (v) which it occupied when the process of 'production', as Clément and Desormes called it, was completed. It was naturally a simple matter to choose the units in such a way that the numerical result obtained in each case was the same (34).

Clément and Desormes were convinced that the quantity pv did not represent all the mechanical power which could be obtained from a vapour or gas. They pointed out that if the bubble was now allowed to rise in the vessel, experiencing an ever-decreasing pressure, it would expand still further, so causing more water to overflow and more mechanical power to be expended. In this clear distinction between the initial process of 'production' and what Clément and Desormes termed the process of 'expansion' (*détente*), the influence of Watt's views concerning the expansive use of steam is not hard to identify (35), and it is certainly no coincidence that, as I show in the next section, high-pressure engines incorporating the expansive principle had arrived in France only some four years earlier. Then Watt's principle and the machines in which it was applied had been something of a revelation to French engineers, who, primarily as a result of the Napoleonic wars, knew little of the important advances in power technology made by their English counterparts in the first decade and a half of the century. Needless to say,

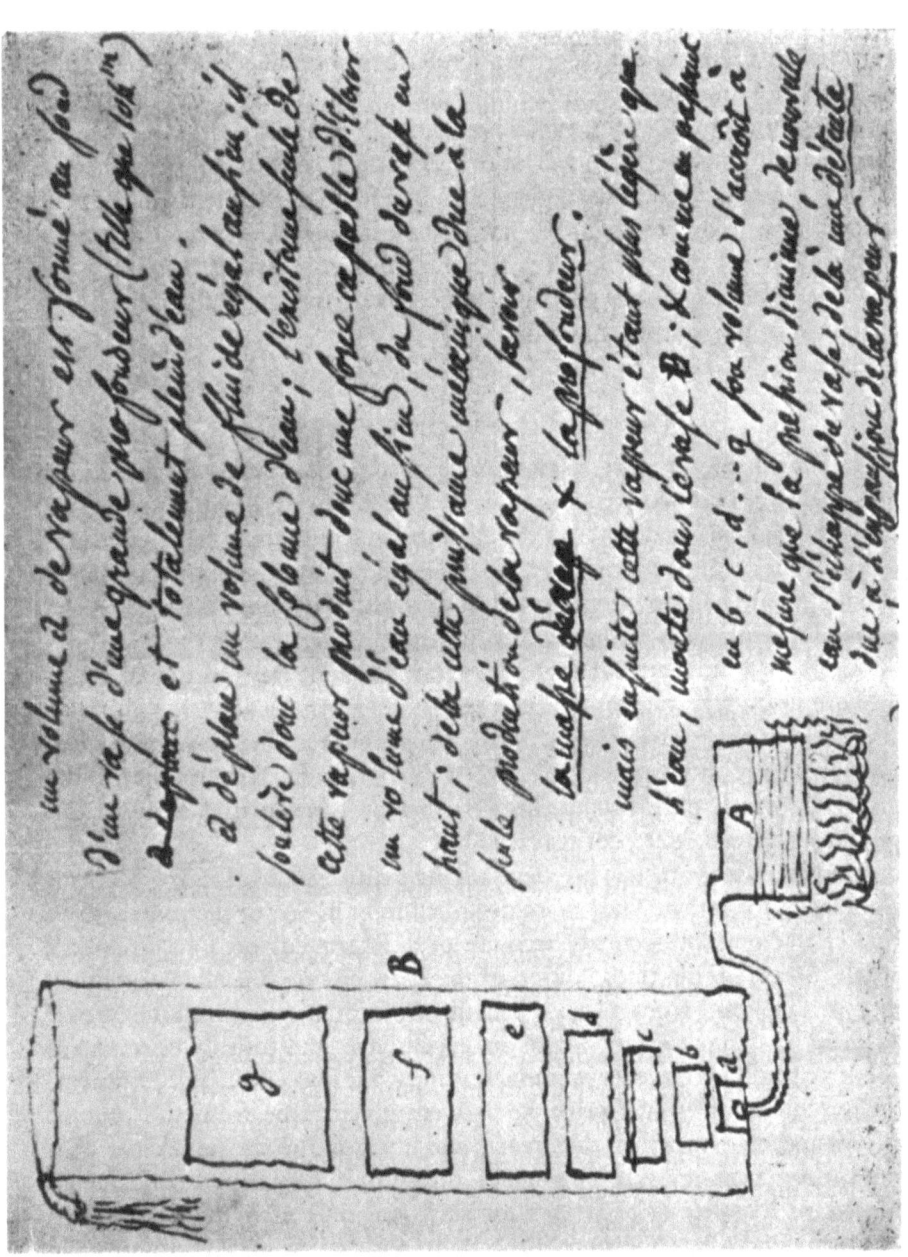

Clément and Desormes's thought experiment, sketched by Francoeur in notes taken at Clément's *Conservatoire* lectures in 1823–1824. Reproduced by kind permission of the Director of the École Nationale Supérieure des Beaux-Arts, Paris.

both the principle and the machines were still arousing intense interest in 1819, when Clément and Desormes wrote their joint paper.

Now it is fundamental to the argument of this paper that Clément and Desormes did not treat the expansive principle in exactly the same way as their contemporaries, but even so I believe that their work, and that of Carnot himself, cannot be understood except in the context of the discussions among power engineers which were taking place in the early years of the Bourbon restoration. It is therefore these discussions, in which the newly arrived expansive principle figured very prominently, which are examined in the following section.

The expansive principle in France

It is well known that Watt himself made little use of his expansive principle (36). He had taken good care to patent it in 1782 and in the specification he had enlarged both on the advantages which could be obtained (in terms of fuel economy) by allowing an engine to operate expansively and also on the mechanism which would be needed if the output throughout the expansion stroke was to be regular and not greater during the admission of steam than after cut off (37). Yet Watt's fear of the dangers of high-pressure steam was so great that he applied his idea in very few engines (38), and even in these cases the fact that the pressure of the steam used was scarcely above atmospheric pressure meant that the true merits of the principle, which only become really apparent when considerably higher pressures are used, were not realized.

Despite Watt's attitude his views on the expansive use of steam were not completely forgotten. They were described fully in one of the most widely read of all eighteenth-century accounts of the steam engine, John Robison's article 'Steam' in the third edition of the *Encyclopaedia Britannica*, published in 1797 (39); and from 1804, when the Cornish engineer Arthur Woolf patented his design for a two-cylinder expansive engine (usually operating at about 3 atmospheres), they were increasingly put into effect (40). However, such accounts of Watt's principle as were given came from the pens of British authors rather than the French, and it was in Britain, not France, that the earliest high-pressure engines of the Woolf type were constructed. Indeed, as has already been mentioned, it was only after the downfall of Napoleon and the restoration of the Bourbons that the expansive principle began to be applied or even widely discussed in France (41). The man chiefly responsible for this important development in French steam-power technology was Humphrey Edwards, an Englishman and a former partner of

Woolf, who gained a monopoly on the import of Woolf engines into France (by means of a *brevet d'importation*) for the ten years from May 1815 (42). It is a measure of Edwards's great success as an importer and erector of the engines that the advantages of high-pressure steam were recognized almost immediately in France and that there the theoretical discussions of the steam engine, which before 1815 had been based on the assumption that the pressure on the piston remained constant throughout the power stroke (43), began very quickly to reflect the innovation. Articles in which expansive working was discussed and advocated for the first time began to be published in some of the leading French journals of the day, notably the *Annales de chimie et de physique* and the *Bulletin de la Société d'Encouragement* (44); and significant changes were to be found also in textbooks, such as J. N. P. Hachette's authoritative and influential *Traité élémentaire des machines* (45). In the first edition of this book, published in 1811, Hachette, as we should expect at that time, did not even mention the expansive principle, but in the second edition, published in 1819, he expounded it fully in the course of his description of the Woolf engine (46). By this time, like most French engineers, he was quite convinced that the maximum effect could only be obtained when steam was used expansively and he even calculated the magnitude of the advantage which could be effected by the application of Watt's principle in an engine.

To a large extent the principles of Clément and Desormes's thought experiment, as described in their paper of August 1819, reflect this interest in high-pressure expansive engines which was already widespread in France, and also the then generally held conviction that such engines were advantageous. However, there were significant ways in which Clément and Desormes's approach was quite original. Of these the most important were:

(i) Their belief that in an ideal engine the expansion of the steam after cut off should take place adiabatically (and not isothermally), i.e. that the bubble in their thought experiment was impermeable to heat; and

(ii) Their use of a new 'law' which enabled them to calculate the mechanical power obtained during the expansion phase.

That Carnot himself incorporated the first of these two ideas in his cycle of operations for an ideal heat engine is too well known to require further comment. But the case for believing that the source of the idea was Clément and Desormes's 1819 paper does not appear to have been argued in detail (47). Any argument to this effect must rest principally on two facts: first, that Carnot had read Clément and Desormes's paper before he wrote the *Réflexions*, and, secondly, that in the early 1820s he was very probably a close friend of Clément. It is also highly relevant to know that, in supposing that the expansion after cut off should take place adiabatically, Clément and Desormes were departing in a most striking way from contemporary opinion. It is true that certain engineers, notably Hachette (48), were aware that some cooling might well occur as a result of the expansion, but all were agreed that such cooling would be slight enough for the existence of iso-thermal conditions to be assumed for the purpose of calculation (a device which was seen as a great simplification since it allowed Boyle's law to be applied). No one, except Clément and Desormes, even hinted that the expansion *had* to take place adiabatically if maximum effect was to be obtained (49).

Until recently the case for believing that Carnot did owe a significant debt to Clément and Desormes would have rested on this evidence alone, but the publication, in 1966, of a previously unknown manuscript paper by Carnot (50) allows us to take the argument a stage further, for in the light of this document it is now possible to show also that Carnot used Clément and Desormes's highly original method for the calculation of the mechanical power obtained in the adiabatic phase. In other words, we can now see that he adopted the second as well as the first of the two assumptions listed above.

THE MECHANICAL POWER OBTAINED IN ADIABATIC EXPANSION

The manuscript paper in question was devoted to a detailed and carefully argued derivation of a general expression for the mechanical power or, in Carnot's words, the motive power (*puissance motrice*) obtained from an ideal heat engine in which saturated steam was the working substance (51). As Dr W. A. Gabbey and Mr J. W. Herivel point out in their commentary, it is impossible to offer an exact date for the document, but from evidence which I discuss in the appendix to this paper we may conclude that it was written between November 1819 and March 1827, though in all probability much closer to the latter date and therefore after the publication of the *Réflexions*.

The cycle of operations described in Carnot's paper was precisely that given in the *Réflexions* for the case of an engine employing steam as the working substance (52). Thus it consisted of only three stages, no mention being made of the final stage of adiabatic compression, which was included in the cycle only when a gas was considered (53). The three stages were:

(i) An isothermal increase in volume (corresponding to the injection of steam at constant pressure into the empty cylinder).

(ii) The adiabatic expansion of steam in the cylinder, after the cut off of supply. In this stage conditions of complete saturation were supposed to be maintained, for reasons which are discussed later in this section.

(iii) Condensation, or a decrease in volume, during which the temperature of the steam was equal to the temperature of the water in the condenser (54).

A further point of similarity between the manuscript paper and the *Réflexions* is that the same methods were used in both cases to calculate the motive power obtained in the first isothermal stage and that which was lost during condensation (work being done *on* and not *by* the steam in this last stage) (55). And in fact it is only in the treatment of the phase of adiabatic expansion that marked differences are to be found. These differences, however, are important ones and, in so far as they provide strong evidence of Carnot's indebtedness to Clément and Desormes, they call for some discussion.

In the very brief examination of the adiabatic phase which appeared in the *Réflexions* Carnot made the quite unrealistic assumption that his ideal heat engine operated between two temperatures which differed by only 1 °C (56). This assumption greatly simplified the calculation, since it implied that the quantity of motive power obtained in the adiabatic phase was so small that it could be ignored. In the manuscript paper, however, Carnot deliberately considered the general case of a steam engine operating between *any* two temperatures and he faced up to the problem, so carefully avoided in the *Réflexions* but already tackled by Clément and Desormes, of calculating the motive power made available in adiabatic expansion.

In their discussion of this same problem in their paper of August 1819 Clément and Desormes made full use of a 'law' which they were then announcing publicly for the first time (although Clément had communicated the result privately to the *Société d'Encouragement pour l'Industrie Nationale* some four months earlier (57)). According to the 'law', which had been arrived at empirically by experiments with an ice calorimeter (58), the total quantity of heat in a unit mass of any given saturated vapour was always the same, being independent of both its temperature and pressure, provided that conditions of complete saturation were maintained. The 'law', of course, is quite incorrect, but the quantity of heat lost by a saturated vapour in condensing and cooling to a fixed final temperature (0 °C in the case of the ice calorimeter) is so nearly independent of its initial temperature and corresponding vapour pressure that Clément and Desormes's mistake is understandable. Until about 1827, when it was disproved experimentally by César-Mansuète Despretz (59), the 'law' won widespread acceptance, although the discoverers themselves do not seem to have received all the credit which was due to them. Their paper of 1819 was never published in full, as has already been noted, and doubts concerning the originality of their discovery were raised almost immediately by an anonymous writer in the *Bulletin de la Société d'Encouragement*, who maintained that Watt had already arrived at a similar conclusion and that evidence to support the claim would soon appear in John Robison's *A system of mechanical philosophy* (60). As it happened, the point was not made, since the letter of 26 March 1814 from John Southern to Watt, which was reproduced in the *System* in 1822 (61) and which presumably constituted the evidence referred to, contained no statement recognizable as Clément and Desormes's 'law'. Certainly Southern considered the possibility that the total heat content of saturated steam was independent of its temperature and pressure, but, as he admitted, his experiments, conducted in 1803, had given him no reason to suppose that this was in fact the case. Watt himself added an undated footnote to Southern's letter stating that 'for many years' he too had 'entertained a similar hypothesis', although he confessed: 'I know of no experiment whereby the truth of it can be demonstrated conclusively' (62). So even Watt, we must assume, would not have disputed Clément and Desormes's priority, had he lived to know of their paper (63). The belief which, in 1819, was to be expressed as a 'law' was clearly very much under discussion in the early years of the century, but there is no evidence that anyone succeeded in confirming it experimentally before Clément and Desormes (64).

These doubts concerning priority do not appear to have prevented most of Clément and Desormes's contemporaries from attributing the discovery of the 'law' unhesitatingly to them (65). In this respect Carnot was no exception. Moreover, despite some reservations towards Clément and Desormes's experimental evidence (66), he followed prevailing opinion in assuming the truth of the 'law' both in the *Réflexions* (67) and in the manuscript paper; and, what is even more important for this study, in the latter document he used the 'law', as Clément and Desormes themselves had used it in 1819, to calculate the motive power of steam expanding adiabatically.

In the period with which we are concerned, some two decades before the discovery of the principle of the conservation of energy, and before the interconvertibility of heat and work was accepted, it was an obvious and necessary consequence of Clément and Desormes's 'law' that the adiabatic expansion or compression of a saturated vapour could not bring about any departure from the initial state of complete saturation. For, in the absence of heat exchange with the surroundings, the total heat content of the vapour remained constant—or so it appeared in the 1820s—and hence, according to the 'law', the temperature and pressure of the vapour always varied in such a way that conditions of saturation were maintained. It followed from this that the temperature and pressure of any saturated vapour undergoing adiabatic volume changes were related in accordance with the well known tables of saturated vapour pressure of which the most widely used seems to have been the one drawn up by John Dalton in the early years of the century (68).

In the light of this conclusion it was possible to give a complete description of the state of the vapour at any stage in the adiabatic compression or expansion, provided that one additional assumption was made: namely, that the vapour could be treated as a perfect gas from the point of view of its obedience to the laws of Boyle and Gay-Lussac (69). Not surprisingly, Carnot followed Clément and Desormes in making this assumption and both in the *Réflexions* and in his manuscript paper he went so far as to set up a general expression of the type

$$\frac{pv}{267 + t} = \text{constant} \qquad [1]$$

for a saturated vapour undergoing a change in volume under adiabatic conditions (70).

Once these basic principles were adopted, it was a simple matter to calculate the effect, in terms of motive or mechanical power, of the adiabatic phase of the expansion stroke of the steam engine, and both Clément and Desormes and Carnot duly performed the calculation.

Clément and Desormes tackled the problem by considering specific cases. To take one example, they imagined a kilogramme of saturated steam at a pressure of 10 atmospheres, or approximately 100 metres of water, being introduced at the bottom of their vessel (71). In this case, of course, the height of the vessel was also 100 metres and the mechanical power of the 'production' stage, corresponding to the injection phase in the real engine, was 100 × 0.210 *dynamies*, the *dynamie* being, at least for Clément (72), the mechanical power required to raise a cubic metre of water through 1 metre, and 210 litres being the volume of 1 kg of steam at a pressure of 10 atmospheres. The calculation, as we see, was a simple one, but the determination of the mechanical power obtained in the adiabatic expansion after cut off was far more complex, involving as it did some laborious numerical work. It was carried out in several stages, consideration first being given to the first 10 metres of the bubble's rise in the vessel, in which process the pressure of the steam inside fell from 10 to 9 atmospheres. At the same time there was known to be an increase in volume from 210 to 228 litres (by using the tables of vapour pressure together with equation [1]), so that 18 litres of water were expelled from the vessel and therefore, taking the mean pressure during the expansion to be 10 atmospheres, 100 × 0.018 or 1.8 *dynamies* of mechanical power were expended. A similar calculation was performed for the fall in pressure from 9 to 8 atmospheres, and so on, the total mechanical power of the expansion phase being simply the sum of the results arrived at for each stage. It is hardly necessary to point out that such a calculation could be performed for steam between *any* two specified pressures.

In terms of the basic assumptions made and the style of his argument, Carnot's treatment of the same problem, as given in his manuscript paper, differed little from that of Clément and Desormes, although the fact that he was seeking a general expression and not a numerical result for a particular case meant that it was somewhat more elegant. For example, he gave the mechanical power made available during the injection of steam simply as pv or $N(267+t)$, where

$$N = \frac{p_{100} \, v_{100}}{367}$$

and t was the temperature of the steam arriving from the boiler (in °C). Moreover, in calculating the total effect obtained in the expansion phase, he used an empirical relationship between the saturated vapour pressure π and temperature θ of steam:

$$\theta = a \, ln\pi + b$$

where a and b were unknown constants (73). From Clément and Desormes's 'law' it followed that this relationship also governed the pressure and temperature of the saturated steam as it underwent adiabatic expansion, so that the derivation of an expression for the over-all effect of expansion (the motive power was $\int\pi\delta\phi$, where ϕ represented volume) thereby became easy. After a good deal of mathematical manipulation Carnot concluded that where the pressure and temperature of the saturated steam changed from p and t to p' and t', the power obtained during a process of adiabatic expansion was:

$$N \left(267+\frac{t+t'}{2}\right) ln \, \frac{p}{p'} - N(t-t')$$

and hence that the output in the complete expansion stroke was

$$N\left(267+\frac{t+t'}{2}\right) ln \, \frac{p}{p'} - N(t-t') + N(267+t) \qquad [2]$$

For Carnot it remained now to calculate the motive power *lost* in the process of condensation, i.e. when work was done *on* the steam and not *by* it. Assuming, in the manuscript paper as in the *Réflexions*, that the temperature after expansion, t', was equal to the temperature of the condenser (74), he arrived at the expression $N(267+t')$ for this, which when subtracted from the quantity given in expression [2], yielded:

$$F = N\left(267+\frac{t+t'}{2}\right) \ln \frac{p}{p'}$$

for the total motive power of the three stages of the steam engine cycle.

In introducing consideration of the third stage in his cycle Carnot was very probably departing from the treatment originally given by Clément and Desormes, who in their 1819 paper seem to have adhered to the custom of the time in that they took account only of the expansion stroke (75). By the winter of 1823-1824, however, Clément had almost certainly adopted the practice of subtracting the effect of the condensation stroke, as Carnot did in both the manuscript paper and the *Réflexions*. The evidence for this appears in a book of notes taken at Clément's *Conservatoire* lectures by Louis Benjamin Francoeur (76) who, as a distinguished mathematician and professor at *Faculté des Sciences* in Paris, worked so devotedly in the interests of French industry and also of popular technical education after the Bourbon restoration (77). Since an inscription in the front of the book indicates that the notes were taken at the course for 1823-1824 (78), it is clear that the idea of considering the effect of the third stage in the cycle *could* have originated, some time between 1819 and late in 1823, with Clément, and it could then have been passed on by him to Carnot. However, we must remember that by the winter of 1823-1824 the *Réflexions* was nearing publication, so that it seems rather more reasonable to continue to suppose that Carnot was the source of the idea and hence that it was he who communicated it to Clément. Clément certainly read the *Réflexions* and he was sufficiently impressed by it to commend it to his audience at the *Conservatoire*, referring to it, in a lecture on 20 January 1825, as 'un ouvrage fort remarquable' (79). Yet despite this evidence, the point must remain open to doubt. Clément was such an original thinker that it is conceivable, although personally I consider it unlikely, that he, and not Carnot, first took the effect of the condensation stroke into account.

CONCLUSION

In a manner which is common enough in the history of science, Carnot was at once an innovator, an isolated, neglected figure, and yet also a man of his time. In attempting to identify the debt which he owed to contemporary interest in steam engines incorporating Watt's expansive principle and, more specifically, to the views on the subject held by Clément and Desormes, I have naturally tended to emphasize the second of these two characteristics.

But it would be wrong to conclude without some reminder of all that was undoubtedly original in the *Réflexions*. For example, neither in the work of Clément and Desormes nor in other early nineteenth-century writings on the steam engine do we find any account of a completely reversible cycle of operations like the one described by Carnot, or any clear analogy, such as Carnot drew, between the 'fall' of caloric in a steam engine and the fall of water in a waterfall or column-of-water engine (80); nor is there any suggestion, even by Clément and Desormes, that for maximum effect the temperature of the steam in the phase of adiabatic expansion must be allowed to fall to that of the condenser *before* condensation. There is no doubt, of course, that precedents for these and most of Carnot's other innovations can be identified in the engineering literature and practices of the day, and, as I mentioned in the introduction, I therefore find it hard to accept that Carnot 'had no forerunners', as one writer has recently maintained (81). Yet often these precedents do seem somewhat remote and they must certainly have needed considerable adaptation and elaboration before they could be usefully applied. In short, the fact that the sources from which Carnot culled his ideas are now more easily identifiable than they were only a decade ago can in no way obscure the genius of the man who synthesized them and so created a theory of the steam engine which, when viewed as a whole, still appears uniquely his own.

ACKNOWLEDGEMENTS

I wish to express my thanks to Professor E. Mendoza for first drawing my attention to the manuscript notes at the *École des Beaux-Arts* and the *Conservatoire National des Arts et Métiers* which are discussed in this paper. I am also grateful to Madame Bouleau-Rabaud, librarian at the *École des Beaux-Arts*, and to Monsieur Jacques Payen, *Chef de travaux* in the *École Pratique des Hautes Études*, *VIe section* (at the *Conservatoire*), who facilitated my access to the manuscripts and kindly arranged for microfilm copies to be made. Finally, I am indebted to Dr A. J. Pacey for the valuable criticisms which he made when he kindly read the paper in manuscript.

APPENDIX

THE DATING OF CARNOT'S MS. PAPER

It seems certain that the paper by Carnot, cited in notes (50) and (51), dates from the period between November 1819 and March 1827. With regard to the first of these two dates, a passing reference to Clément as 'professeur au

Conservatoire des arts et métiers' (Gabbey and Herivel, *op. cit.* (50), p. 156) indicates that the paper must have been written after Clément was appointed to his Chair (on 25 November 1819). The evidence for the second date appears not in the paper itself but in the notes taken by J. M. Baudot at one of Clément's lectures on 8 March 1827. On that day (*Conservatoire* note-books, volume 3, *cahier* 2, pp. 41-43) Clément dictated precisely the result which Carnot arrived at in his paper, even to the point of using the same nomenclature. He did not acknowledge his debt to Carnot by name but said that the result had been given to him by 'un mathématicien distingué'. Baudot added: 'La formule algébrique [i.e. Carnot's] n'est ici que comme sujet d'exercice pour ceux qui voudront l'employer; toutefois, le Professeur avoue qu'il n'en a jamais fait usage; il préfère le calcul arithmétique'.

The fact that the result does not appear in Baudot's notes for the years 1824-1825 and 1825-1826, or in Francoeur's notes for 1823-1824, strongly suggests, although it clearly does not prove, that Clément did not learn of Carnot's paper until 1826 or 1827. Since Carnot and Clément were certainly acquainted by then, and probably even close friends, this in turn suggests that the paper was written about that time and hence that in it Carnot was seeking to remedy an obvious weakness in the quite unrealistic treatment of the expansion stroke which he had already given in the *Réflexions*.

NOTES

(1) J. P. Muirhead, *The origin and progress of the mechanical inventions of James Watt*, 3 vols. London, 1854, **I**, p. 62.

(2) *Ibid.* **3**, p. 60.

(3) T. S. Kuhn, 'Carnot's version of Carnot's cycle', *Am. J. Phys.* **23**, 94n (1955).

(4) T. S. Kuhn, 'Engineering precedent for the work of Sadi Carnot', *Actes du IXe Congrès International d'Histoire des Sciences (Barcelona-Madrid, 1959)*, Barcelona and Paris, 1960, pp. 530-535.

(5) D. S. L. Cardwell, 'Power technologies and the advance of science, 1700-1825', *Technology cult.* **6**, 188-207 (1965).

(6) The view that Carnot's *problem* in the *Réflexions* was one of engineering and that he was there addressing engineers rather than physicists is now of course a common-place. See, for example, J. T. Merz, *A history of European thought in the nineteenth century*, 4 vols. Edinburgh and London, 1896-1914, **2**, p. 117; L. Rosenfeld, 'La genèse des principes de la thermodynamique', *Bull. Soc. R. Sci. Liège*, **10**, 197-212 (1941); S. Lilley, 'Social aspects of the history of science', *Archs int. Hist. Sci.* **2**, 392-394 (1948-1949).

(7) See Kuhn, *op. cit.* (4), pp. 533-534; and Cardwell, *op. cit.* (5), p. 202.

(8) N. L. S. Carnot, *Réflexions sur la puissance motrice du feu et sur les machines propres à développer cette puissance*, Paris, 1824. Throughout this paper I refer to this, the first, edition of the *Réflexions*.

(9) See the passage, in a letter of 28 May 1769 from Watt to Dr William Small, cited at the head of this paper.

(10) In a letter of 19 November 1781 to Matthew Boulton (Muirhead, *op. cit.* (1), **2**, p. 135), Watt wrote: 'It is fourteen years since I thought of the double cylinder and expansive engine.' A comment by John Robison, quoted in J. P. Muirhead, *The life of James Watt*, 2nd edition, London, 1859, p. 71, suggests that this statement may have been true, but in his *A system of mechanical philosophy*, 4 vols. Edinburgh, 1822, **2**, p. 127, Robison stated categorically that the principle first occurred to Watt in 1769.

(11) Muirhead, *op. cit.* (1), **3**, pp. 60-73.

(12) For a description of the engine, based on Woolf's patent specification, see *Phil. Mag.* **19**, 133-137 (1804).

(13) Clément first met Desormes about 1800, when the latter was an assistant in Guyton de Morveau's chemical laboratory at the *École Polytechnique*. Later he married Desormes's daughter and adopted the surname of Clément-Desormes, probably in the mid-1820s. This change of name has since caused some confusion. For example, in C. L. Louandre and F. Bourquelot, *La littérature française contemporaine . . . ,* 6 vols. Paris, 1842-1857, **3**, pp. 16-17, it is stated (incorrectly) that it is wrong to distinguish Clément from Desormes and that there was just one person, Clément-Desormes. Unfortunately this error is repeated in F. Hoefer (ed.), *Nouvelle biographie générale*, 46 vols. Paris, 1855-1866, **10**, p. 793.

(14) R. Fox, 'The intellectual environment of Sadi Carnot: a new look', a paper read in August 1968 at the 12th International Congress on the History of Science (Paris) and to be published shortly in the *Actes* of the Congress.

(15) For biographical details of Clément (1778-1841) see his obituary, by Charles Dunoyer, in *J. économistes*, **1**, 208-213 (1842). On Desormes (1777-1862) see *Dictionnaire de biographie française*, Paris, 1933- in progress, **10**, p. 1501.

(16) See the Royal Society's *Catalogue of scientific papers (1800-1863)*, 6 vols. London, 1867-1872, **1**, p. 950, and **2**, p. 269.

(17) Their entry was reproduced, with the title 'Détermination expérimentale du zéro absolu de la chaleur et du calorique spécifique des gaz', in *J. Phys.* **89**, 321-346 (1819). For a discussion of this paper and its reception see P. Costabel, 'Le "calorique du vide" de Clément et Desormes (1812-1819)', *Archs int. Hist. Sci.* **21**, 3-14 (1968).

(18) Clément and Desormes seem to have thought of themselves as being primarily industrialists. For example, in their paper cited in note (17), they described themselves as 'manufacturiers'.

(19) See note (30).

(20) The publication probably took place in the spring of 1824, since it was on 14 June that the book was presented before the *Académie des Sciences* by the engineer P. S. Girard, on behalf of Carnot. See *Académie des Sciences. Procès-verbaux hebdomadaires des séances de l'Académie tenues depuis la fondation de l'Institut jusqu'au mois d'août 1835*, 10 vols. Hendaye, 1910-1922, **8**, p. 101, referred to hereafter as *Procès-verbaux*.

(21) In his lecture on 20 January 1825 Clément referred to Carnot as 'un des auditeurs de ce cours' and also recommended the *Réflexions* to his audience. The relevant passage appears on p. 19 of the second of two books of notes bound in volume 1 of MS. 8°Fa40(2) in the library of the *Conservatoire National des Arts et Métiers*, Paris. The notes, which were taken at Clément's lectures by J. M. Baudot, are bound in three ·volumes and entitled: 'Conservatoire des Arts et Métiers—Chimie industrielle,

Professeur: M. Clément-Desormes—Journal des Cours de 1825 à 1830'. For a full description of these volumes see J. Payen, 'Une source de la pensée de Sadi Carnot', *Archs int. Hist. Sci.* **21**, 18-32 (1968). The reference to Carnot is quoted in full in E. Mendoza, 'Contributions to the study of Sadi Carnot and his work', *ibid.* **12**, 394 (1959).

(22) See Hippolyte's biographical sketch appended to the Gauthier Villars edition of the *Réflexions* (Paris, 1878), p. 77. Reference to the friendship was also made in a short biography written by Sadi Carnot, Hippolyte's son, and communicated to the Italian Count Paolo de San Roberto (see *Atti Accad. Sci. Torino*, **4**, 157 (1868)).

(23) I suggest this date, since it was not until 1819 that Carnot obtained the rank of staff officer and so returned to reside in Paris, having spent most of the time since his graduation from the *École Polytechnique* (in 1814) in the provinces, first at the *École de l'Artillerie et du Génie* at Metz and then, as a young lieutenant, in various army garrisons throughout France (see H. Carnot, *op. cit.* (22), pp. 74-75).

(24) Although the *Conservatoire* had existed since 1794, it was not until 1819 that public instruction was offered. On the establishment of the new chairs see A. de Monzie, *Le Conservatoire du Peuple*, Paris, 1948, pp. 69-80.

(25) Fox, *op, cit.* (14).

(26) The fact that Carnot was never even a candidate for membership of the *Académie* or for a post at the *École Polytechnique* serves to emphasize that he was working on the fringe of the scientific community of Paris—an important point in understanding why his ideas attracted so little attention.

(27) On this episode see Joseph Bertrand's *éloge* of Dupin in his *Éloges académiques*, Paris, 1890, pp. 234-236. It is interesting to note that the third professor appointed in 1819, the economist Jean-Baptiste Say, was no friend of the restored royalist government either (see de Monzie, *op. cit.* (24), p. 78). The hostility of the Bourbon governments towards the *Conservatoire* seems to have been really relaxed only during the ministry of the duc Decazes (December 1818 to February 1820), who, acting on Dupin's advice, was chiefly responsible for the establishment of the public lecture-courses in November 1819 (see de Monzie, *op. cit.* (24), p. 70).

(28) *Procès-verbaux*, **6**, pp. 480 and 481 (16 and 23 August 1819). The paper, read by Clément, is described simply as 'un mémoire sur les machines à vapeur'. *Cf.* the title given in note (29).

(29) C. B. Desormes and N. Clément, 'Mémoire sur la théorie des machines à feu (extrait)', *Bull. Sci. Soc. Philomat. Paris.* 1819, pp. 115-118.

(30) *Réflexions*, p. 98n, where Carnot says of the paper: '. . . j'en ai dû la connaissance à la complaisance de l'auteur [Clément]'. The paper was by no means unknown in the scientific community of Paris. It was probably also read in its complete form by the engineer C. L. M. H. Navier (see *Annls Chim. Phys.* **17**, 357 (1821)) as well as by the referees appointed by the *Académie*, Fourier, Arago, Thenard and Gay-Lussac.

(31) Among others who were tackling similar problems at about the same time were A. T. Petit (*Annls Chim. Phys.* **8**, 287-305 (1818)), J. N. P. Hachette (*Traité élémentaire des machines*, 2nd edition, Paris, 1819, pp. 213-214 and 223-225) and Navier (*op. cit.* (30), pp. 357-372).

(32) This thought experiment has already been described by Professor E. Mendoza in Mendoza, *op. cit.* (21), pp. 394-395, and in his 'Sadi Carnot and the Cagnard engine', *Isis*, **54**, 262-263 (1963); also in Fox, *op. cit.* (14). The present description is a recon-

struction from the *Société Philomathique* extract of 1819 viewed in the light of Clément's accounts as recorded in the sets of lecture notes cited in notes (21) and (76).

(33) At the risk of causing some confusion, I have decided to use the terms 'mechanical power' and 'motive power' throughout this paper in preference to 'work', which was introduced with its precise modern meaning (as *travail*) by G. G. Coriolis only in 1829 (see D. S. L. Cardwell, 'Some factors in the early development of the concepts of power, work and energy', *Br. J. Hist. Sci.* **3**, 218 (1966-1967)).

(34) Clément and Desormes naturally used an engineer's unit (as did Hachette and Coriolis, among others): a cubic metre of water, or 1000 kg, raised through 1 metre. Carnot used the same unit in the *Réflexions*, although in his MS. paper, discussed later, his unit was much smaller: 1 kg raised through 1 metre (what he termed the *dyname*). This latter unit (though not its name) was chosen by G. J. Christian, Navier and J. V. Poncelet. Dupin's *dyname* was 10^6 times the size of Carnot's—a huge unit.

(35) The two processes correspond clearly to the injection and expansion phases of a steam engine operating expansively.

(36) See, for example, S. Smiles, *Lives of Boulton and Watt*, 2nd edition, London, 1866, pp. 228-229 and 304; and H. W. Dickinson, *James Watt, craftsman and engineer*, Cambridge, 1936, p. 132.

(37) Muirhead, *op. cit.* (1), **3**, pp, 60-73.

(38) Another factor which led Watt to neglect the expansive engines—the shortage of men with the skill to operate them—is mentioned in Smiles, *loc. cit.* (36). Dr A. J. Pacey has pointed out to me that yet another factor was the difficulty of applying the expansive principle in rotative cotton-mill engines. For such an application it was essential to have a governor capable of varying the point in the stroke at which cut off occurred, and the design of such a governor appears to have presented very serious problems.

(39) *Encyclopaedia Britannica*, 3rd edition, 18 vols. Edinburgh, 1797, **17**, pp. 765-766. In fact expansive operation attracted a good deal of attention in the last two decades of the eighteenth century, although Watt did not have the field to himself. It was in 1781 that Jonathan Hornblower constructed and patented his compound engine incorporating the expansive principle and in the 1790s that Boulton and Watt sought so actively to suppress his activities.

(40) On the early history of the Woolf engine see R. Jenkins, 'A Cornish engineer: Arthur Woolf, 1766-1837', *Trans. Newcomen Soc.* **13**, 55-68 (1932-1933).

(41) There is hardly any evidence of French interest in the Woolf engine during the Napoleonic period, although one account by the engineer Michel-Ange Lancret, based on the *Phil. Mag.* article of 1804 (cited in note (12)), did appear in the *Bull. Soc. d'Encouragement*, IIIe année, brumaire an 13 (1804), pp. 108-112. Also two engines operating expansively appear to have been built by an inspector of the French Mint, Gengembre, in 1808 and 1812. I am grateful to Dr D. S. L. Cardwell for drawing my attention to Gengembre's brief description of these engines in *Annls Chim. Phys.* **4**, 190-191 (1817).

(42) Jenkins, *op. cit.* (40), p. 59.

(43) See, for example, the first edition of Hachette's *Traité élémentaire des machines*, Paris and St Petersburg, 1811, pp. 123-124.

(44) See, for example, the articles by A. R. Bouvier and A. T. Petit in, respectively, *Annls Chim. Phys.* **3**, 177-192 (1816), and *ibid.* **8**, 287-305 (1818). Among the numerous references in the *Bulletin* at this time the most important are: *Bull. Soc. d'Encouragement*, XVIᵉ année, 1817, pp. 267-270 (by C. P. Molard); XVIIᵉ année, 1818, pp. 169-174 (Hachette) and pp. 365-384 (Hoyau).

(45) Also in a work which Carnot is known to have read: Héron de Villefosse, *De la richesse minérale*, 3 vols. Paris, 1810-1819, **3**, pp. 87-108.

(46) Hachette, *op. cit.* (31), pp. 210-216. See also *Bull. Soc. d'Encouragement*, XVIIᵉ année, 1818, pp. 169-174, where the treatment had already appeared.

(47) Although, in his article in *Isis*, **54**, 262-263 (1963) (cited in note (32)), Professor E. Mendoza has noted Carnot's praise for Clément and Desormes's treatment of the expansive principle (*Réflexions*, p. 98n). The point has been taken up again very recently in Payen, *op. cit.* (21), pp. 32-37.

(48) Hachette, *op. cit.* (31), pp. 198 and 214-215. For a similar view see Davies Gilbert's paper in *Phil. Trans.* 1827, p. 34, and also J. Farey, *A treatise on the steam engine, historical, practical and descriptive*, **2**, p. 53n. This second volume of Farey's book was never published and the information is taken from the proof sheets in the Patent Office library, London, which probably date from about 1840. In the first volume (London, 1827) Farey clearly used Boyle's law in his calculation (see p. 368).

(49) Boyle's law continued to be used in calculations of the effect obtained after cut off throughout the first half of the nineteenth century. See, among some of the better known works of the time, T. Tredgold, *The steam engine*, London, 1827, p. 161; J. V. Poncelet, *Introduction à la physique industrielle, physique ou expérimentale*, 2nd edition, Metz and Paris, 1839, pp. 204-207; H. V. Regnault, *Mém. Acad. Roy. Sci.* **21**, 8-9 (1847).

(50) W. A. Gabbey and J. W. Herivel, 'Un manuscrit inédit de Sadi Carnot', *Revue Hist. Sci. Applic.* **19**, 151-166 (1966).

(51) The manuscript, which is in the library of the École Polytechnique, Paris, is entitled 'Recherche d'une formule propre à représenter la puissance motrice de la vapeur d'eau'.

(52) *Réflexions*, pp. 17-18.

(53) In fact, by beginning the adiabatic compression before condensation is complete, it is possible to complete the cycle even when steam is the working substance. But his acceptance of Clément and Desormes's 'law' (discussed later in this section) meant that Carnot could not envisage such a possibility, since he believed that no change of state could occur during adiabatic compression or expansion.

(54) Though in the manuscript paper Carnot first considered the general case in which the temperature of the steam after expansion and that of the condenser are not equal, and only then showed how his result could be simplified by assuming that the two temperatures were in fact the same.

(55) *Cf. Réflexions*, pp. 79-88, and Gabbey and Herivel, *op. cit.* (50), pp. 154-156 and 161.

(56) *Réflexions*, pp. 81-88.

(57) *Bull. Soc. d'Encouragement*, XVIIIᵉ année, 1819, p. 255.

(58) For the experiments which were described in April, however, Clément had used a boiler (*chaudière*), presumably from a steam engine, at the Oberkampf factory at Jouy.

(59) On the discrediting of the 'law' see R. Fox, 'The background to the discovery of Dulong and Petit's law', *Br. J. Hist. Sci.* **4**, 16n (1968-1969).

(60) *Loc. cit.* (57). The book is referred to simply as 'un ouvrage de Robison, revu par M. Watt, qui paraîtra bientot par les soins de M. Brewster'. It is unmistakably Robison's *System*, which appeared, under Brewster's supervision, in Edinburgh in 1822 (see note 10).

(61) Robison, *op. cit.* (10), **2**, pp. 160-175.

(62) *Ibid.* p. 167n. For further evidence that Watt worked on this subject and at least gave serious consideration to the possibility that the heat content of saturated steam did not vary with temperature, see: J. Black, *Lectures on the elements of chemistry*, ed. J. Robison, 2 vols. Edinburgh, 1803, **1**, p. 190, and J. Dalton, *A new system of chemical philosophy*, Manchester, 1808, **1** (part 1), pp. 131-132n. In the latter work Dalton records that Watt's suggestion was taken up by Peter Ewart, who believed that it was probably true.

(63) He died on 19 August 1819.

(64) This point is true also of Hachette, who seems to have assumed the truth of the 'law' in a paper as early as June 1817, though (on his own admission) without any experimental justification. See *Nouveau bulletin des sciences, par la Société Philomatique de Paris*, 1826, p. 53.

(65) Yet they may well have been sufficient to prevent the publication of the 1819 paper, since Robison's views on Watt's priority had already been circulated to Berthollet and Prony, among others (*loc. cit.* (57)).

(66) Gabbey and Herivel, *op. cit.* (50), p. 156. The fact that no such doubts were expressed in the *Réflexions* is additional evidence that the manuscript paper was written after 1824.

(67) *Réflexions*, pp. 65-66.

(68) In both the *Réflexions* and the manuscript paper Carnot used the tables in J. B. Biot, *Traité de physique expérimentale et mathématique*, 4 vols. Paris, 1816, **1**, pp. 272 and 531, which were based on data given by John Dalton in *Mem. Manchr. Lit. Phil. Soc.* **5**, 550-563 (1802).

(69) i.e. Gay-Lussac's law governing the regular expansion of gases by heat.

(70) *Réflexions*, p. 67n, and Gabbey and Herivel, *op, cit* (50), pp. 155-158. $\frac{1}{267}$ per °C was Gay-Lussac's expansion coefficient for gases.

(71) Calculations for an engine generating at 10 atmospheres appear in the *École des Beaux-Arts* manuscript (see note (76)), pp. 50-52, and on pp. 38-43 of the *Conservatoire* notebook cited in note (21).

(72) This definition of the *dynamie* is repeated several times in the notes of Clément's lectures. Although the unit was widely used (see note (34)), the term *dynamie* in this sense seems to be Clément's own.

(73) It should be noted that Carnot did not use the now more familiar $pv^{\gamma} = $ constant and $p^{1-\gamma}(t+267)^{\gamma} = $ constant relationships for adiabatic volume changes, although these had been derived by S. D. Poisson and published in *Annls Chim. Phys.* **23**, 15 (1823).

(74) However, see note (54).

(75) The effect of the condensation stroke was not so much as mentioned even by authorities of the stature of Petit (see *op. cit.* (31)) and Hachette (*ibid.*). It continued to be ignored until Carnot's work was rediscovered in the late 1840s.

(76) The notebook, which is now MS. 407 in the library of the *École Nationale Supérieure des Beaux-Arts*, Paris, is one of forty-one volumes of manuscripts that Francoeur left to the *École*. Although the notebook does not bear the name of Francoeur, the handwriting is unmistakably his, as Madame Bouleau-Rabaud has pointed out to me. See also note (78).

(77) See L. G. Michaud (ed.), *Biographie universelle* . . . , new edition, 45 vols. Paris, 1842–1865, **14**, pp. 639–640.

(78) The inscription reads: 'Notes prises aux leçons de Clément au Conservatoire des Arts et Métiers rue St. Martin. Cours de 1823 à 1824.' The notes of the individual lectures are undated.

(79) The reference is in the *Conservatoire* notebooks at *loc. cit.* (21).

(80) Clément's calculations, as reproduced in his lectures, showed clearly that the output of an engine increased as the pressure of the steam was raised (though by no means in proportion to the pressure). Yet, feeling that the advantages of high-pressure steam did not outweigh the practical difficulties involved, he advocated the use of moderate pressures, of the order of just a few atmospheres. Such pressures were sufficient to allow the benefits of expansive working to be obtained—an important point, since, as Clément observed in a lecture on 24 January 1825, 'La puissance mécanique de la détente est beaucoup plus précieuse que celle de la production'. See the *Conservatoire* notebook cited in note (21), pp. 43–46.

(81) B. Cimbleris, 'Reflections on the motive power of a mind', *Physis*, **9**, 417 (1967).

INDEX

Abbeville: V 128
Abbotsford: VI 314
Abria, Jérémie Joseph Benoît: III 549n
Académie de Médecine: II 242
Académie des Sciences, Paris: I 442;
   II 242, 249, 266–7, 271, 279, 281; IV 1,
   4, 10–11; VI 312; VII 333n, 341; VIII 2,
   14, 16n; IX 90, 112–15, 132; XI 155;
   XII 236 (*see also* Prizes and prize
   competitions)
— age of members: II 269
— *éloges*: I 456; IX 122
— permanent secretaries: I 452; IX 114n,
   126–7
— publications: I 456; II 267, 278
Academies: II 241–82; III 543–64
— Amiens: II 248–9
— Arras: II 246n; III 544n
— Bordeaux: II 246n, 274; III 544–7, 555n,
   562–4
— Caen: II 246n, 248–9; III 544–7, 562–4
— Cherbourg: 246n; III 544n
— Dijon: II 248; III 545
— Ionian: VI 306–10
— Lyon: II 246–9; III 544n, 546n; X 361–4
— Marseille: II 246n; III 544n
— Nancy: II 246n; III 544n, 562–4
— Nîmes: II 246n; III 544n
— Rouen: II 246n; III 544n, 546n
— Toulouse: II 246n, 247, 253; III 544n,
   547n, 558
Adiabatic heating and cooling: VII 334–6;
   VIII 13–15; X 355–70; XI 150; XII 239–45
Adiabatic phase in Carnot cycle: XII 234,
   237–46
Aepinus, Franz Ulrich Theodor: IX 93–4
Affinity, chemical: I 448, 451; IX 95–9
Agassiz, Louis: V 154
Agrégation: I 472; II 242
Agulhon, Maurice: III 555
Alesia: VI 311
Ali Pasha: VI 308
Alsace: I 464; II 257, 271; V 127–68
Altkirch: V 152
Amiens: II 248–9, 256n
Amontons, Guillaume: VII 333–4

Ampère, André Marie: I 454; IX 116–17,
   119, 128–31
Ampère, Jean-Jacques: I 443n, 472n
Andersonian Institution: VI 315, 317
André Koechlin et Cie: V 143, 161n
Andrews, Thomas: I 469
Angers: V 151n
Arago, Dominique François Jean: I 447,
   448n, 451–2, 456, 463, 470, 472n; II 269;
   VIII 2–3, 5, 21–2; IX 102–35
Archiac, E.J.A. Desmier de Saint-Simon,
   vicomte d': II 267
Arcueil: I 448, 451–2, 458, 464, 472–3;
   VIII 5, 11, 16–17; IX 90, 98, 107–9, 123,
   127, 133–6
Arnold, Johann Christian: VII 334–5
Arras: II 246n; III 544n
Art galleries: III 554–5
Association Française pour l'Avancement
   des Sciences: II 243, 272–6, 278, 280;
   III 552
Association Normande: II 259n; III 547
Association Scientifique de France:
   II 272–4
Athénée, Paris: I 463
Athénée de Lyon: II 246
Atomic theory: I 451–2; VIII 1, 7–9, 16–22;
   IX 116, 119, 129–30
Augoyat, Antoine Marie: VI 309–10
Augsburg: V 139–40
Austria: V 150
Auxerre: II 258

Babbage, Charles: I 447; IX 90
Baccalauréat: I 462; II 242
Baillaud, Benjamin: II 253
Balard, Antoine Jérôme: I 462; II 271
Balbiani, Edouard Gérard: II 277n
Bâle: V 143
Balfour, Henry: X 355n, 365–7
Ballot, Charles: V 141
Balzac, Honoré de: I 458
Banks, Sir Joseph: VI 315
Barreswil, Charles Louis: II 271
Barrow, John: VI 316